Clean Energy, Climate and Carbon

Dedication

For my children and grandchildren

Clean Energy, Climate and Carbon

Peter J Cook

National Library of Australia Cataloguing-in-Publication entry

Cook, P. J. (Peter John), 1938–

Clean energy, climate and carbon/Peter J. Cook.

9780643094857 (pbk.)
9780643106826 (epdf)
9780643106833 (epub)

Includes bibliographical references and index.

Carbon sequestration.
Geological carbon sequestration.
Greenhouse gas mitigation.

628.532

Published exclusively in Australia and New Zealand by

CSIRO PUBLISHING
150 Oxford Street (PO Box 1139)
Collingwood VIC 3066
Australia

Telephone: +61 3 9662 7666
Local call: 1300 788 000 (Australia only)
Fax: +61 3 9662 7555
Email: publishing.sales@csiro.au
Website: www.publish.csiro.au

Published exclusively throughout the rest of the world (excluding Australia and New Zealand) by CRC Press/Balkema, with ISBN 978-0-415-62106-9

CRC Press/Balkema
P.O. Box 447
2300 AK Leiden
The Netherlands
Tel: +31 71 524 3080
Website: www.crcpress.com

Front cover image by iStockphoto

Set in 10.5/14 Adobe ITC New Baskerville
Edited by Michele Sabto
Cover and text design by James Kelly
Typeset by Desktop Concepts Pty Ltd, Melbourne
Printed in China by 1010 Printing International Ltd

CSIRO PUBLISHING publishes and distributes scientific, technical and health science books, magazines and journals from Australia to a worldwide audience and conducts these activities autonomously from the research activities of the Commonwealth Scientific and Industrial Research Organisation (CSIRO). The views expressed in this publication are those of the author(s) and do not necessarily represent those of, and should not be attributed to, the publisher or CSIRO.

Original print edition:
The paper this book is printed on is in accordance with the rules of the Forest Stewardship Council®. The FSC® promotes environmentally responsible, socially beneficial and economically viable management of the world's forests.

CONTENTS

PREFACE

As a boy growing up in England I was fascinated by weather and by the extremes of weather. I developed an interest in science and thought it might be interesting to be a meteorologist, spending months in lonely places taking weather readings. With this vague thought in mind, I enrolled at Durham University in mathematics, physics and chemistry, with geology as a subsidiary. But my plans all changed when I attended the inspirational first year lectures of Professor Kingsley Dunham. Geology was for me! These lectures also triggered an interest in past climate change and ancient environments. An early career with the Australian Bureau of Mineral Resources (BMR – now Geoscience Australia) provided outstanding opportunities to work on some wonderful geology, including a spell in Antarctica, where I was fascinated by the dry valleys near the Davis Antarctic base and the abundant evidence of relative sea level change there. A PhD at the University of Colorado and some years working on sedimentary phosphate deposits, provided career variety but still kept me in touch with past climate change. Direct involvement in the record of recent sea-level change arose when Dr Norman Fisher, an outstanding Director of BMR, agreed to my suggestion to initiate a study of estuaries starting with Broad Sound on the central Queensland coast. This was (and probably still is) an area with marvellous depositional systems and a somewhat daunting fauna of large snakes, larger crocodiles and very large sharks. A series of beach deposits or cheniers made it possible to plot the course of relative sea level change in the area over the past 5000 years. This was followed by the opportunity in the mid 1970s, to work on an exceptional series of ancient beach ridges in southern Australia, extending more than 100 kilometres inland, first described by Reg Sprigg, one of the giants of Australian geology. Using new dating techniques we were able to show that the rise in sea level evident from the beach ridges could be correlated with orbital changes as first proposed by Sprigg.

It was probably Marchetti who first suggested in the late 1970s the idea of long term storage of CO_2 as a way of combating the increasing emission of CO_2 arising from human activity. His idea was to inject the CO_2 into the deep ocean, but at that time there was no widespread concern regarding climate change and indeed, at about that time, there was a view that any increase in CO_2 was likely to be a good thing because it would increase plant growth and prevent the onset of a new ice age. The concept of storage was taken up again about 1990, but with the focus on geological storage. Researchers at the Alberta Research Council were amongst the pioneers, driven in part by the need to consider options for disposal of waste gas (acid gas containing high CO_2) from the petroleum industry.

In 1990 I was appointed Director of the British Geological Survey (BGS) and about that same time, supported by the Joule program of the European Commission, researchers at the Survey along with EU colleagues, started to consider the feasibility of storing CO_2 in rocks underlying the North Sea and adjacent areas. This was an outstanding program of research which for the first time showed the potential regional importance of CO_2 storage. Undoubtedly one of the most significant events in taking geological storage forward occurred in 1996 when Statoil commenced the Sleipner Project in the North Sea, injecting approximately a million tonnes of CO_2 a year into a geological formation located at a depth of about 800 metres below the sea floor. It was also during this period of the 1990s that there was an awakening of interest and increasing concern regarding climate change. Despite the UN Framework Convention on Climate Change and the Kyoto Protocol, it was still difficult to engage the power industry or UK government departments on the issue (as I found as I walked the corridors of Whitehall) throughout much of the 1990s. Nonetheless the science progressed in the UK, Canada, USA, Norway and the Netherlands in particular during this time.

In 1998, I returned to Australia and took up a position as Director of the Petroleum Cooperative Research Centre (APCRC). At that time, some very preliminary work on CO_2 was underway in Australia by the Gorgon partners on the possibility of CO_2 storage under Barrow Island in Western Australia, but there was no public discussion on the potential of CO_2 storage and no research into the topic. A critical question that needed to be asked was – is geological storage a significant mitigation opportunity for Australia or is it likely to be just 'cottage industry'? To answer this question in 1998 I established the GEODISC (Geological Disposal of CO_2) Program within the APCRC, with the financial backing of six natural gas companies and the Australian Greenhouse Office, together with research support from Geoscience Australia, CSIRO and several universities. Over the subsequent four years, this Program was very successful in confirming that Australia has a high potential for the large scale application of geological storage of CO_2.

By 2002, Australian governments and industry were becoming more conscious of the issue of climate change and of the opportunity that CO_2 storage might offer, so in 2002, I took a proposal to Government and industry for a greatly expanded program of research and development encompassing not only storage but also capture of CO_2. This provided the basis of the establishment of the Cooperative Research Centre for Greenhouse Gas Technologies (CO2CRC) in 2003. It brought together not only the gas industry but also the coal industry and the power sector,as well as various governments and a number of universities and research bodies, establishing CO2CRC as one of the leading collaborative CO_2 research organisations in the world. Also at about that same time, the Intergovernmental Panel on Climate Change (IPCC) initiated a Special Volume on Carbon Capture and Storage. This was an important step in confirming the international recognition of carbon capture and storage (now known as CCS) and the part it could play in global abatement of emissions. I was the Co-ordinating Lead Author (with Sally Benson) for the IPCC chapter on geological storage – a difficult but ultimately satisfying role in that the Special Volume did pull together all of the key information available at 2004–2005.

Since then, awareness of climate change has risen in all countries and this has been reflected in an increased level of interest in clean energy technology activity generally. Much of this has been driven by a range of government initiatives with some measures quite effective and others not. In Australia, as in many OECD countries, the measures taken have included the signing of the Kyoto Protocol, mandatory renewable energy targets and support for R&D. For many people, these new initiatives have been puzzling and the clean technologies being developed, unfamiliar.

It was for this reason that I first started thinking about a book to discuss some of these technologies in a way that was accessible to the broader public. The initial intention was to focus entirely on one technology, CCS, because it was the technology with which people were least familiar and it was the one on which CO2CRC focused its research attention. However, it soon became clear that CCS needed to be put into a much broader climate change context. But nor was this sufficient and the scope was then further extended, with CCS as a key component in the broader clean energy family.

In all, this book has taken far longer to write than I anticipated. Part of the problem was that throughout the period of its writing I was Chief Executive of CO2CRC, a large, diverse and geographically dispersed organisation involving more than 200 researchers. As a result, much of the initial writing was done on planes, in hotels and in airports and my wife Norma showed exceptional forbearance as yet more weekends were sacrificed to yet more writing and rewriting. The difficulty in bringing the book to a close was that not only was clean technology evolving, but climate change policies around the world were also evolving. In Australia, the political setting for clean energy was especially difficult and governments changed in no small measure due to energy and climate issues. Clearly this book could not wait until everything was resolved! Therefore, this book reflects the status of clean energy and related issues up to August 2011 with the expectation that these matters will further evolve. However, I believe the book also contributes to the debate regarding how best to address climate change through the introduction of clean energy technologies not only in Australia but also in many other countries, by enabling people to be better informed about the various technologies and the at-times complex issues relating to their deployment, and the contribution that they can make to reducing emissions not just now but in decades to come.

Elsewhere in this book there is a comprehensive acknowledgement list but there are some people that deserve to be acknowledged here, particularly Anni Bartlett, Michael Soroka and Rhonda Evans for their roles in contributing to this book in many ways, and Rebecca Jones and Anna Nguyen for their graphic design skills. Finally, yet first and foremost, I particularly thank my wife Norma for her ongoing support, companionship and understanding.

Professor Peter J Cook, CBE, FTSE

ACKNOWLEDGEMENTS

A book such as this which seeks to cover such a broad range of science and technology is highly dependent on a diversity of inputs and contributions from many people, for nobody can claim to be an expert in all the fields covered. One of the most important inputs is the time and effort involved in seeking out and compiling large amounts of information from a great diversity of sources. Here the work of Michael Soroka and Anni Bartlett, both of CO2CRC, was absolutely outstanding and I am forever grateful to them. Michael has also been a tower of strength in handling the formatting of material for the publisher. When I started planning this book I felt that having good graphics to explain complex ideas was critical to its success and I have been very fortunate to be able to work with some highly talented graphic designers who have been able to turn vague ideas and concepts into a clear and concise illustration. In this regard I particularly thank Rebecca Jones, Anna Nygen, Roslyn Paonin and Lee-Anne Shepherd of CO2CRC for their contributions, aided by Anni Bartlett using her skills as an educator.

The quality of this book has been very much dependent on the generosity of many people who freely gave of their time to provide information or advice or peer review. In a diverse and highly technical field such as clean energy, such input is absolutely critical to the accuracy and objectivity of the information presented. People I should particularly acknowledge in this regard include Alex Thorman (CO2CRC), Dr Alexandra Golab (Digital Core Labs), Mr Barry Hooper (CO2CRC), Dr Charles Jenkins (CSIRO), Dr Dennis van Puyvelde (ACALET), Professor Dianne Wiley (University of New South Wales), Dr Jonathon Ennis-King (CSIRO), Professor John Kaldi (University of Adelaide), Dr Lincoln Paterson (CSIRO), Dr David Etheridge (CSIRO), Dr Graeme Pearman (Graeme Pearman Consulting), Dr John Soderbaum (Acil Tasman), Dr Karl Gerdes (Chevron), Professor Paul Webley (Monash University) and Associate Professor Sandra Kentish (University of Melbourne). There are also 200 researchers within CO2CRC and its many collaborating organisations who have contributed over the years to developing the collective knowledge represented in this book, and to whom I extend my thanks for their dedication and their outstanding science, which I hope I have been able to faithfully represent here.

The aim of this volume is overwhelmingly to present facts and information. Nonetheless at times opinions are presented, particularly in the final chapter. I should make it absolutely clear that these are my opinions and do not necessarily reflect the views of other CO2CRC participants or contributors. Within CO2CRC I have benefitted greatly from excellent and supportive relationships with Mr Tim Besley AC and Mr David Borthwick AM and the CO2CRC Board and I thank them for that support.

At CSIRO Publishing, John Manger was particularly forgiving as various deadlines were missed and the timescale become evermore extended. Michele Sabto and Tracey Millen were excellent (and tolerant) as editors.

Finally this book would not have been possible without the ongoing support and understanding of my family, not just during the writing of this book (though this was undoubtedly a period when even more understanding and tolerance than ever was required!) but for the many years prior, with particular thanks to my wife Norma.

1 THE CONTEXT

It is impossible to pick up a newspaper, read a blog, or watch the news on TV without being aware of the 'climate debate' and of the range of views that accompany it. The debate is complex because it covers a wide range of intersecting issues. For many people the complexity is increased by the range of acronyms, terms and concepts used, which are often not well understood: UNFCCC, IPCC, anthropogenic, parts per million, clean energy technologies; the list goes on. People talk about 'carbon' but in some cases they really mean 'carbon dioxide', which further adds to the confusion. It is not the intent of this book to explain every term used, every concept developed or every solution proposed, but many are dealt with in sufficient detail to provide a background to the development and deployment of clean energy technologies and why they are critical to addressing climate concerns. But at the same time, technology issues cannot be considered in a vacuum; they have to be understood within the context of the climate debate and of energy more broadly. This book provides that context, enabling clean energy technology to be considered in an informed and practical manner, before focusing on one of the least known, but very significant, clean energy technologies, carbon capture and storage or CCS.

What the book does not discuss to any extent is the other form of carbon storage, namely soil carbon or biocarbon storage. Appropriately managed soils can store large amounts of carbon and bring benefits to agricultural productivity. As stated in the *Australian Government Climate Commission Report* released in 2011:

> *'for many reasons increasing carbon storage in land ecosystems is a necessary and desirable component of a comprehensive approach to greenhouse carbon in a secure geological formation, locked away from the influences of climate variability and change or from the direct impacts of human management'.*

The reader should turn to other texts on soil carbon. Nor does this book deal with what is called 'geo-engineering', such as putting sulphur dioxide or other particulates into the atmosphere to decrease the amount of radiation reaching the earth.

This book is concerned with how to decrease CO_2 emissions to the atmosphere resulting from human activity, and focuses on clean energy, or low emission, technologies. While there is vigorous debate about global warming and climate change, the weight of scientific evidence indicates that climate change/global warming is occurring and that it is very likely those changes correlate with the increasing concentration of atmospheric CO_2 resulting from human activity. Some of the more dire predictions of imminent catastrophe and pestilence are not helpful to rational debate although the concept of a 'tipping point' with irreversible changes cannot be dismissed. These issues are convincingly discussed in some considerable detail by the Intergovernmental Panel on Climate Change (IPCC). The conclusions of the IPCC process, based on the work of many reputable scientists from all over the world, are for the most part accepted as a starting point for this book.

It is not the intention to revisit here the massive amount of evidence supporting anthropogenic climate change as this is presented in the IPCC reports. There is of course some controversy surrounding aspects of the IPCC process and here I should declare my involvement as a Co-ordinating Lead Author of the IPCC Special Report on Carbon Capture and Storage, published in 2005. For me, that involvement exposed the strengths and weaknesses of the IPCC process, but left me with the overwhelming view that whatever the shortcomings, overall the IPCC process was rigorous and balanced.

Climate change science: the controversies

There has been some controversy around the IPCC, culminating in what has been called 'Climategate'. Subsequently, a number of reviews of the IPCC have been conducted, including those by the Netherlands Environmental Assessment Agency, the Royal Society and the US National Research Council. All concluded that the IPCC's key findings were valid. Most recently, in 2010, the Inter Academy Council conducted an independent review of IPCC processes and procedures in response to criticism of the IPCC Fourth Assessment report. It too found that '*the IPCC process has been successful overall and has served society well*', and while it makes a number of important recommendations for improving IPCC procedures, it leaves the reader in no doubt that IPCC processes are appropriate and its findings credible. The scientists responsible for triggering the 'Climategate' controversy engaged in some questionable practices, but the review of the affair by the Royal Society concluded that while some of their behaviour was not appropriate, this did not invalidate their overall conclusions regarding climate change. There have also been many attacks on the science of climate change, questioning the data and interpretation of the data. It is healthy to question the views of experts and to listen to opposing views. But it is also appropriate to weigh this against the strength of the data and the standing of the scientists offering the opinion. On that basis, a jury would rule in favour of the validity of anthropogenic (human induced) climate change. But as Harold Shapiro has put it:

> *'The intersection of climate science and public policy is certain to remain a controversial arena for some time as so many competing interests are at stake including the interests of future generations and the diverse interests of different nations, regions and other sectors of society around the world. Moreover, thoughtful controversy will remain a critical ingredient in stimulating further developments on the scientific frontier of climate change'.*

Amen to that!

The debate on what action should be taken to address climate change, and when, has also stirred up controversy. Some, such as Bjorn Lomborg, argue that it would be unwise to take climate actions that would impose a massive financial burden on national economies, or on the global economy. Nicholas Stern in the UK, Ross Garnaut in Australia and others have looked at the economics of climate change and have argued convincingly for action now. According to Stern and Garnaut, it would be cheaper to act now than be forced to take more drastic (and more expensive) action in the future.

Global and national efforts to take action on climate change

What does 'taking action' on climate change really mean? The aim of most governments and international organisations is to ensure that 'dangerous climate change' does not occur. We do not actually know what 'dangerous' means, but it is often taken to require that the rise in the global temperature should be kept to a maximum of two degrees Celsius. Translating this target into a maximum acceptable concentration of atmospheric CO_2 is by no means straight forward. Estimates for this figure range from as little as 350 parts per million (ppm) of CO_2 (in other words we are already over the limit) to 450 ppm or 550 ppm CO_2, compared to the current atmospheric concentration of 395 ppm.

A general view appears to be that 450 ppm CO_2 equivalent is an appropriate target. But by what date should this target (or maximum) be reached to avoid 'dangerous climate change'? Some argue that 2100 is the appropriate date, but many others point out that our emissions must decrease long before then and suggest 2050 instead, with intermediate emission reduction targets at 2020, 2030 and 2040. The fact that there is as yet no international agreement on targets is used by some to argue against taking any action at all at this time. But there is a strong case for the view that if we don't act now, and we subsequently discover that climate change and its impact is much more severe than forecast, it may be too late to take remedial action. There are also convincing arguments that countries should take individual action to decrease emissions, despite the absence of a collective agreement.

The view taken in this book is that the course of action must be balanced, practical and must begin soon. We must seek to decrease CO_2 emissions to the atmosphere as quickly and as cost effectively as possible, without foregoing the important social and economic benefits that access to affordable and reliable energy provides.

But what about the billion or more people in developing countries that have yet to receive those benefits? Surely in our efforts to decrease global emissions we cannot deny developing countries access to low cost energy. Yet if every developing country were to have the same level of fossil energy use as, say, the United States, we would be faced with a massive increase in global emissions – far in excess of current projections and well into 'dangerous climate change' territory. Recognition by developed countries of the need to make cuts in their emissions to make 'head space' for developing countries may be part of the answer, so that over time there is a convergence of average per capita emissions between developed and developing countries.

To provide context to discussions on clean energy technology, it is worth summarising the status of international climate negotiations. The United Nations Framework Convention on Climate Change (UNFCCC), the outcome of the so-called Earth Summit held in Rio de Janeiro in 1992, had as its objective, the stabilisation of atmospheric greenhouse gases at or below the concentration at which there would be dangerous anthropogenic consequences to the earth's climate system. One hundred and ninety-four countries have now signed this Framework Convention and it has been in force since March 1994.

However, the UNFCCC did not offer guidance on a specific concentration of greenhouse gases that was seen as 'dangerous', nor did it provide an agreed mechanism for decreasing emissions. That was left to the Kyoto Protocol on Climate Change which was developed in 1997 to establish legally binding reductions in emissions in relation to 1990 emissions, with reduction generally of the order of 5–8% by 2008–2012. However, the Protocol did not come into force until 2005 when there were finally a sufficient number of signatory countries (84) accounting for at least 55% of total global emissions. Most of the cuts in emissions were to be made by developed (Annex 1) countries, with no formal requirement being placed on Non Annex (developing) countries to limit their emissions. In any event, the United States, then the world's largest CO_2 emitter, did not ratify the Protocol, and since 1990 emissions from many Annex 1 countries have grown enormously.

Therefore it has to be said that while there was (and continues to be) important symbolism in the Protocol, at the global level, it has not been successful in decreasing global emissions of CO_2. Prior to, and subsequent to, Kyoto, there have been a range of Conferences of the Parties (COP) meetings, but these have had only limited success in developing an international agreement. This disappointing trajectory for the COP meetings culminated in the 2009 COP meeting in Copenhangen, which, despite attempts to depict it otherwise, was spectacularly unsuccessful.

I was a participant in the Copenhagen scientific meeting that preceded the COP and was disappointed by the outcome. Since then, a ministerial level meeting in Cancun in 2010 also proved to be of limited value in taking matters forward. The subsequent meeting in Bonn reached no decisions on the form of any future climate agreement, although there was agreement reached to include CCS in the Clean Development Mechanism (and the Green Climate Fund). Not everybody will agree with me on this rather bleak assessment of the achievements of climate negotiations over the past decade and more. But it is telling that expectations for the COP meeting in Durban in 2011 are modest. There is no escaping the fact that the Kyoto Protocol expires at the end of 2012 and that there is little prospect that a new international treaty will be in place by then.

Some countries may work towards their post-Copenhagen aspirational emission targets, but many are setting their own targets (qualified to varying degrees), to decrease their emissions, whether in absolute terms or in terms of carbon intensity. What is clear for the moment, is that it would be unwise to assume that international emission targets will be agreed upon and implemented in the foreseeable future. This is no reason for individual countries to delay taking action where they can, but obviously the nature of those actions will be influenced by expectations of when (and if) a new international treaty will be in place. Despite this uncertainty, one thing is clear: there are no grounds for complacency. According to the International Energy Agency, in 2010 emissions rose to 30.6 gigatonnes CO_2, a 5% increase from the previous maximum of 29.3 Gigatonnes of CO_2 reached in 2008.

About this book

This book makes no assumptions whatsoever regarding future agreements. It proceeds on the basis that action is required now to limit emissions even in the absence of a binding international agreement. It focuses on clean energy technologies. This alone is a massive topic and it is necessary to be realistic about what can be presented here, given that there have been thousands of papers, articles, books and films on the issue. In fact the book also attempts to cover not only clean energy but also climate and carbon. It is worth taking some time to explain why and how these topics are covered.

Climate, and more specifically climate change, is the context for all the discussion in this book so Chapter 2 starts by reminding the reader of the nature of, and the evidence for, the changes in the composition of our atmosphere in the past. It outlines what is happening now and considers the implications of those changes to the future of our planet. However, this is not a book about climate change. Climate change is an extra-ordinarily broad and complex topic that is concerned not only with global warming, but also encompasses ocean acidification, extreme events, ecological change, sea level rise and many other issues. The approach taken in Chapter 2 is to focus not on climate change per se, but on the issue of rising levels of atmospheric CO_2, and to examine both the geological and the recent record of CO_2 concentration in the atmosphere. This sets the scene for the discussion that follows about the steps that can be taken to decrease CO_2 emissions.

The starting point for considering the increasing concentration of atmospheric CO_2 is the increasing use of energy. In 2008 the International Energy Agency (IEA), pointed out that:

> *'Approximately 69% of all CO_2 emissions are energy related. The IEA World Energy Outlook 2010 (IEA, 2010a) projects that, without changes in current and already planned policies, global energy related CO_2 emissions will be 49% higher in 2035 than in 2007, with fossil fuel demand increasing by more than 40% and remaining the dominant source of energy'.*

In other words, much of the extra CO_2 entering the atmosphere is a consequence of energy use, especially the production of electricity (Figure 2.1). But other activities, including manufacturing, heating, cooling, domestic activities, transportation and construction also contribute. We use energy in every aspect of our daily lives. Chapter 3 summarises the disparate sources of energy and CO_2. Energy is used as the solution to many of the everyday problems at work, in the home, or at play. Too cold? Turn the

Figure 1.1 This now-classic image of the world at night clearly illustrates the global nature of the energy-greenhouse problem, with massive concentrations of light (and energy use) in eastern China, northern India, Europe and the eastern United States. It also serves to show that there are large areas of the world, such as large parts of western China, southern India and most of Africa, where there is only limited access to light and power. (Image courtesy of C. Mayhew and R. Simmons (NASAGSFC), NOAA NGDC, DMSP Digital Archive)

heating up! Too far to walk? Get in the car! Too much like hard work? Get a machine to do it! But as much as energy is used as a solution, it also creates a problem, whether through overuse or because of the nature of the energy source.

In fact, energy is not solely responsible for the increased concentrations of CO_2 entering the atmosphere. There are some large industrial sources of CO_2 resulting from chemical reactions (such as CO_2 released from limestones during the manufacture of cement) which should not be ignored and which are discussed in Chapter 3 and elsewhere. Land clearance and changed farming practices also play a significant role. Subsequent discussion briefly deals with these, but unless otherwise stated, further reference to CO_2 emissions should be taken to mean 'non-farm' anthropogenic CO_2 emissions.

Energy is the dominant source of the problem of increasing atmospheric CO_2 and therefore it is necessary to consider the range of options for decreasing energy-related emissions. Chapter 4 discusses a wide range of energy technologies. Some of these technologies are familiar and others not. Each have their advantages and disadvantages, and these are outlined to allow the reader to come to their own view on the pluses and minuses of each technology. There is no question that the public (and politicians) favour renewable energy for a range of reasons, notably of course that they do not (for the most part) emit CO_2 and (for the most part) are 'sustainable'.

Chapter 5 highlights one of the key principles underpinning our range of responses to increasing CO_2 emissions, namely that there is no single technology solution but rather a portfolio of energy technologies. It also questions the sustainability of some apparently 'sustainable' technologies, from the point of view of:

- the full life cycle of the solution: does it produce more or less energy than it uses; does it unsustainably impact on another resource, such as water or soils?
- the time scale within which 'sustainability' should be considered: for a hundred years, or a thousand years, or does it just need to be sustainable for the next few decades until something else turns up?

The practicality of the various technology solutions is obviously a critical issue and Chapter 5 summarises

- the level of maturity of various technologies: can they be deployed now, or will it be many decades?
- the degree to which they can meet our energy needs: can they provide power all the time, or just some of the time?
- the extent to which they impact on our natural or social environment.

In a thoughtful book by David McKay, the practical extent to which the various clean energy technologies can be deployed is considered in some detail. He reaches the conclusion that renewables cannot possibly meet all the energy needs of the United Kingdom. This is perhaps not surprising for a small country with a relatively large population. A study of Australian clean energy technology opportunities, by Peter Seligman, applying much the same approach used by McKay, reaches a rather different conclusion regarding the opportunities for deploying renewable energy technologies. This again is not surprising, given that Australia is a large, hot, dry, sparsely populated country. This analysis does not extend to consideration of how Australia would replace its export income of A$70–80 billion per annum if fossil fuels were no longer needed. Clearly this would require a profound restructuring of the Australian economy, and many other economies.

A considered approach must be taken regarding any claims for a particular technology, especially claims that a particular technology is 'The

Answer'. As will be seen, there is in fact no single answer but rather a portfolio of technologies to decrease CO_2 emissions. The clean energy portfolio will vary from country to country and from region to region. But there are two unavoidable facts common to all countries:

- Globally most electricity is currently generated from fossil fuels.
- We are faced with the 'inconvenient truth', confirmed by the International Energy Agency, that use of fossil fuels is increasing rather than decreasing and that it is likely to continue to increase for some years to come.

The only technology we have that can decrease emissions from ongoing use of fossil fuels is carbon, capture and storage (CCS). But it is not a well known technology: a study conducted by Massachusetts Institute of Technology (MIT) in 2009, indicated that only 17% of Americans, or one person in six, had heard of CCS. This is despite the fact that virtually 100% of the population of the United States (and most other developed countries) is addicted to the use of fossil fuels! Therefore part of the aim of this book is to make CCS an understood technology.

Chapters 6 to 9 very specifically address key aspects of CCS. Chapter 6 is concerned with the range of technologies that can be used to separate and capture CO_2 from major stationary sources, Chapter 7 discusses how we can transport large quantities of CO_2. Chapter 8 identifies the opportunities for geological storage of CO_2, and the focus of Chapter 9 is the safety and effectiveness of CCS.

The question of the cost of CCS compared to other clean energy technologies is addressed in Chapter 10. It is a difficult question to answer, both because it is highly dependent on the location and nature of the particular CCS project, and also because each clean energy technology seems to have a different way of calculating costs. The result is that all too often we are left comparing apples and pears!

Recent attempts to base costs of the various technologies on what is termed levelised costs of electricity (LCOE) have been helpful in providing a common base for comparison of costs. The conclusion reached through consideration of LCOE is that CCS will be part of the clean energy mix for as long as people choose to use fossil fuels to meet much or even just some of their energy requirements.

Given our currently high level of dependency on fossil fuels, we face a challenge in decreasing CO_2 emissions. Leaving CCS out of the clean energy portfolio makes this challenge even greater. Chapter 11 seeks to address this question, and here it is important to acknowledge that it is impossible to avoid becoming drawn into political and policy issues. However, in this context it is important to remember that this book does not take the view that CCS is 'the answer'. Rather its perspective is that CCS is part of the answer – along with renewables, energy efficiency and switching to lower emission fuels.

This book is firm in its view that CCS must not be left out of the portfolio, that we have no realistic solutions to climate change if we leave it out. This is a matter of urgency because CCS is not moving forward at the speed or at the scale required. The reasons are perhaps less technical and more political, financial and economic; perhaps there is a lack of policy drivers or the existing policy drivers are ineffective or costly, or all of the above.

In contrast to the largely impersonal approach in Chapters 1–11, there is some use of the personal pronoun in Chapter 11. This serves to emphasise the nature of the views expressed and ensure that no one else is blamed for them! Whether the views and proposals are accepted or not, hopefully they will be recognised as carefully considered.

They have been developed over the past 20 years from a scientific perspective, but often within, or on the margins of a policy perspective.

The IPCC prides itself on providing technical advice that is policy relevant not policy prescriptive and this description is certainly appropriate for the first 10 chapters of this book. In contrast, while Chapter 11 is most certainly not 'policy prescriptive', it does identify key policy and other issues and suggests ways in which they might be addressed.

Chapter 11 may raise some eyebrows among those who believe that scientists should not venture into consideration of economics and policy. But economists and policy makers frequently venture into inherently scientific and technical issues (such as climate change and technology options) with impunity. Hopefully the same level of indulgence will be extended to a scientist venturing outside the confines of a rigid scientific discipline! Indeed if we are to resolve the extraordinarily complex issues of climate change, and the role of clean energy technologies in them, it is essential that we take a multidisciplinary approach that transcends traditional discipline barriers. This book is part of that effort.

2 CO_2 AND CLIMATE CHANGE

Greenhouse gases

The introductory chapter to this book touched on a range of issues regarding greenhouse gas emissions. Paramount amongst these is the increasing level of atmospheric CO_2 and consequential climate change. But what are greenhouse gases (GHGs) and how do they influence global warming and climate?

Most people have heard of carbon dioxide and know it is a 'greenhouse gas'. But it is just one of many. A few common GHGs are water vapour (H_2O), carbon dioxide (CO_2), methane (CH_4) and nitrous oxide (N_2O). These occur naturally but the rate at which some of them are being added to the atmosphere by man's activities is of increasing concern. In addition, there are GHGs (also recognised in the Kyoto Protocol) which are solely the product of human activity, such as perfluorocarbons, hydrofluorocarbons and sulphur hexafluoride (SF_6). These are used as refrigerants, solvents and cleaning fluids and are very powerful greenhouse gases. But as they are present in only small quantities in the atmosphere, their total greenhouse effect is presently fairly modest, compared to the impact of the more abundant GHGs. Some fluorocarbons, or CFCs, are also the major contributor to the development of the ozone hole in the uppermost parts of the atmosphere.

Energy in the form of solar radiation is emitted from the sun. The majority passes through our atmosphere and heats up the earth's surface. It is then reflected back into the atmosphere mostly as infrared radiation (similar to that emitted by domestic infrared heaters). Much of that reflected solar radiation passes out into space, but GHGs 'soak up' infrared radiation and re-emit it (including back to Earth), thereby further warming the atmosphere (Figure 2.1).

It is the presence of natural GHGs which warm up the atmosphere sufficiently to make the earth habitable; without greenhouse gases, life as we know it would not be possible on earth. So what then is the problem with GHGs? The Earth's temperature is controlled by the difference between the total amount of solar radiation coming into the earth system and the total going out. If more is coming in than is leaving the earth system, then the earth warms. How much a particular factor influences this balance is known as its radiative forcing and each molecule of greenhouse gas contributes to that radiative forcing.The amount contributed depends on the composition of the molecule and the abundance of the molecule in the atmosphere. Each greenhouse gas has a different warming potential, as shown in Table 2.1. Using the warming potential of CO_2 as the baseline, other GHGs have much stronger warming potentials, but are generally present in very much smaller quantities than CO_2. Therefore despite its relatively modest warming potential CO_2 is, after water vapour, the single most important greenhouse gas.

The greenhouse effect

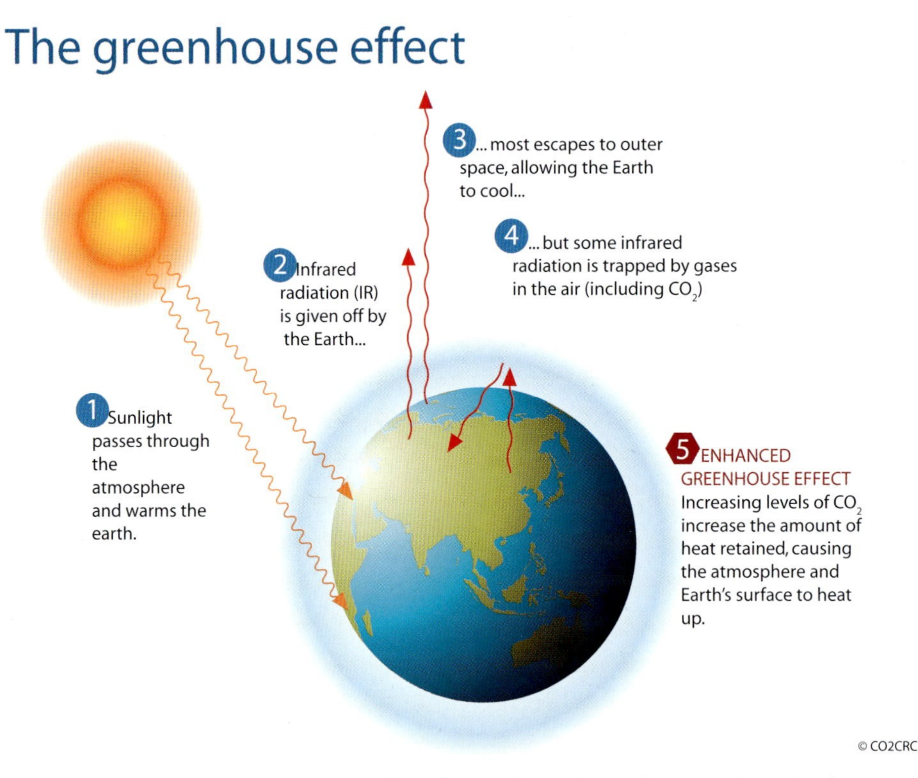

Figure 2.1 The greenhouse effect. Solar energy reflected back from the earth's surface (as infra red radiation) is trapped by greenhouse gases and warms the atmosphere, resulting in the greenhouse effect and global warming.

Molecule for molecule, sulphur hexafluoride (SF_6) is a much more powerful GHG than methane (CH_4), which in turn is a much more powerful GHG than CO_2 (Table 2.1). However, their atmospheric concentrations are in the reverse order and because of this, the total radiative forcing of CO_2 is far greater than that of CH_4, which in turn is far greater than that of SF_6 (Table 2.1).

Table 2.1: Global warming potential over 100 years on a per molecule basis*

Carbon dioxide (CO_2)	1
Methane CH_4 (ppb)	25
Nitrous oxide N_2O (ppb)	298
HFC-134a CF_3CH_2F (refrigerant)	1430
Carbon tetrachloride CCl_4	1400
CFC-12 CF_2Cl_2 (refrigerant)	10 900
Sulphur hexafluoride SF_6	22 800

Data source: IPCC 2007
*Units are Global Warming Potential (GWP), a unit relative to the amount of warming from the same mass of CO_2.

The total radiative impact of water vapour in the atmosphere is greater than that of any other GHG component and this is sometimes used as an argument for ignoring the increase in atmospheric CO_2. But there is no evidence to suggest that man is significantly changing the amount of water vapour in the atmosphere whereas over the past 200 years, carbon dioxide has increased by almost one third compared to natural levels. Future climate change will impact on water vapour and cloud distribution but the current atmospheric models do not allow those impacts to be accurately assessed.

Methane and nitrous oxide concentration in the atmosphere have also measurably increased but their overall concentration is still low (Figure 2.2).

Some components, such as aerosols and fine particles, reflect back some of the sun's radiation

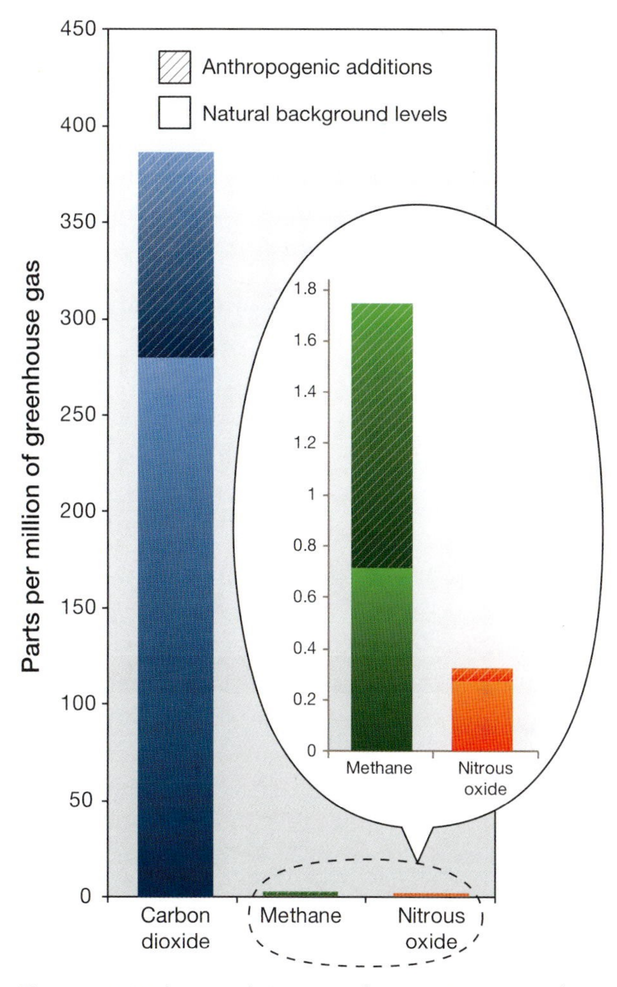

Figure 2.2 Methane and nitrous oxide concentrations in the atmosphere have measurably increased but their overall concentration is still low. (Data source: Blasing 2001)

and therefore tend to cool the earth. Some of these are emitted during volcanic eruptions; others are produced by biological processes and some are emitted during the combustion of coal. In other words and somewhat ironically, combustion of coal can potentially produce agents of both global warming (CO_2) and global cooling (particulate matter), but it appears the impact of the former far outweighs the impact of the latter.

It has been suggested that geoengineering of the upper atmosphere (by shooting aerosols into it) may be one solution to global warming. However, overall man's record of major engineering feats to deal with the exigencies of climate, weather, erosion, or desertification (the list goes on) is at best mixed and at this stage it would seem more sensible to treat the cause (by decreasing GHG emissions to the atmosphere) rather than just dealing with the symptoms (by adding something else to the atmosphere).

The nature of carbon dioxide

Carbon dioxide (CO_2), the single most important anthropogenic greenhouse gas in terms of total radiative forcing, is a molecule comprised of one carbon atom and two oxygen atoms (Figure 2.3). One of the ongoing greenhouse 'confusions' is that people sometimes refer to carbon (which has an atomic weight of 12) and at other times they talk about carbon dioxide (which has a molecular weight of 44). If emissions are expressed as tonnes of carbon, then the value has to be multiplied by 44/12 (3.67 times) to convert it to tonnes of carbon dioxide. The stability of the CO_2 molecule and the strength of the bonds within the molecule is the reason why we cannot just break down CO_2 into its 'benign' constituent parts of carbon and oxygen. To do this requires a great deal of energy, which in turn generally means that yet more CO_2 is made! Many schemes

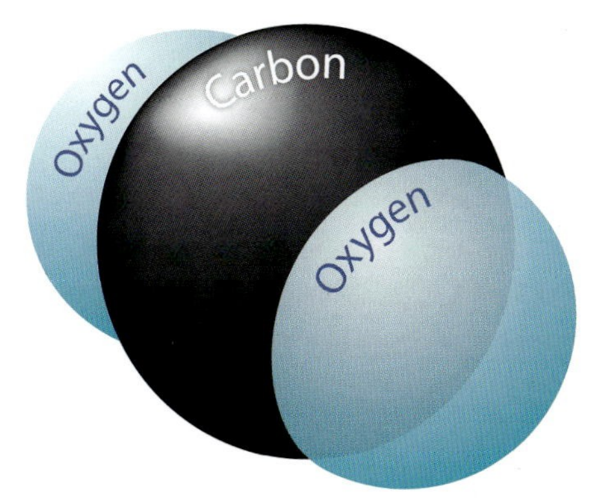

Figure 2.3 The carbon dioxide molecule, consisting of one carbon atom flanked by two oxygen atoms, has a very stable structure. Depending on the temperature and pressure conditions, it can be a gas, a liquid or a solid.

have been proposed for converting the CO_2 molecule to longer chain carbon molecules (such as plastics), but the reality is that the energy needed to break down the molecule makes this a difficult and energy intensive industrial process, though of course plants do it very successfully through photosynthesis (Figure 2.4).

At normal temperatures and pressures, CO_2 is a colourless and odourless gas. It has an atmospheric concentration of approximately 0.039% or 390 parts per million (ppm) and is an essential component of most forms of life on earth. The energy in sunlight helps to convert the water plus carbon dioxide (from the atmosphere) plus various nutrients in plants, into carbohydrates (which is stored energy) and oxygen (Figure 2.4). Within the plant cells, chloroplasts trap sunlight at the blue and red ends of the spectrum and reflect green light (hence the green colour of most plants). Plants also release CO_2 during respiration when they convert the carbohydrates into energy for growth. The plants, with their stored carbon, may also be consumed by animals and ultimately this carbon is returned to the earth as waste, or exhaled as CO_2 to the atmosphere.

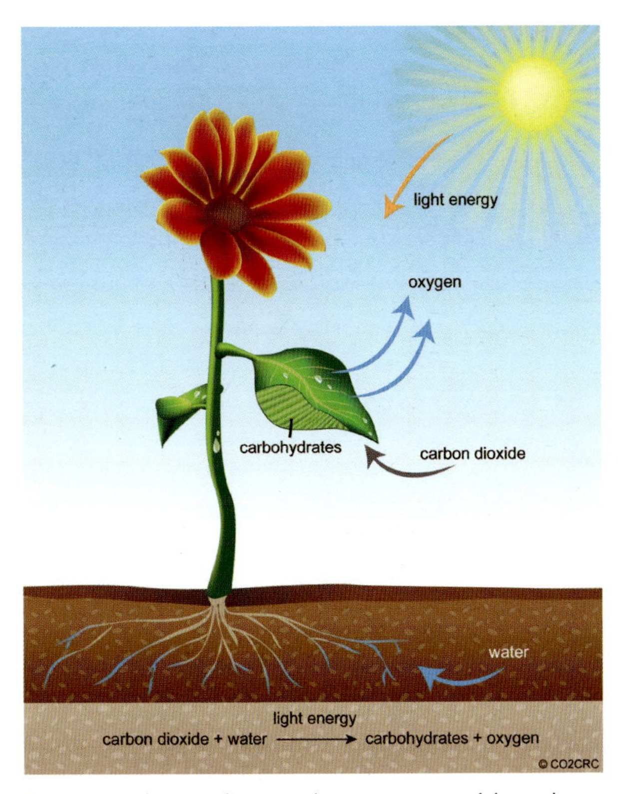

Figure 2.4 Photosynthesis is a key component of the carbon cycle. Plants convert atmospheric CO_2 into carbohydrates and oxygen and also release CO_2 during respiration.

Carbon (and CO_2) is an integral part of our lives. Descriptions of CO_2 as 'poisonous' or 'toxic' or a 'pollutant' are misleading, given that it is an essential part of virtually all forms of life. The carbonation (the bubbles) in soft drinks and champagne is CO_2. Our breath contains several per cent CO_2. In the office or classroom at the start of the day the CO_2 content is around 400 ppm (parts per million), but by the end of the day it can be in excess of 500 ppm (0.05% CO_2) which is about the same as some projections for likely atmospheric concentrations by the year 2050.

None of the projections for atmospheric CO_2 to 2050 are likely to impact directly on human health, though there may be indirect effects such as the spread of harmful insects due to rising temperatures or more frequent and more severe bushfires or heat waves. At high concentrations, CO_2 is an asphyxiant gas and can be hazardous. Prolonged exposure to CO_2 concentrations of 3–5% in air causes respiratory problems and headaches. At 8–15% CO_2 and above, nausea is followed by unconsciousness and death if the person is not moved to the open air and/or given oxygen.

CO_2 is chemically stable in the atmosphere, but in the presence of water it gradually dissolves to form an unstable and quite corrosive acid known as carbonic acid. Under normal atmospheric temperatures and pressures CO_2 is a gas. At high pressures, such as those encountered in natural gas fields, or in some industrial processes, the CO_2 can be present in a relatively dense form, with properties of a gas and a liquid. Conversely at low temperatures (minus 78°C), CO_2 becomes solid 'dry ice', a commodity used in the food industry. Dry ice will also form spontaneously if high pressure CO_2 is suddenly released, with rapid expansion of the gas resulting in a drop in temperature to below the point at which dry ice forms as fine particles. This, together with the resultant cooled air, produces a mist, sometimes used to dramatic effect in theatres and at rock concerts!

Carbon dioxide is an integral part of the global carbon cycle (Figure 2.5) and the sources and the

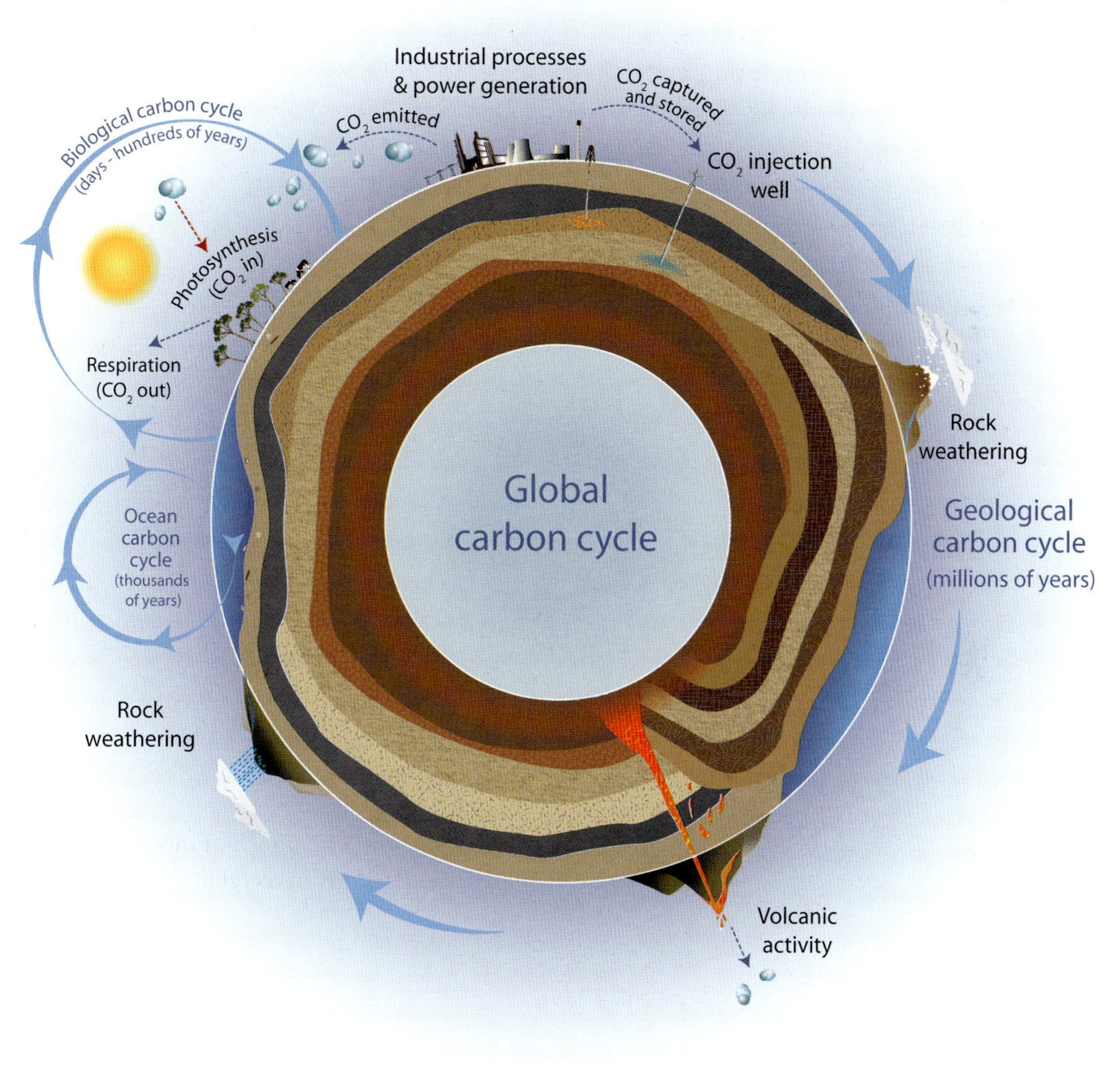

Figure 2.5 The global carbon cycle. This cycle involves a complex interplay of geological processes such as volcanic activity, plate movement and weathering; biological processes such as animal activity, photosynthesis and respiration; and human (anthropogenic) activity such as industrial processes, heat and power generation. The scale of human activity is now so enormous that it is perturbing some of these natural cycles.

concentrations of carbon and carbon dioxide have varied through time in response to both the geological carbon cycle (including such geological processes as rock weathering, plate tectonics, volcanic activity) and the oceanic and biological carbon cycle. The manner in which the concentrations have varied through time, and the drivers of these changes, can be seen by studying the isotopes of carbon in the geologic record.

There are three naturally occurring carbon isotopes:

- ^{12}**C** is the lightest and most common isotope and is preferentially used by plants during photosynthesis
- ^{13}**C** is a somewhat heavier isotope, is more common in deep geological sources of carbon such as volcanoes
- ^{14}**C** is the heaviest, rarest and least stable of the naturally occurring isotopes and is formed as a consequence of the impact of cosmic rays (and more recently fallout from nuclear testing) on the CO_2 molecule.

The ratio between ^{13}C and ^{12}C in the atmosphere, referred to as delta C13 ($\delta^{13}C$) (Figure 2.6), varies in response to increased or decreased periods of photosynthesis (as more or less ^{12}C is stored in plants), or in response to changes in the amount of carbon that is stored in geological carbon 'sinks' or (in modern times) by the release of CO_2 from fossil fuels (derived from ancient plants). The oxygen isotopes ^{18}O and ^{16}O also vary through time, in response to changes temperature rather than to the carbon cycle. Therefore, by studying the changes in the oxygen and carbon isotopes, we can potentially see how both temperatures and carbon sources have varied over time.

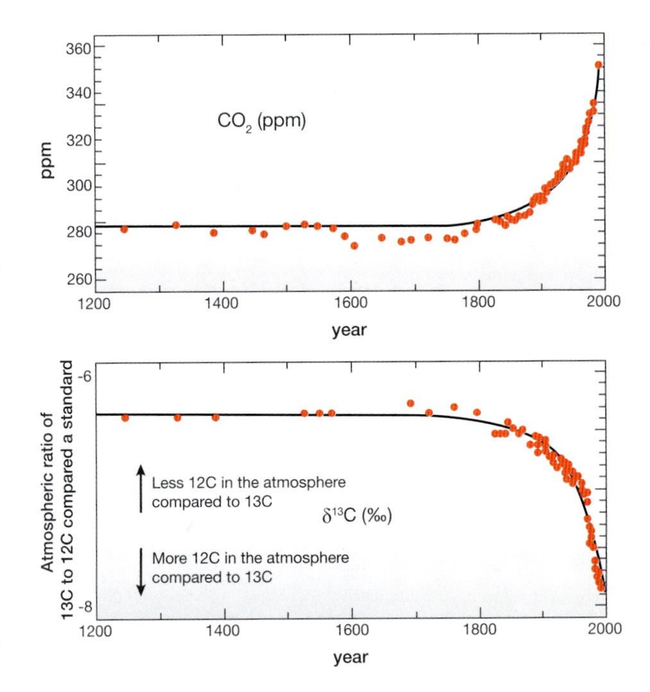

Figure 2.6 Compared to previous centuries, over the past 200 years there has been a marked increase in the atmospheric concentration of CO_2. This is reflected in a marked change in the isotopic carbon ratio which can be directly attributed to the burning of fossil fuels. (Adapted from Trudinger *et al.* 2005)

Carbon dioxide and earth's history

The earth is approximately 4.5 billion years old. Primitive forms of life have been around for at least 3 billion years and by 2 billion years or earlier there was probably a significant global biomass made up first of primitive unicellular organisms and then of multicellular organisms. More complex invertebrates occurred first during the Ediacaran period (about 600 million years ago) and then at the Precambrian-Cambrian boundary, about 530 million years ago, when a massive explosion of invertebrate life occurred, perhaps related to a major input of nutrients, particularly phosphate, at this time. From that time on, the biota and the carbon cycle became progressively more complex, with constant cycling of carbon between the geosphere, the hydrosphere, the biosphere and the atmosphere.

That carbon cycle has varied through time in terms of the amount of carbon that is held in one or other of the carbon stores (the carbon

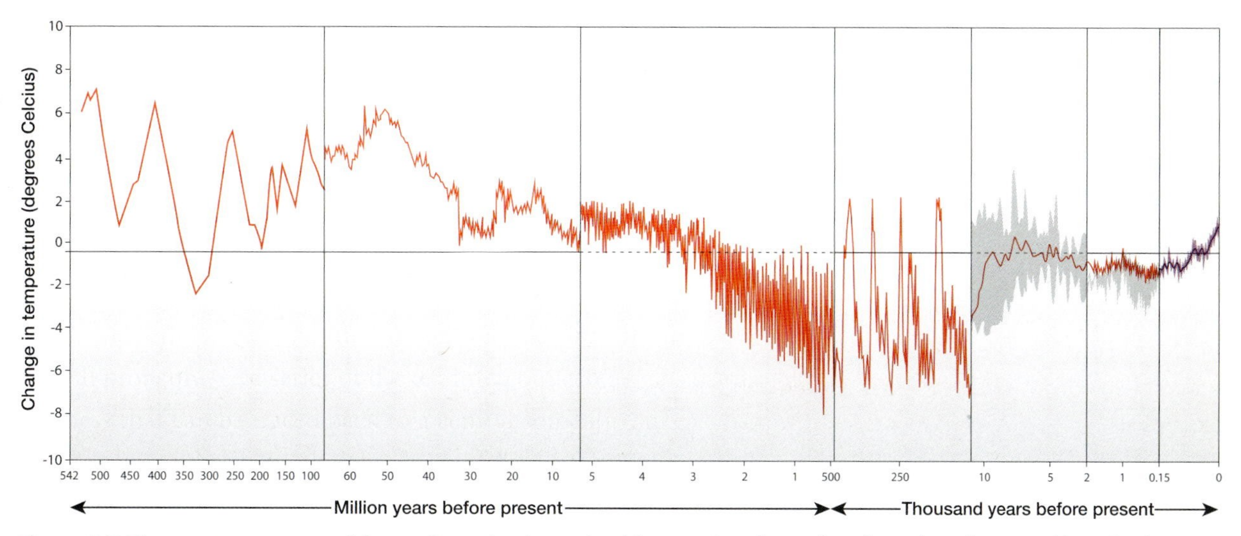

Figure 2.7 The past temperature of the earth can be determined from various lines of geological evidence and has clearly varied considerably over the past 500 million years or so (the Phanerozoic period). The further back in time, the less is the certainty of the time and magnitude of the temperature change. In the last 10 000 years, we are able to see a great many variations in temperature using a variety of 'proxies' such as ice core, coral banding and tree rings. (Data source: Royer *et al.* 2004, Zachos *et al.* 2001, Lisiecki *et al.* 2005, Petit *et al.* 1999, Jouzel *et al.* 2007. Image adapted from Rhodes)

sink) and the speed at which it is recycled between the sinks (the carbon flux). There have been times when massive quantities of carbon have been stored in the geosphere: for example during the Carboniferous (300–350 million years ago) when thick coals were deposited in deltas in many parts of the world, or during the Cretaceous (65–145 million years ago) when thick organic-rich marine shales were deposited in the world's oceans, perhaps as a result of decreased deep oceanic circulation coupled with high marine organic production.

One of the most profound changes occurred with the evolution of land plants about 400 million years ago and this can be seen in the geological record. The other profound and much more recent change resulted from the action of man starting perhaps tens to hundreds of thousands of years ago, first with fire, then with agriculture. William Ruddiman, a paleoclimatologist at the University of Virginia, has suggested that a rise in CO_2 and methane, which occurred about 8000 years ago is a reflection of the 'expansion of agriculture at that time.

Land clearance for agriculture and more recently for urban development as well as changes in land use – for example, massive increases in rice production – are responsible for some of the increase in atmospheric CO_2 concentration. Increases in livestock (especially ruminants) have also produced increases in GHGs, especially methane, although as pointed out earlier, the total greenhouse effect arising from increased methane is modest compared to that arising from CO_2.

Strong evidence that much of the increase in atmospheric CO_2 is the consequence of the burning of fossil fuels is provided by the carbon isotope record, which convincingly mirrors the rise in atmospheric CO_2 concentration. It is highly likely that global temperatures have increased as a result of man-made (anthropogenic) changes in the concentration of GHGs, but there are complexities in the correlation because of natural variations in climate acting on a variety of time scales, making it difficult to separate the natural changes from anthropogenic changes with absolute certainty.

There is also confusion in many people's minds regarding weather and climate and their variability over time.

Weather versus climate

What is weather and what is climate? The difference between the two is largely a reflection of the period of time and the size of the area under consideration. In its discussion on climate and weather, NASA explains the difference between the two as follows:

> *'In most places, weather can change from minute-to-minute, hour-to-hour, day-to-day and season-to-season. Climate however, is the average of weather over time and space. An easy way to remember the difference is that climate is what you expect, like a very hot summer and weather is what you get, like a hot day with pop-up thunderstorms . . . Weather is what conditions of the atmosphere are over a short period of time, and climate is how the atmosphere "behaves" over relatively long periods of time.'*

So, weather is short-term; climate is long-term.

There is also a spatial component, in that weather relates to a location or a region or a zone whereas it would be somewhat meaningless to try to apply the term globally, because the weather is so locally and regionally variable. Climate, on the other hand, can be applied at all spatial scales including, of course, climate change on a global scale. The phrase 'climate change' refers to the fact that, when averaged out over sufficient time, there are trends of warming or cooling, moistening or drying, at a global scale.

Large scale global climate trends are obvious in the geological record. The occurrence of glacially-derived sediments at particular times in earth's history offers compelling evidence of very widespread glaciations, perhaps extending some 650 million years ago to near-equatorial latitudes. Similarly there have been times when some types of limestones, indicators of warm tropical seas, were much more extensive than the present day, extending to near-polar latitudes. Some of these apparent changes are the result of plate tectonics – the plate drifted into a different climatic zone. Others cannot be fully explained on this basis and appear to be the result of the global climate as a whole being warmer or colder.

So, at time scales measured in hundreds of thousands to hundreds of millions of years, it is clear that the earth's climate has changed over time, although the further we go back, the coarser the time scale within which we recognise those changes and therefore the greater the difficulty in saying exactly when they took place and whether they were random or cyclical events. Changes in climate are often mirrored by rising and falling sea levels, with changes of metres to hundreds of metres in sea level, and over timescales ranging from thousands to millions of years. The evidence for this comes from the sedimentary record such as is provided by seismic surveys, deep drilling and geological mapping. Records of sea level change also come from exposed rock sequences and from the occurrence of stranded (and uplifted) coral reefs – such as those found in Papua New Guinea and Timor – or beach ridges, such as those found in southern Australia (Figure 2.9). More detailed information on climate change can be obtained from the sediments themselves, from the isotopic record found in sediment cores in lakes and oceans and from coral cores and ice cores as well as from tree rings. All of these provide 'proxy' indicators of temperature and/or sea level rise and fall.

Causes of pre-human climate change

What caused these past changes when clearly humans had no hand in them? In some cases, changes that occurred in the geological past may reflect external events such as variations in solar radiation or large meteoritic impacts. Others, as outlined earlier, may be the consequence of terrestrial events such as increased volcanic activity and large quantities of aerosols, water vapour and particulates being injected into the upper atmosphere, or by continental drift and the repositioning of plates.

Changes in climate are reflected in the global carbon cycle and in organic evolution. But organic evolution may itself have produced climate change at times. For example the first large scale appearance of photosynthetic organisms would have had an impact on the production of carbon dioxide, initially by marine plants and then by terrestrial plants. The processes which control the carbon cycle in the atmosphere, the ocean and the terrestrial biosphere are affected by temperature and, according to David Frank and others, are likely to provide positive feedback that leads to further global warming. In other words, warming of the earth's climate could cause increased biological activity and net release of CO_2 into the atmosphere that could in turn amplify global warming.

Changes in the earth's orbit almost certainly account for some of the climatic changes evident in the geological record (Figure 2.8). An understanding of the combined effect of the earth's various orbital movements on climate was first brought together in an elegant fashion by the Serbian mathematician Milutin Milankovitch. He related climate change to variations in the earth's axis of rotation (the precession of the equinoxes), which has a period of 23 000 years, the variation in the tilt of the

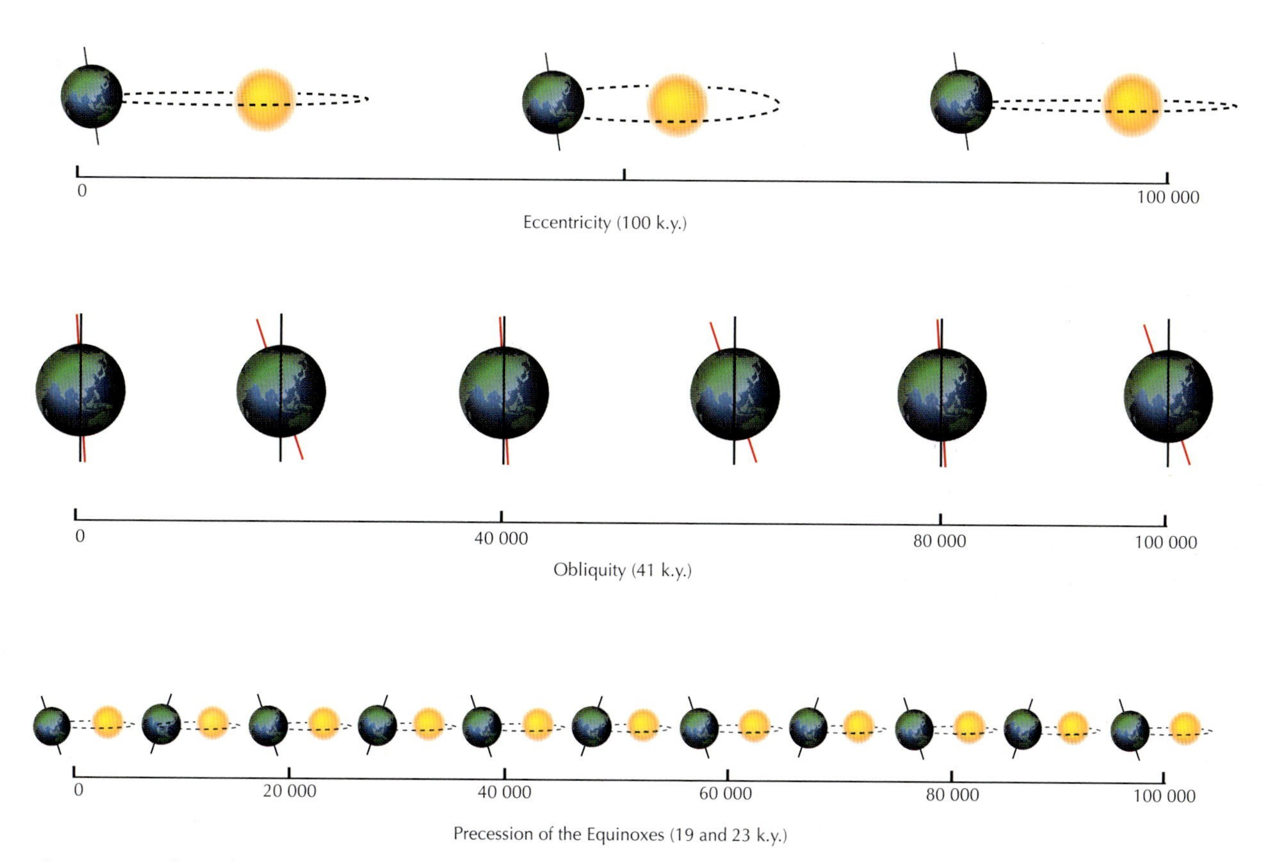

Figure 2.8 In the 19th century the mathematician Milutin Milankovitch suggested that orbital and axial movements of the earth produce cyclical changes in solar radiation and climate, with periodicities of 100 000, 41 000, 23 000 and 19 000 years.

earth's axis (the obliquity) which has a periodicity of around 41 000 years, and the shape of the earth's orbit (the eccentricity) which has a periodicity of approximately 100 000 years. The interaction between these various changes, and between these and others, produces cycles of climate and sea level change.

One of the first to apply this theory to the geological record was the Australian geologist Reg Sprigg who, in 1948, proposed that a well preserved series of beach ridges in south-eastern South Australia could be related to sea level change and the Milankovitch cycles (Figure 2.9). More recent work has tended to support the view of Sprigg that the record of sea level change is indeed related to orbital forcing and that there is cyclicity evident in the changes. Interestingly, the Milankovitch Cycles have also been used by Hays, Imbrie and Shackleton to suggest that without the anthropogenic greenhouse effect, 'the long term trend over the next 7000 years is towards extensive Northern Hemisphere glaciations'.

There are also more random terrestrial changes which have had a profound impact on past climates. Continental drift has been a major long term climate driver; as large land masses drifted into high or low latitudes. In some instances the climate changed as continental plates separated and new seaways formed or existing seaways closed, producing major disruptions in global ocean current patterns and climate. The collision of the Indian and Asian plates around 50 million years ago produced the Himalayas, profoundly changing the climate of the Asian region, initiating the monsoonal system and the central Asian deserts.

So there is abundant evidence of past (pre-human) climate change and it is reasonable to assume that non-anthropogenic climate change is underway now and will continue in the future on geological time scales. The geological evidence of a direct link between ancient climates and atmospheric CO_2 concentration is less clear. The over-riding influence on the long term climate cycles (over millions of years) may be related more to changes in solar radiation, orbital changes, plate tectonics and a combination of these, than to natural variations in carbon dioxide. Indeed, at times, changes in the atmospheric concentration of CO_2 might

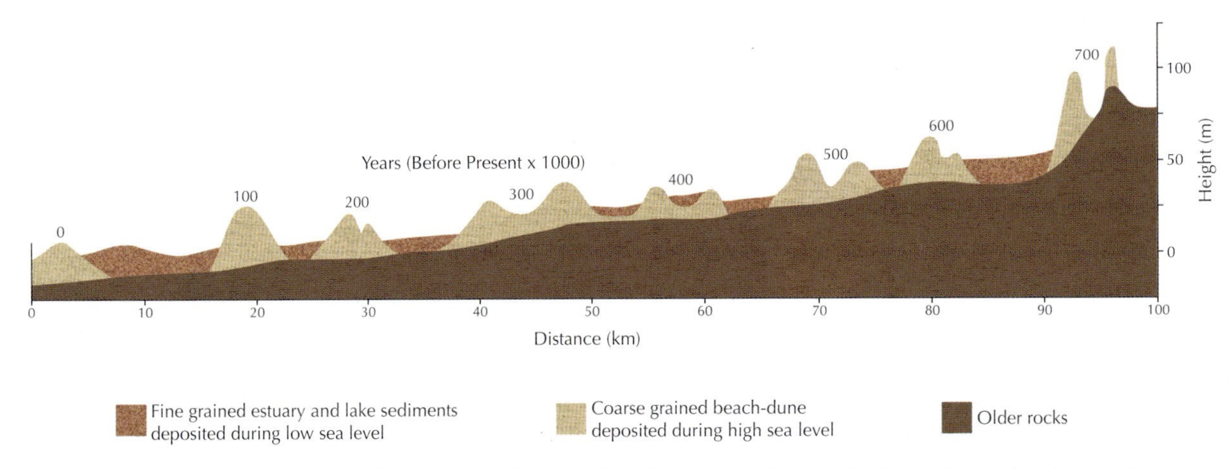

Figure 2.9 One of the regions where the Milankovitch curves have been 'ground truthed' is in south-east South Australia. There, a series of beach ridges extend across the coastal plane. The oldest beach ridge (approximately 700 000 years old) occurs 100 km inland and at a height of almost 100 metres above present day sea level. Seaward, the beach ridges are progressively younger and correspond to sea level change at 100 000-year intervals, which coincides with changes in the eccentricity of the earth's orbit. (Adapted from Idnurm and Cook 1980)

themselves have been the consequence of global warming or cooling and the attendant impact on the biosphere and the carbon cycle, rather than the CO_2 itself being the primary driver for such change.

But it is important to stress that past changes that are evident from the geological record do not invalidate the concept of recent anthropogenic climate change. They merely complicate it, in that changes over say the past 200 years have been imposed on the much longer term natural variation in the earth's climate over say the past 2 million years.

Distinguishing natural climate change from anthropogenic climate change

So how do we separate natural climate change from anthropogenic climate change? The rate of increase in atmospheric CO_2 concentration over the past 200 years appears to have been much faster and more marked than the rate of change evident in the longer term ice, tree ring or geological record. Similarly, although perhaps more controversially, the rise in temperature has also been more marked in the past 200 years than in previous millennia.

Tree rings (Figure 2.10) provide a record of change of temperature and of the carbon budget extending back several thousand years. Coral banding (Figure 2.11) can also provide an indication of coastal climate change extending back hundreds and perhaps thousands of years. Lake and nearshore sediments offer a record of climate change extending back for hundreds to thousands of years with varying levels of confidence.

One of the best long term records of temperature change comes from ice cores (Figure 2.12). Bubbles of air within the core provide us with samples of 'fossil air', and by determining the isotopic composition of the air and of the ice, we can establish the temperature and the CO_2 content of the atmosphere at the

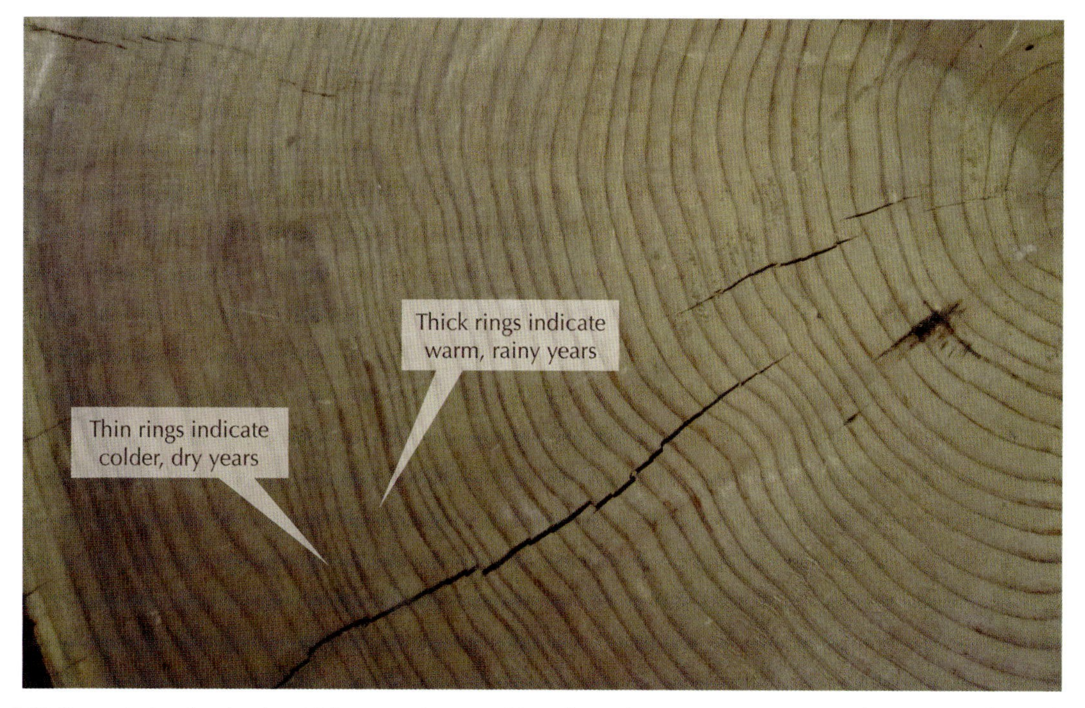

Figure 2.10 By analysing the density, thickness and composition of tree rings, we can see a very clear pattern of past climate change, which can be extended over thousands of years. (Image courtesy of T. Bartlett)

Figure 2.11 Coral banding can be a useful indicator of changes in coastal conditions, including increased coastal runoff during the wet period and decreased runoff during dryer hotter periods. (Image courtesy of E. Matson, Australian Institute of Marine Sciences)

time the ice was formed. The ice cores show a record of temperature and CO_2 going back 800 000 years, with work underway to extract core as old as 1 million years.

Is a CO_2 climate correlation reflected in the ice record? A strong general correlation is evident, with the change in temperature (obtained from the oxygen isotope record) following the same general trend as the atmospheric CO_2 record (Figure 2.13). Fine-scale correlation is not always evident and on occasions the change in temperature apparently preceded the CO_2 change. But it has to be remembered that a surrogate measure of temperature is used, which can lead to some uncertainties. However, the uncertainties appear to diminish, the more we look into the historical record over the past 200 years.

Figure 2.12 Antarctic ice core contains abundant 'fossil' air bubbles, which can be analysed for their CO_2 methane and isotopic composition to determine past atmospheric conditions. (Image courtesy of Tas Van Ommen)

Up to around 1800 AD, the carbon produced from decomposition and respiration was approximately in balance with that taken up by plants. However, the advent of the industrial revolution and the massive increase in the use of fossil fuels, particularly coal, had a major impact on that balance. Human activity, whether agriculture, resource exploitation, urbanisation, deforestation – the list goes on – has had a global impact on the biosphere and on the physical nature of the earth's surface. So much so that Crutzen has proposed the term '*Anthropocene*' to mark this unique period in the earth's history, which is taken to have commenced at the start of the Industrial Revolution some 200 years ago.

The prospect of global warming resulting from the rising atmospheric concentration of CO_2 associated with the Industrial Revolution was first suggested in 1896 by the Swedish chemist Arrhenius, although it is only in the past 30 years or so that the concept has come into prominence. So, what is the evidence of CO_2 increase and related climate change over the past 200 years? Again, the record offered by isotopic and related studies of ice cores, lake sediments and tree rings, provides a record of temperature over the past 200 years at a number of sites. By compiling such information, a global picture of climate change emerges, underpinned by the historical record.

Reliable instrumental measurement of temperature and other weather conditions has been available since about 1850. It was also at about this time that national meteorological offices started to be established, providing for the first time, a global record of air

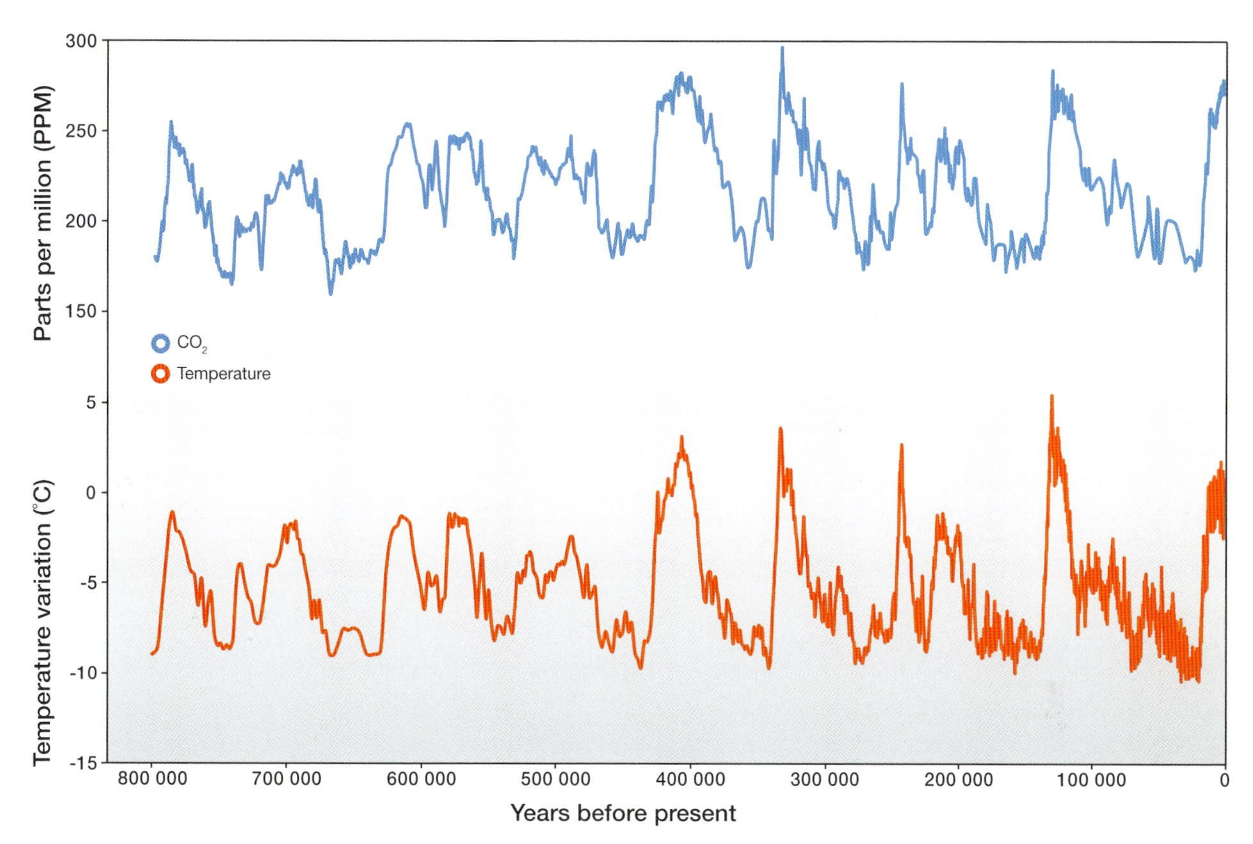

Figure 2.13 Analysis of Antarctic ice core has provided an extraordinarily detailed picture of variation in the CO_2 content of the atmosphere and of the average temperature (determined from the oxygen isotopes) under which the ice formed. There is clearly a strong correlation between the temperature curve and the CO_2 content over the past 800 000 years. (Data source: US Global Change Research Program)

temperatures, sea surface temperatures, temperature profiles for the atmosphere and oceans and, most recently, satellite data. These records indicate an overall global warming trend, although the extent of global warming over the past 200 years and the degree to which any rise in global temperature is directly related to the increase in CO_2 concentration, is for some people still a matter of uncertainty.

The consensus of the scientific community is that there is an overall warming of planet Earth, and the reports of the Intergovernmental Panel on Climate Change (IPCC) provide useful summaries of the evidence. There are some discrepancies between the temperature records of different parts of the earth's surface and between different parts of the atmosphere. This may be a reflection of imperfect data or computer models rather than evidence contrary to global warming, but nonetheless it is important to continue to collect data to resolve these apparent discrepancies. 'Heat islands' associated with the megacities that have grown up in the past 50 years, also complicate the record. Nevertheless the overall global trend appears to be one of warming.

The work of Mann and his co-workers, first published in 1998 and refined in subsequent publications, shows a marked increase in global temperature over the last 200 years, particularly over the last century (Figure 2.14). This reconstruction of global temperatures, based largely on tree ring studies, has been challenged by a number of people, most notably McIntyre and McKitrick,

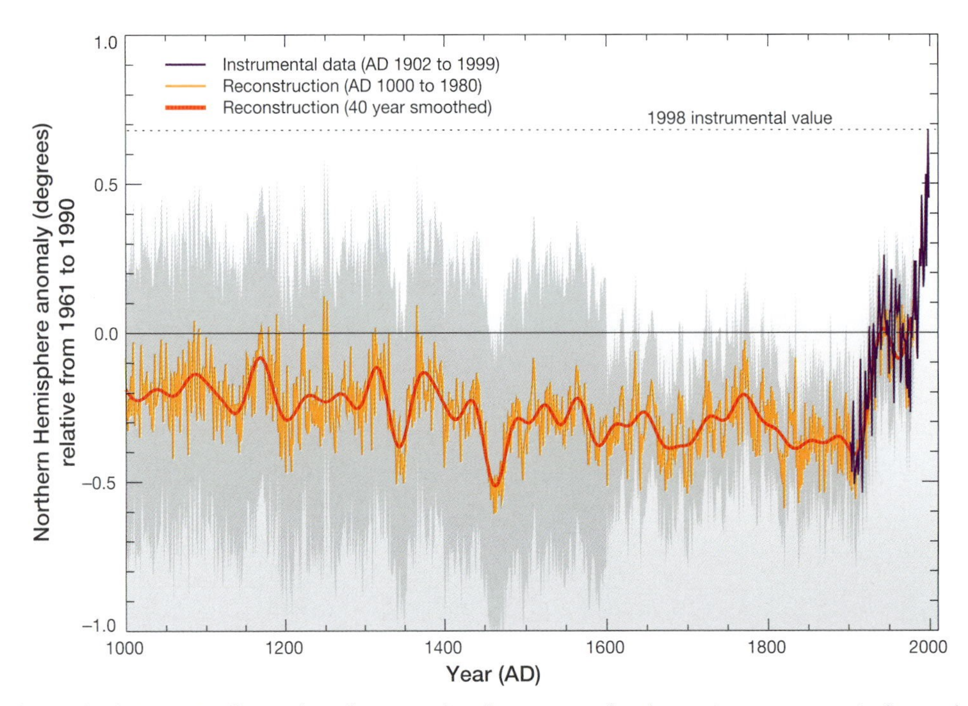

Figure 2.14 The work of Mann, Bradley and Hughes is used to demonstrate the change in temperature in the northern hemisphere, particularly over the past 100 years or so. The temperatures have been reconstructed using a combination of proxy measures such as tree ring, ice core and isotopic data, with more recent values obtained from the use of satellite and other instrumental data. The reconstructed temperature values have been attacked by critics, but have been able to withstand close scrutiny and the overall warming trend has been generally accepted. (Data source: Mann *et al.* 2009)

on the basis that the data are so 'noisy' (i.e. so variable), that there is no discernable upward trend in temperature and/or that some of the tree ring data should not have been included. It would be no exaggeration to say that literally hundreds of climate scientists, mathematicians, statisticians, modellers, and scientistis from a range of other disciplines, have analysed, evaluated and assessed the hockey stick reconstruction. While there are certainly some who contest the validity of the reconstruction (and the science is all the better for being questioned), the overwhelming majority conclude that the hockey stick reconstruction is valid and that the rise in global temperature evident over the past 200 years is indeed anomalous compared to any other period not only in the past 1000–2000 years, but also, many would argue, in the past 100 000 years or more.

If then it is accepted that there has been a rise in temperature of the order of 0.6 degrees Celsius over the past century, which is beyond the normal range of variability, what will happen over the coming century if we take no action? Again, there are many models and projections employed to determine this and the IPCC Assessment Reports are a ready source of these. There is controversy over the trajectory of the projections, but the consensus appears to be that there will be two degrees Celsius (or more) of warming over this century, if we continue to emit CO_2 to the atmosphere at current rates.

Sea level change as evidence for global warming

Evidence of global warming is also suggested by the retreat of many of the world's glaciers (Figure 2.15), the shrinking of the Greenland

Figure 2.15 The Franz Josef Glacier in New Zealand, like many glaciers around the world, has been retreating, but with some short-term advances in the first half of the 20th century.

icecap and the loss of Arctic Ocean sea ice (Figure 2.16). For the present, the evidence for any significant decrease in the Antarctic Icecap or a major global rise in sea level is not compelling. However, there appears to be a rise in sea level of around 3 millimetres resulting from the melting of glaciers and thermal expansion of the ocean over recent decades. Separating absolute sea level change from relative sea level change can be difficult, as the continents and the continental shelves themselves can be uplifted as a result of the movement of the continental plates producing a relative rise or fall in sea level. They can also undergo 'glacial rebound': i.e. they bounce up, when the weight of an ice sheet is removed as a result of melting (Fig 2.17). Scandinavia shows abundant evidence of this sort of rebound.

The earth's surface can also subside as a result of compaction of sediments, for example, in areas of urban growth, or as a consequence of extraction of groundwater. Major coastal cities such as Jakarta, Bangkok and Shanghai all provide graphic illustrations of land subsidence. In fact many areas of coastal subsidence represent a more immediate challenge to governments than sea level rise related to global warming (Figure 2.18).

This is not to say we should ignore the potential consequences of global warming–related sea level rise, particularly to some of the small island nations or to some of the world's major deltas. Where an area is subsiding due to relative sea level rise, and this is coupled with a global rise in sea level, the area will need

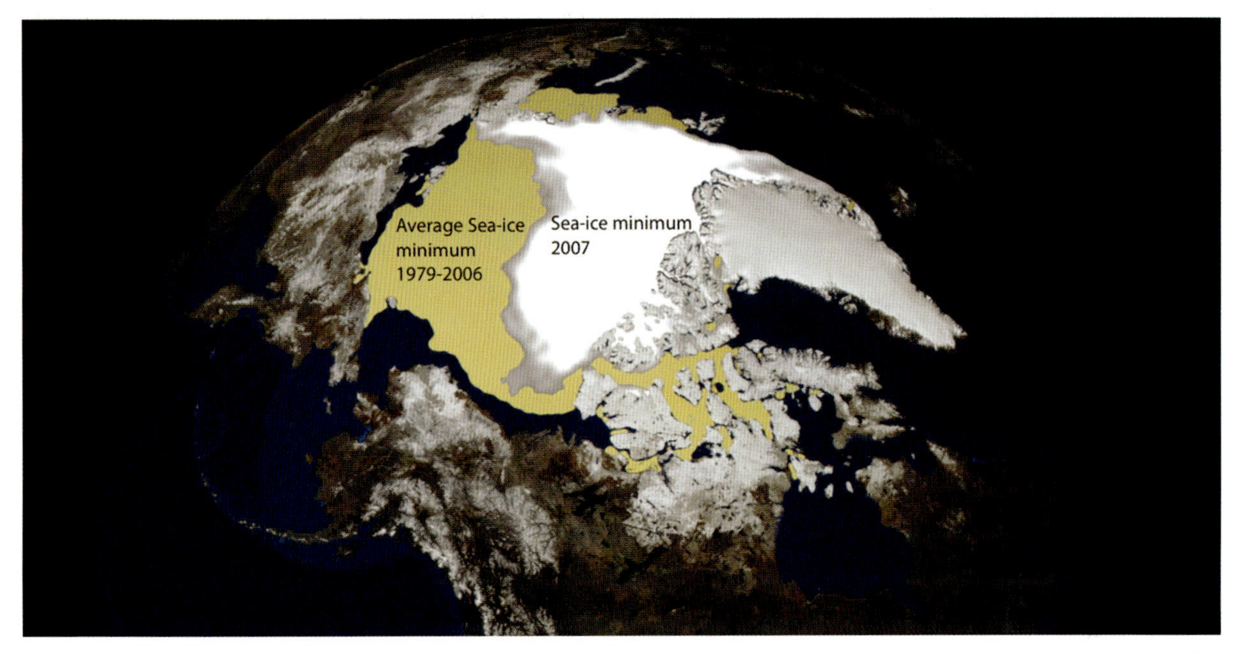

Figure 2.16 The rapid loss of Arctic sea ice is used as evidence of global warming. However, it is important to bear in mind that the extent of the sea ice varies greatly from year to year and is a reflection of a number of factors, only one of which is temperature. (Images from NASA/Goddard Space Flight Center Scientific Visualization Studio)

Figure 2.17 When large ice sheets advance over land, the weight of the ice pushes the land down. Upon warming and retreat of the ice, the weight is removed and the land rebounds producing a relative fall in sea level. Scandinavia is still rising due to this glacial rebound effect. (Adapted from Flint 1971)

improved coastal defences and vulnerable populations may need to be moved to higher ground. In contrast, in an area where the rate of uplift is greater than the projected global rise in sea level, there is less cause for local concern, although no grounds for global complacency! So, the impact of any global sea level rise on coastal communities and infrastructure will vary from negligible to profound, depending on the stability (uplift or subsidence) and the topography of the coastal zone.

Global warming and extreme weather events

Some climate models suggest that global warming will produce an increase in the intensity of tropical storms and many hold up Hurricane Katrina, which devastated New Orleans, as an example of this. However, for the present, the evidence for this correlation is uncertain – some violent storms may have more to do with 'weather' than 'climate change': a particularly hot summer cannot be taken as clear

Figure 2.18 Shanghai is one of the cities most under threat from sea level rise because it is also undergoing subsidence due to urban development and withdrawal of groundwater.

evidence of global warming. Nonetheless it is reasonable to assume that if the oceans warm, there will be more energy available, leading to more intense and more violent storms, though there is some uncertainty about whether the frequency will also increase.

Consideration of possible biological/ecological change and the spread of tropical diseases as a result of global warming is outside the scope of this book, but these may also be issues that could have a profound impact on some countries in the future. Desertification is also an unwelcome and likely consequence of climate change.

Act now or later?

It would be foolish to defer action until we are absolutely certain that global warming and climate change are happening as a result of rising CO_2 concentrations. Over the past 200 years, human activities have released a total of approximately 500 billion tonnes of CO_2, a massive contribution to atmospheric CO_2 compared to any other time in the past million years.

This represents a major perturbation of the natural carbon cycle and present day carbon fluxes and sinks (Figure 2.19). Sedimentary rocks are by far the greatest carbon sink (a carbon sink is something that absorbs more carbon than it releases), and are the source of our fossil fuels. Sedimentary rocks store as much as 100 000 000 billion tonnes of carbon – an inconceivable amount. But that geological carbon remains locked up – unless we choose to burn it.

The oceans store around 1000 billion tonnes of carbon, the atmosphere around 750 billion and terrestrial vegetation about 600 billion. Human activity is today adding in excess of 30 billion tonnes of CO_2 each year to the atmosphere. If we are to stay below dangerous levels of CO_2, no more than an additional 400-500 billion tonnes can be emitted to the atmosphere by 2050 and we are currently on track to add far more than this. While not all emitted CO_2 remains in the atmosphere – some is taken up by the ocean – the majority does remain an active greenhouse gas in the atmosphere for hundreds of years. In other words, there is no 'quick fix' – we are locked into whatever changes result from the

Figure 2.19 There are a number of pathways (fluxes) by which CO_2 moves from a source (such as vegetation) to a sink (such as the ocean). The natural flows of carbon between the various sources and sinks is now being significantly changed by human activity, notably deforestation, land use change and fossil fuel use. Some of the largest carbon sinks are fossil fuel (coal, oil, gas) deposits, which accumulated over hundreds of millions of years and which are now being burned, thereby circulating ancient carbon back into the atmosphere as CO_2.

excess anthropogenic greenhouse gases already present in the atmosphere.

Just because there is no 'quick fix' does not mean we should not take action, even if the nature, magnitude, timing and cost effectiveness of that action is still a matter of debate. There can be no dispute that the CO_2 content of the atmosphere is increasing, as revealed by:

- the tree ring and ice core record
- the record of atmospheric CO_2 concentration measured at the Mauna Loa Observatory in Hawaii (even more convincing than the tree ring and ice core record)
- the southern hemisphere record obtained at the Cape Grim Observatory in Tasmania.

These last two show a steady climb in CO_2 content of about 2 ppm per annum (Figure 2.20). The annual wiggle in the graph, most marked in the Northern Hemisphere record, is a reflection of seasonal plant growth and variations in photosynthesis. The difference between the patterns for the northern and southern hemisphere is also a consequence of the Southern Ocean taking up large quantities of CO_2.

The correlation between increasing CO_2 concentration and the carbon isotope ratio provides convincing evidence that fossil fuels are the source of most of the increased carbon dioxide, although as pointed out earlier, land use changes are also a contributor. There is a related

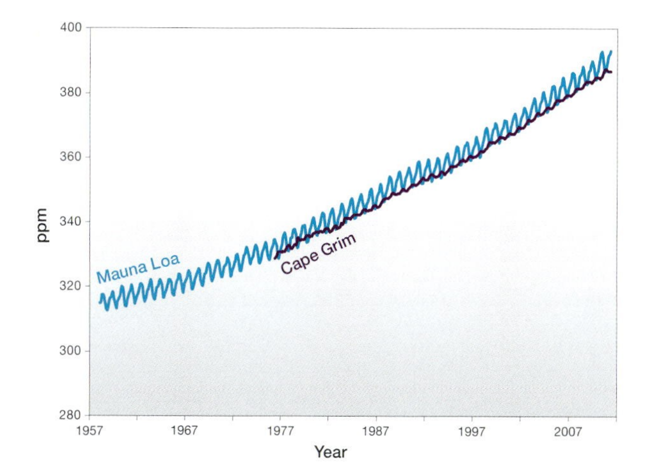

Figure 2.20 The atmospheric record at both the Mauna Loa Observatory, Hawaii, and the CSIRO Cape Grim Observatory, Tasmania, provide clear evidence of increasing concentrations of CO_2 over time. Not surprisingly, the two graphs show a high degree of correlation but there are also some interesting differences. For example the northern hemisphere curve shows a significant annual variation in CO_2 content compared to the southern hemisphere. This is due to the abundance of land plants in the northern hemisphere and the relative paucity of land plants in the southern hemisphere. In addition, the southern ocean removes about 2 Gtonnes of CO_2 from the atmosphere every year, which is responsible in part for the northern hemisphere having a higher CO_2 concentration than the southern hemisphere. (Data source: NOAA/CSIRO)

consequence of increasing atmospheric CO_2 that cannot be ignored. As mentioned earlier, not all the CO_2 emitted from human activity remains in the atmosphere: one third is taken up by the world's oceans. Whilst this is a good thing from the perspective of global warming, it does lead to another adverse effect – acidification of the oceans. Acidification is cause for concern as it affects calcifying organisms (organisms that build their external skeletal material out of calcium) such as corals and shellfish. Ocean acidification also, and perhaps even more significantly, is likely to affect the tiny calcifying micro organisms or plankton (such as foraminifera) that make up a very significant proportion of the total marine biomass. While some argue that there may be modest potential benefits from increased atmospheric CO_2, such as increased terrestrial plant growth, there are no obvious benefits from ocean acidification – only adverse impacts on the marine biota.

Conclusions

Perhaps not all the forecasts of extreme climatic events will come to pass and perhaps not all the dire social and ecological consequences that are predicted, will happen. But even the most hardened sceptics would surely accept that just as we take out home insurance, so it is appropriate to take out global climate insurance by decreasing CO_2 emissions, rather than delaying action until all the uncertainties are addressed. Put another way, if we don't know all the consequences of increasing CO_2 with certainty – i.e. we don't know precisely what we are doing – then logic suggests we should stop doing it! This is often referred to as the Precautionary Principle.

How soon should action be taken to decrease CO_2 emissions? The easy answer is as soon as possible, but at the same time we cannot cease to use of fossil fuels overnight, for this would stop much of the world's food production, make all of the world's major cities uninhabitable and inaccessible and send the human race back to a darker, hungrier future than anyone would wish. Therefore the actions must be appropriate and feasible and, as discussed later in this book, it is necessary to have a portfolio of actions to decrease emissions of CO_2 arising from the use of fossil fuels.

3 WHERE AND WHY ARE WE PRODUCING SO MUCH CO_2?

The production and use of energy and its impacts on CO_2 emissions: an overview

Carbon dioxide is produced through a range of natural and anthropogenic processes and activities. Paramount among these is the production and use of energy. Energy is used in homes, businesses and farms for heating and cooling, cooking and washing, appliances and lighting. It is used for motorised transport (cars, buses, motorcycles, aeroplanes, trains, trams and ships) and for industrial processes, including manufacturing the goods we buy and harvesting and processing the food we eat. Energy allows us to transform our environment by building houses, businesses and infrastructure. We use it to pump water to our homes, farms and cities, to run our hospitals and schools, grow our food and keep our work places illuminated, heated and cooled. Access to energy is essential to life as we know it. But providing that energy produces massive quantities of CO_2 (Figure 3.1).

The widespread availability of low cost energy has played an enormously important role in improving the quality of life around the globe, increasing our mobility, comfort, health and life expectancy, as well as offering material benefits. At the same time, the rapid increase in population, particularly in developing countries, has exacerbated many of the drivers of climate change. This will only increase as ever more people aspire to the benefits of developed countries. The demand for energy, especially for electricity, but also for transport and energy-intensive industrial products such as iron, steel, cement and fertiliser, has led to increased emissions of CO_2 to the atmosphere, particularly over the past 30 years (Figure 3.2).

The International Energy Agency (IEA) estimates that in 1971 total global emissions

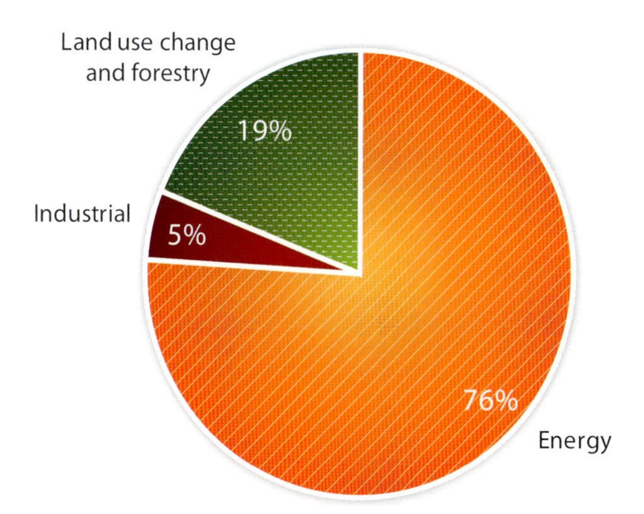

Figure 3.1 Global carbon dioxide emissions directly attributable to human activity are dominated by energy production and use. Emissions from industrial activity shown here exclude energy use and refer only to activities that result in direct CO_2 emissions, such as cement manufacture or iron and steel production. Land use change is obviously also a significant source of CO_2 emissions. (Data source: OECD/IEA 2010)

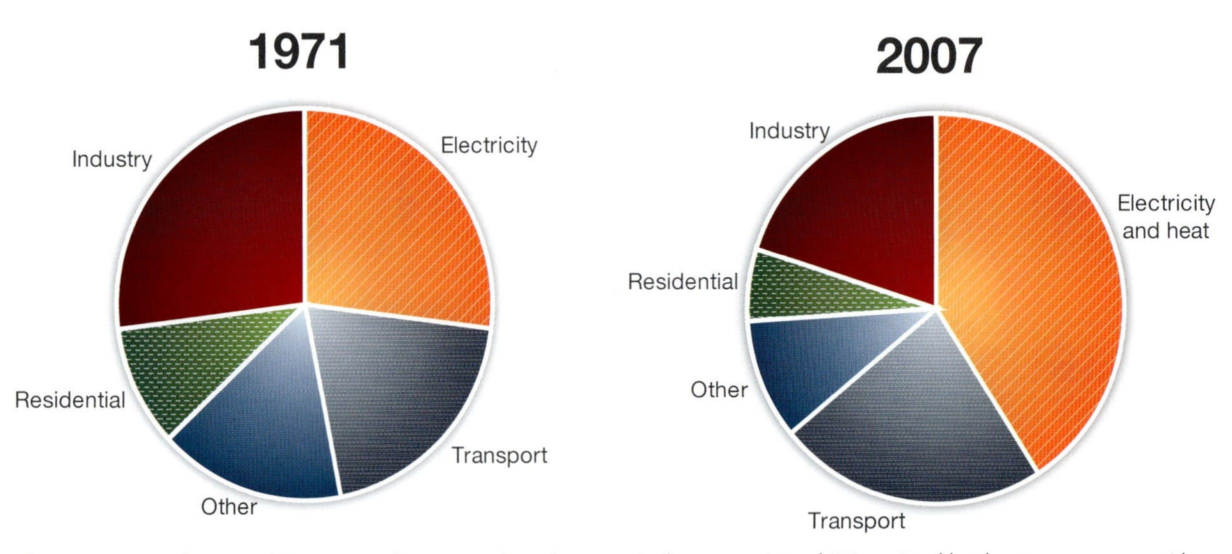

Figure 3.2 Over the period 1971–2007 there was a large increase in the proportion of CO_2 emitted by the energy sector, with transport increasing somewhat and some sectors showing a proportionate decrease. In fact, all sectors increased in absolute terms, with total global CO_2 emissions more than doubling over this period. (Data source: IEA 2009)

from energy were 14.1 G tonnes CO_2; 36 years later, those emissions had more than doubled, to 30 G tonnes CO_2. If we look at the way we use energy, it is evident that over this period, while every sector has increased its CO_2 emissions in absolute terms, proportionally the biggest increases have been to meet growing demand for electricity and heat, and to a lesser extent, transport. The share of emissions from the industrial and residential sectors has declined significantly, though emissions have increased in absolute terms. While some of these changes may result from the manner in which the emissions are attributed to sectors, there can be no doubting the massive increase in emissions from the electricity sector.

Energy is not the only source of CO_2; land clearing is also a significant source. As described in Chapter 2, plants use photosynthesis to capture the energy of the sun and in the process store the carbon in CO_2. Land clearing impacts on the carbon cycle in two ways: first, as plants are cleared and burned, the vegetation releases its stored carbon to the atmosphere as CO_2; second there may be a reduced amount of vegetation to take up the CO_2 already in the atmosphere. According to Rasheed, parts of Africa and central and South America have seen overall deforestation as high as 50% of forested land since 1950 and even today annual rates of deforestation in wooded areas are around 1% per annum.

Farming activities following clearing, such as soil cultivation carried out on newly deforested areas, can release more of the carbon that was stored in the soils. But it is not only land clearing for agriculture or urban development that impact on atmospheric CO_2 and the carbon cycle. More than a billion people worldwide rely on wood as their main source of energy for cooking and heating and this can cause deforestation. This is evident particularly in Africa, where rapidly expanding populations have made the practise of using wood for domestic purposes unsustainable (Figure 3.3). The amount of CO_2 emitted from land use, land use change and forestry (LULUCF) can vary with the climatic conditions and there are no universally agreed methods for measuring and verifying the resulting emissions,

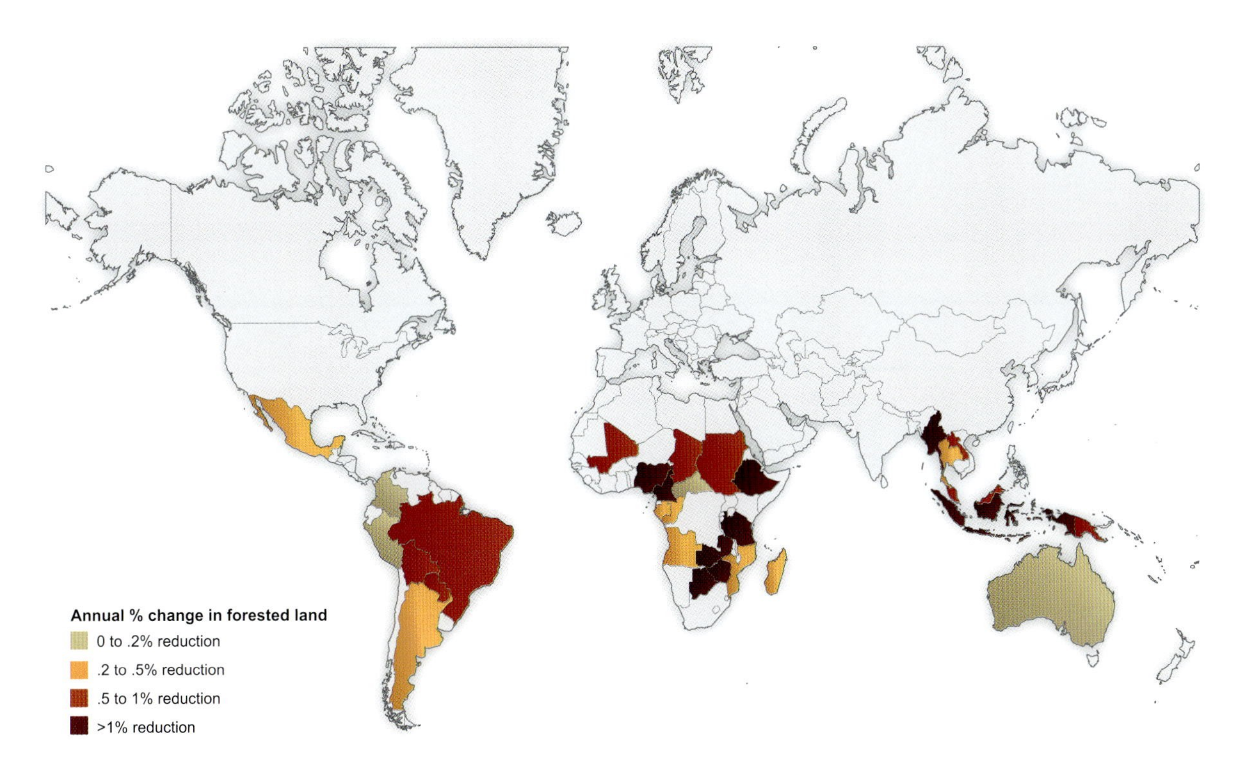

Figure 3.3 Land use change (land clearance, deforestation, changed farming practices), often result in increased emissions of carbon dioxide. Reduction in forested land is a particular feature of South America, equatorial Africa and South-East Asia, with Australia also showing significant land clearance. Whilst some of this has now been brought to a close, the consequences obviously continue in terms of decreased uptake of atmospheric CO_2. (Data source: FAO 2011)

but they are significant contributors of excess CO_2 to the atmosphere.

Along with energy and land change, there are also some industries where CO_2 is emitted as part of the manufacturing or refining process. Oil, gas and coal extraction produces greenhouse gases (methane and CO_2) as a by-product. Iron and steel production uses metallurgical coal to reduce the iron oxide into iron; cement production generates CO_2 from chemical reactions when reducing limestone.

The global picture of CO_2 emission is complicated by the fact that each country takes a different approach to documenting its sources. Some regions such as East Asia, Europe and North America have many stationary emission sources, while other regions, notably Africa and most of South America, have low levels of electricity use and few major emissions sources. The United States and China are dominant as CO_2 producers, but India, Japan and Russia are also significant emitters. There is then a large 'tail' of other countries each producing 300–600 million tonnes of CO_2 (Figure 3.4).

If the per capita production of CO_2 is considered, then the picture becomes rather different (Figure 3.5). For example, China and India have quite low per capita emissions compared with developed countries. Conversely, countries such as Australia, Canada, the Netherlands, Saudi Arabia and Taiwan, modest emitters in terms of absolute amounts of CO_2, are major emitters on a per capita basis. Among the developed countries, the variation in per capita emissions highlights the heavy reliance of some countries on fossil fuels for stationary energy production. There are also

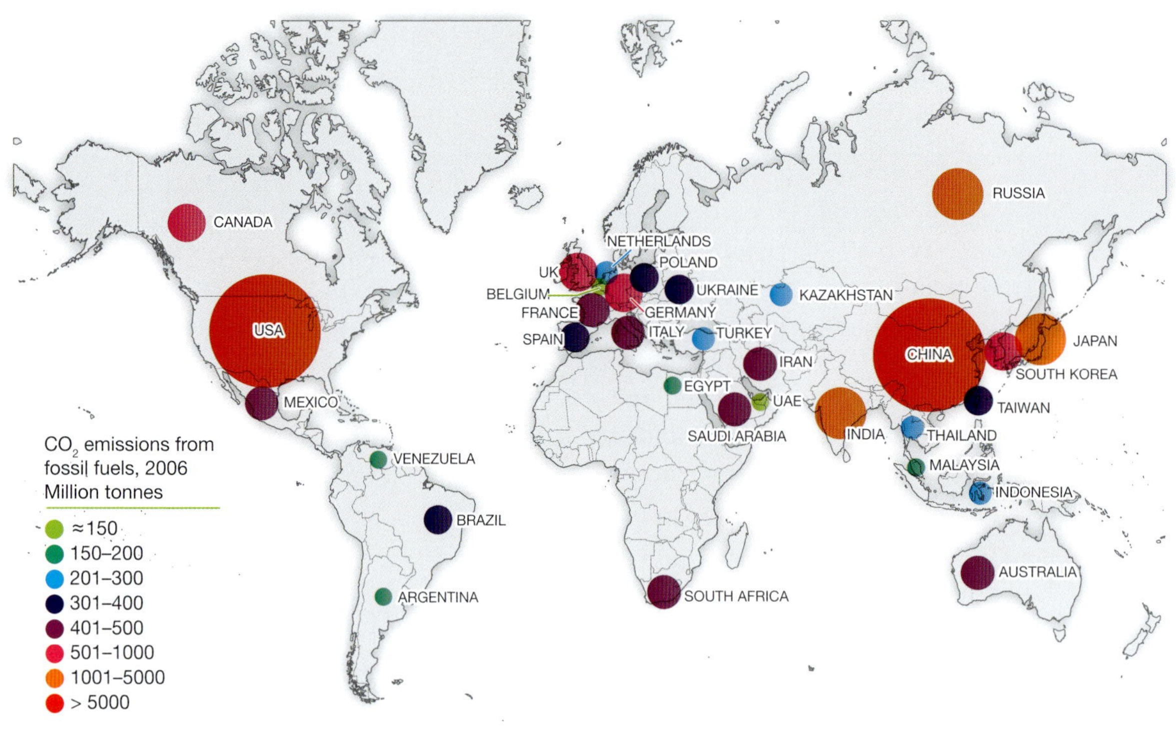

Figure 3.4 The geographic distribution of emissions (much of it from stationary sources) is shown for all major emitting countries. China and the United States are clearly the dominant emitters, re-enforcing the point that if there is to be any global greenhouse agreement, it will not be effective unless it includes both these countries. (Data source: Boden *et al.* 2011)

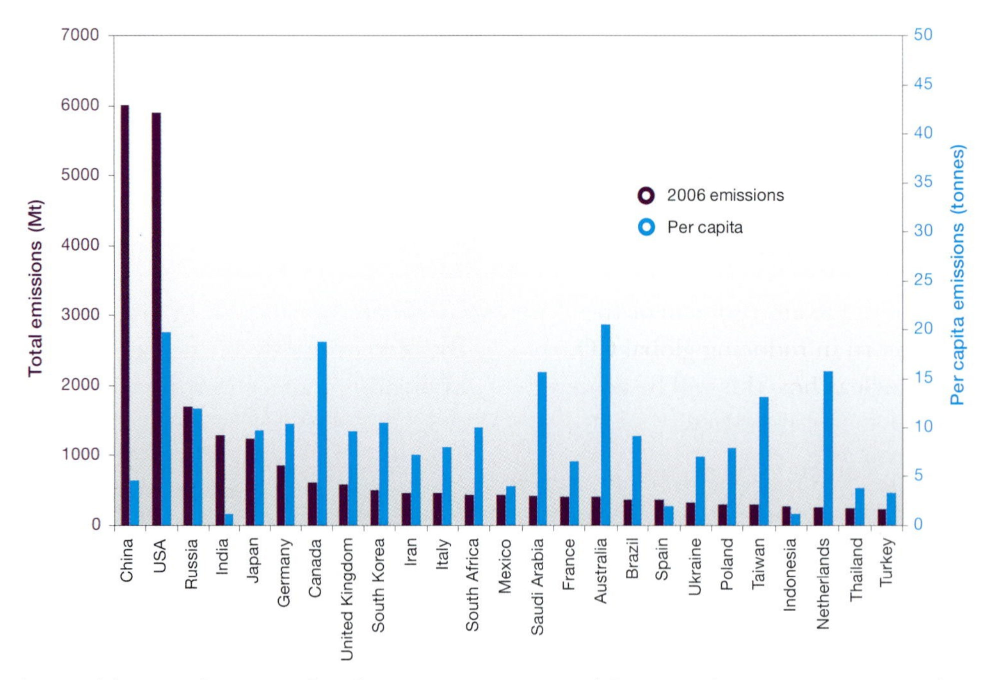

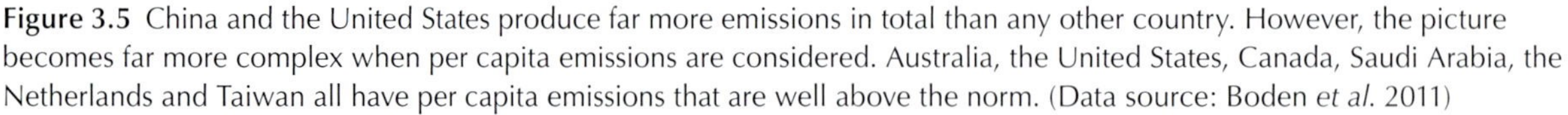

Figure 3.5 China and the United States produce far more emissions in total than any other country. However, the picture becomes far more complex when per capita emissions are considered. Australia, the United States, Canada, Saudi Arabia, the Netherlands and Taiwan all have per capita emissions that are well above the norm. (Data source: Boden *et al.* 2011)

many other factors influencing per capita emissions such as population density and urbanisation, availability of hydroelectricity or nuclear power, the energy intensity of the industrial base, and climate.

So there is great variability in the carbon footprint from country to country and, it could be argued, great inequity in the extent to which the atmosphere, a commons, is used for the 'disposal' of CO_2. But exactly who should reduce their emissions, and by how much, is a complex question. The extreme variability in per capita emissions (Figure 3.5) has a general correlation to level of development, with developed countries having several times the rate of per capita emissions of most developing countries. Developing countries, not unreasonably, seek to have the benefits of access to the reliable, cheap and abundant energy (and access to the atmosphere for emissions) that has been available to the developed world for the past 200 years. Conversely, developed countries are reluctant to lose the economic benefits derived from their energy intensive industries to countries not constrained by international agreements on reducing emissions.

Recognising the finite nature of the capacity of the 'commons', the atmosphere, can we develop an equitable approach to its use? Clearly this has to be a goal, but in the absence of an international agreement and the overall ineffectiveness of the Kyoto Protocol or the Copenhagen Accord in reducing global CO_2 emissions, it is unclear how this will be achieved at this time. Some of the general mitigation issues are discussed in the final chapter of this book.

Combustion has been a primary source of energy since the time of the discovery of fire by early humans who began to use the energy of combustion, burning wood for warmth, cooking and other energy needs. By the third century BC, coal was being used for heating in China. Britain first used coal for domestic heating in the 13th century AD and by the 16th century it had displaced wood as the main domestic energy source in major towns such as London. With the development of the steam engine, humans started to use fossil fuels, predominantly coal, to provide the energy to power first pumps and other machines, and then trains. By the mid 1800s, coal was being converted to gas to provide street lighting. The discovery of oil and natural gas ushered in a new era in the use of fossil fuels for lighting and to power engines particularly for transport. Throughout the 20th century, the use of coal to generate electricity became increasingly important.

Today, transport and electricity production are the sectors that use the most fossil fuel, accounting for around 80% of the world's energy use. They are therefore the dominant sectors in terms of CO_2 emissions. It follows that if we are concerned about CO_2 emissions, we have to be particularly concerned about reducing emissions from electricity production and transport.

The use of fossil fuels

Why are we such enthusiastic users of fossil fuels, particularly in the developed world, but increasingly in the developing world? There are three reasons: cost, convenience and availability. Fossil fuels are extraordinarily cheap for the amount of energy they contain. Transportation fuels vary considerably in price from country to country, primarily as a consequence of the level of taxation that a government chooses to levy. But the underlying price of a barrel of oil is much the same in all countries because it is a globally traded commodity. Even at $100 or more a barrel, it is still cheap for the amount of energy that it provides.

But the real cost of oil is higher if we account for what are usually called 'externalities'. These are indirect costs that attach to all energy options including renewable energy to some degree, but

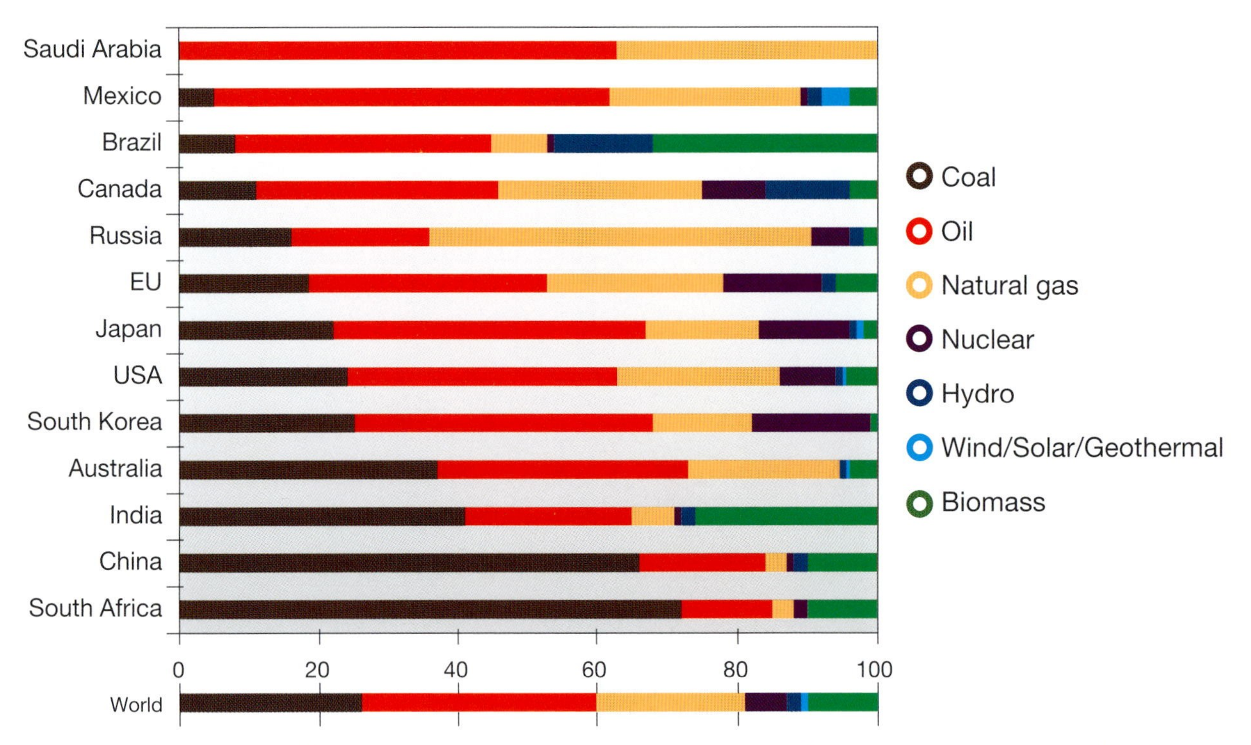

Figure 3.6 The energy mix (where countries get their energy from) varies enormously from country to country and largely reflects the abundance or absence of in-country energy resources. At one end of the spectrum, Saudi Arabia gets its energy from oil and to a lesser extent gas. South Africa (closely followed by China) uses coal as its dominant energy source. Russia has gas as its single most important source of energy and Brazil has a large component of biomass in its energy mix. (Data source: OECD/IEA 2011)

particularly fossil fuels. Examples of externalities include not only CO_2 emissions and the potential consequences and costs of climate change but also a wide range of other negative impacts such as:

- injuries or fatalities associated with mining or drilling
- the health impact of particulates
- the overuse of vehicular transport leading to less physical exercise and greater levels of obesity.

But there is also a positive side to the use of fossil fuels in terms of living standards – a life that is not only richer socially and intellectually but also safer. The lights stay on and people are able to travel; electricity or gas is readily available for cooking; and so on. Increased use of electricity correlates with improved health and increased life expectancy. However, this is not an absolute correlation. For example, a number of European countries have low per capita CO_2 emissions but very high living standards.

Fossil fuels are abundant, convenient, and widely available. The engines they power are easy to use and reliable and they offer the prospect of almost instant heating and cooling, power and movement. Our infrastructure and our way of life is built around fossil fuels; it is not easy to change that dependency because it is convenient to continue to do things the way that we have been doing them for the past 100 to 200 years.

Coal in particular is an extremely widespread resource with massive deposits found in many parts of the world. The size of the global coal resource is enormous compared to oil. At our current rate of coal use, the known resources will

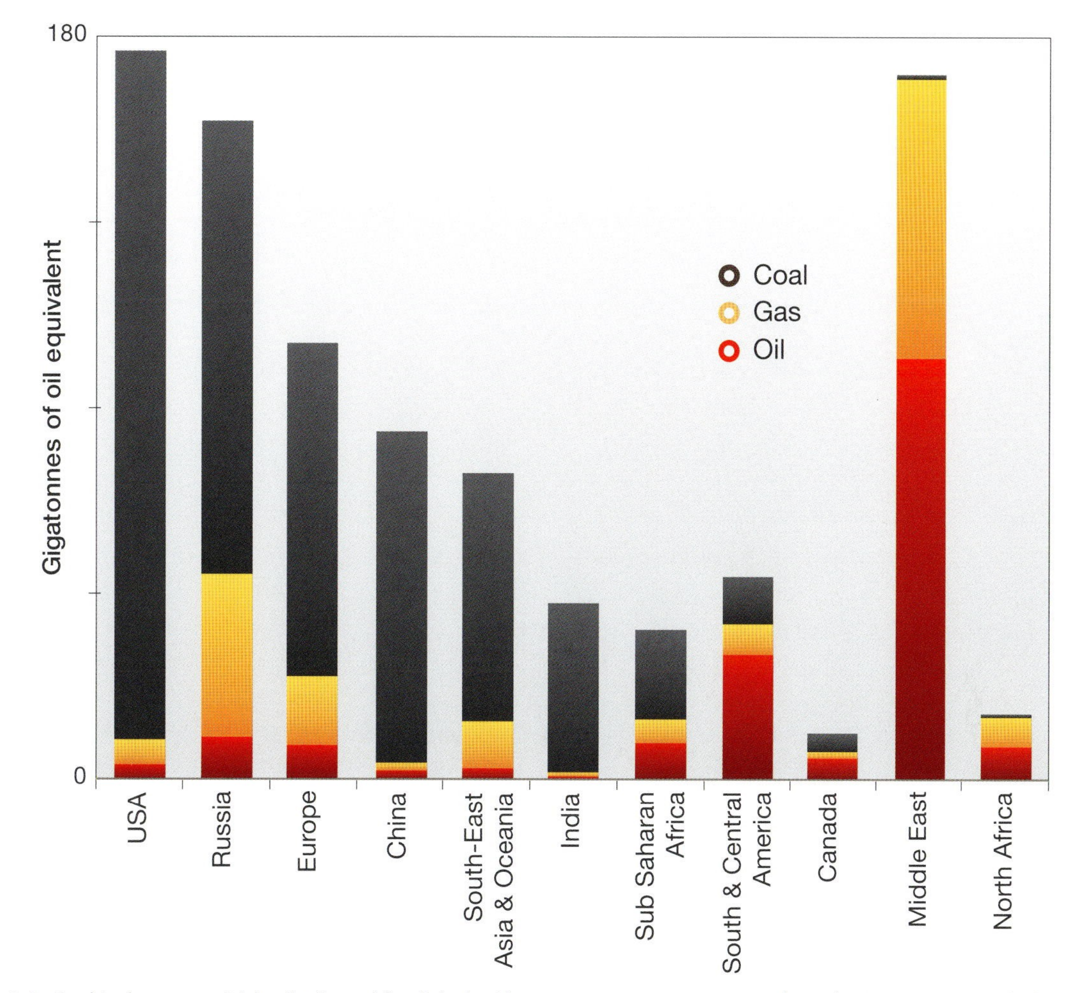

Figure 3.7 Coal is the most widely distributed fossil fuel with massive resources in a number of countries. Gas is fairly widely distributed but total energy reserves are significantly smaller than for coal. Oil is the most geographically restricted fossil energy resource, with the Middle East dominating the resource scene. This distribution has a major impact on energy use, energy trade and energy security. (Data source: BP 2011)

last for centuries, but, if we choose to use that coal in the manner it is currently used, then we will have to live with the potentially profound consequences of its emissions. Oil is much less abundant than coal and major oil resources are found in only a few countries, which means that many countries face security of supply issues (Figure 3.7).

The finite nature of oil, has led to the concept of 'peak oil', first developed by M. King Hubbert some 60 years ago. Hubbert pointed out that as the production of oil continues to increase, the rate of production will become greater than the rate at which new oil fields are discovered. In other words, the rate of production will no longer be sustainable and will decrease until finally there is no oil left to produce.

This very simple concept is at the heart of the idea of resource sustainability, a concept that first received attention in the work of Malthus more than 200 years ago. Malthus believed that world starvation due to increasing population growth and limited world food production was a looming disaster. He did not of course foresee the use of greatly improved agricultural practices, better varieties of plants or improved fertilisers.

Similarly, improved technology gives access to previously inaccessible or unknown resources or provides a cheaper way of producing oil from a deposit that was previously too expensive to develop. As the price of oil increases, the amount of oil considered economically recoverable (the reserves) increases. Any discussion of when peak oil (or peak gas or peak uranium) will occur has to be considered not just from the perspective of how finite the total resource and the economic reserve appears to be, but also bearing in mind how much we might be prepared to pay for it in the future. In the past decade or so, the price of oil has varied from as little as US$20 a barrel to as much as US$150 a barrel and the oil reserves that can be economically exploited have in theory decreased or increased to some extent with that price. In fact, oil companies do not base their investment decisions to develop new oil or gas fields on the highest commodity price, but on the probable realistic (usually assessed quite conservatively) price in 10 or 20 years time, based on historical trends and sensible market predictions.

But some commodities such as oil, show significant price elasticity. In other words, people are prepared to pay a high price for the convenience of oil which in turn means that oil production will continue to increase. Obviously the increase cannot go on forever but it does explain why concepts such as 'peak oil' can appear to be almost a mirage; it always seems to be a few years ahead.

Provided there are alternative energy sources, increasing oil prices will, ultimately, mean that we will cease to use 'conventional oil' (i.e. oil that we pump out of the ground) and then, in the longer term, even 'unconventional oil' (i.e. oil extracted from oil sands or oil shales). In the case of unconventional oil, we may be unable to handle the increased amount of CO_2 emissions resulting from production of unconventional oil. Alternatively, it may become too expensive because of a price on carbon.

Natural gas, like oil, tends to occur in specific gas provinces. While the resource is finite, we are finding more and more natural gas reserves and resources which can be conveniently and cheaply transported by pipeline to the user. The IEA has suggested that global gas reserves are sufficient to last more than 200 years based on current rates of production. Some of the gas fields are far from markets and pipelines are not a viable option. To overcome this, the gas can be turned into liquefied natural gas (LNG) which can be transported to markets by ship. The process of making LNG results in significant quantities of CO_2 being generated because of the energy requirements of the LNG process. But even taking this into account, the amount of CO_2 produced per unit of energy is still less for LNG than for the same energy derived from coal.

In a carbon-constrained world there will be movement from coal to lower carbon intensity gas, but the extent of this will vary from place to place, depending on the availability and cost of gas or LNG compared to coal. An example of the impact of technology on resources is provided by 'shale gas' (Figure 3.8). Until recently, shales (fine grained rocks) were regarded as unprospective for gas. However, development of a new understanding of these rocks, coupled with innovative drilling technologies to allow the drilling of horizontal wells and new techniques for fracturing ('fracking') the rocks to provide permeability, has suddenly made it possible to produce massive quantities of natural gas from these rocks. As a result of recent exploitation of shales, such as the Barnett Shale in the eastern United States, over the past 5–10 years or so, the United States has been transformed from a gas-deficient nation to one with abundant gas reserves.

However, a study by Cornell researcher Robert Howarth has suggested that the fracking process results in the emission of large quantities of methane, decreasing the benefits of using gas rather than coal. More research is needed to

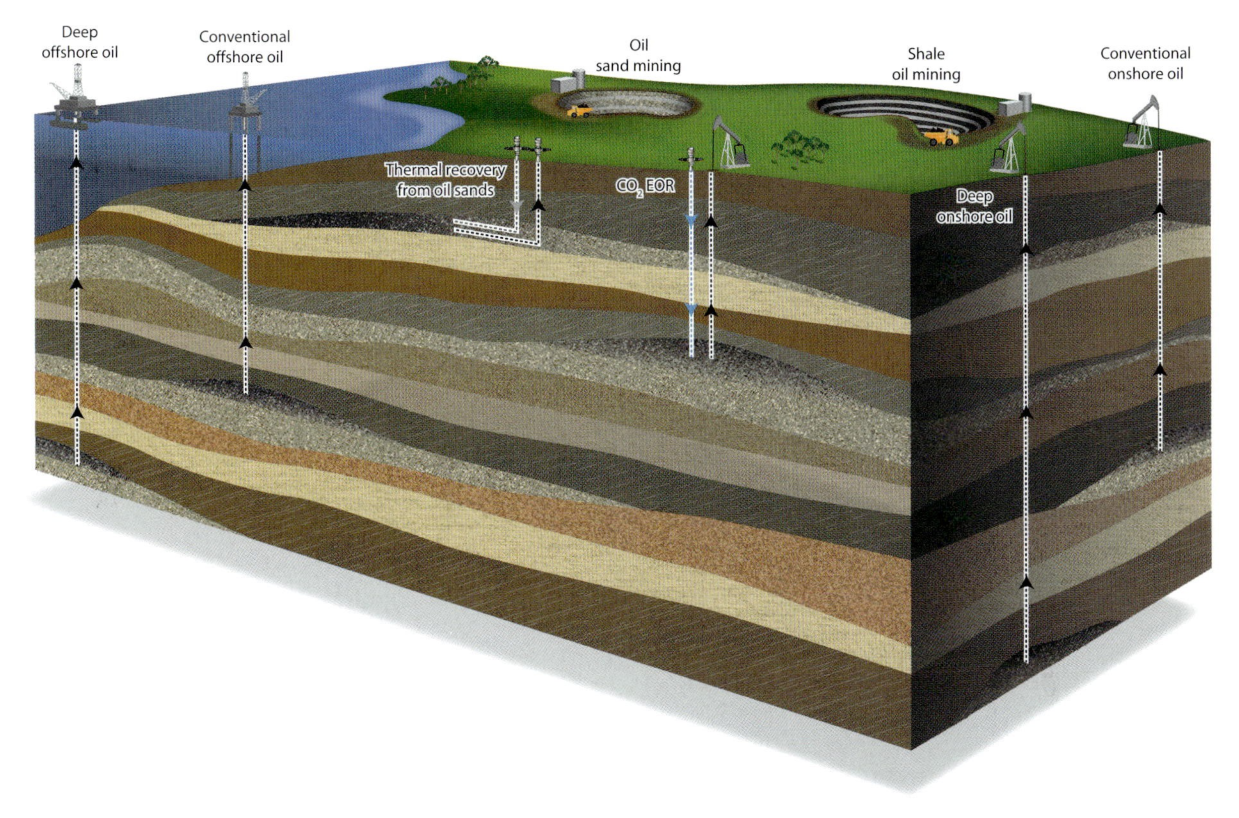

Figure 3.8 Schematic cross section showing different underground sites for various types of oil reserves. Technology development has a major impact on resource exploitation by, for example, bringing down the cost of oil extraction and by making it possible to access oil that was previously too deep to reach. Unconventional oil can be extracted from oil sands and from organic-rich shales. Enhanced oil recovery using CO_2 can also extract additional oil. All of these together have increased identified world oil reserves and resources very significantly in recent decades.

resolve this issue. This technology impacts greatly on the energy picture in the United States and also on the opportunities for decreasing CO_2 emissions by substituting gas for coal. Similarly, eastern Australia, an area until recently regarded as gas deficient, is being transformed into a major gas (LNG) exporter through the development of massive coal seam gas deposits that can be extracted with new production techniques.

Therefore, while we are depleting oil and gas resources at a rapid rate, there is no prospect of them running out in the medium term (at least the next 50 years), though they will become progressively more expensive to exploit and process. They will also become more expensive to use if a price is applied to the CO_2 that is emitted as a result of that use. Coal is abundant and very widespread and will continue to be mined and used in many countries and exported to many others for many decades. Based on the projections of the IEA, for example, it would be prudent for us to plan for ongoing use of fossil fuels for many years to come, in transport, electricity production and industry, but at the same time work to ensure that we use fossil fuels in smarter and cleaner ways during the transition to clean energy technologies such as renewables.

Two key sectors: electricity production and transport

Globally, CO_2 emissions from transport have increased by more than 2% per year since 1970 according to the IEA and they now account for 25% of all energy-related emissions. While there

Figure 3.9 As liquid fuels increase in price there will be a move away from conventional transport fuels to electric cars, such as this electric car at a recharging point in Grosvenor Square, London.

have been significant improvements in the energy efficiency of various forms of transport, these gains have been outstripped by the increase in the total number of cars and trucks. Although personal transportation is the largest component of emissions from transport, 10% is used in the transportation of food and goods and the increasingly global nature of the economy is exacerbating this, with an annual growth rate of more than 3%. As a result, the 'carbon footprint' of food is increasing as our food is obtained from progressively more distant sources.

Figure 3.10 The newly commissioned steam turbine and generator at the Shanghai Waigaoqiao power station, owned by the Shenergy Company. This is one of the most efficient super critical coal-fired power stations in the world.

Transport emissions can be significantly reduced through efficiency improvements such as redesigning vehicles to be more aerodynamic, improving tyres and incorporating light-weight materials. The design and production of hybrid and electric cars has also attracted much attention. But more efficient transport systems do not have to rely solely on improved cars; better roads, congestion management and integrated design of freight systems can have large impacts on reducing the amount of time vehicles spend on the road. Better public transport also has a role to play. Because the source of transport-related CO_2 is mobile and the emissions from individual vehicles are small, direct capture of transport-related CO_2 emissions is not an option. However, some biofuels (but certainly not all) may offer a less carbon-intensive approach. In the future, there will be greater use of low carbon fuels (for example, hydrogen) or less carbon intensive fuels (biofuels), but the biggest shift is likely to be to electrically powered cars that plug into the electricity grid (Figure 3.9). However, unless the battery is recharged from wholly renewable sources (or nuclear), a move to electric cars will transfer CO_2 emissions to a gas or coal-fired power station. We will also need new transport infrastructure systems to deliver new fuels such as hydrogen or charge batteries.

As pointed out previously, at the present time, the single largest source of anthropogenic CO_2 is the production of electricity from fossil fuels. Electricity can be produced in a steam turbine by burning fuel (usually coal) which generates the heat, which then boils water to produce steam. The steam is further heated to produce high pressure superheated steam and the expansion of the pressurised steam turns the turbine which creates electricity (Figure 3.10). In gas turbines it is the rapid expansion of the combusted gas which creates the airflow that turns the turbine.

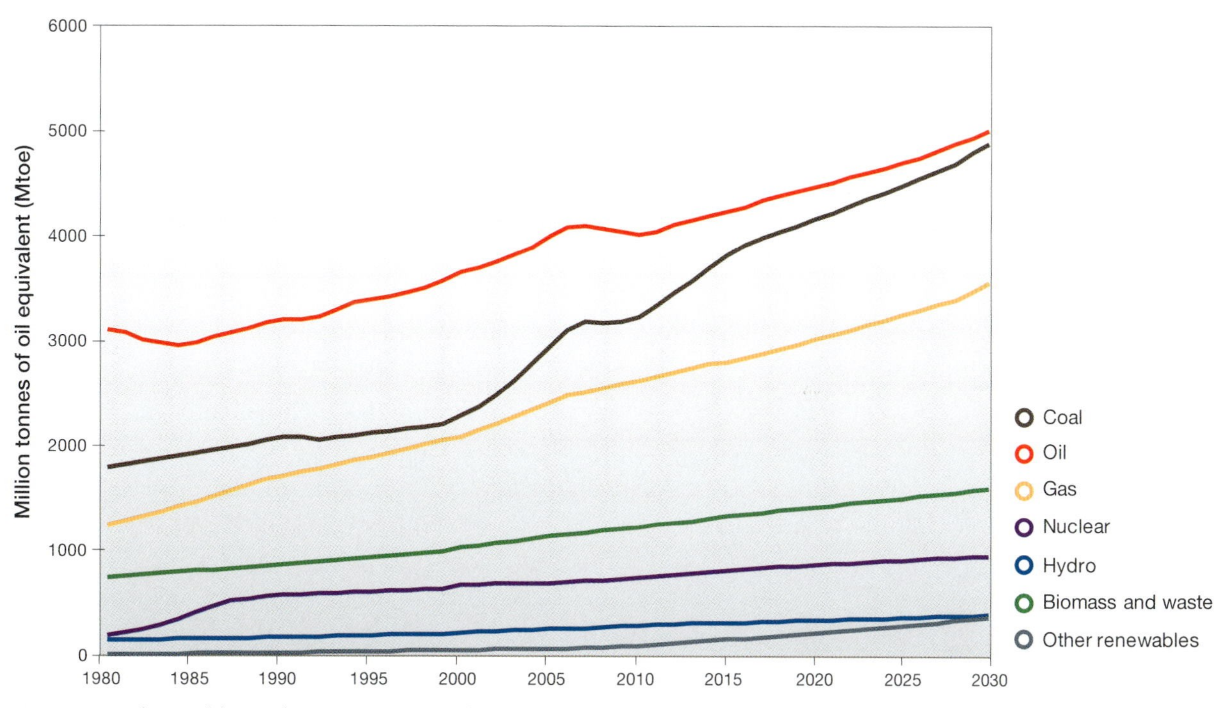

Figure 3.11 The world use of energy is projected to continue to grow significantly to 2030 (and probably beyond). All fossil fuels are expected to show growth, with much of that in developing countries. Nuclear and hydro have the lowest growth trajectory. Renewables show significant growth but from a low base. In all, this illustrates that there is no single clean energy technology that addresses growth in energy demand. (Data source: OECD/IEA 2010)

The rate of growth in electricity consumption has slowed in many developed countries over the past five decades, due in many cases, to a decrease in the manufacturing sector (which has now transferred to the lower-cost developing countries). Greater energy efficiency has also been important. However, electricity consumption is still increasing very significantly in developing countries.

In most areas the global distribution of coal or gas fired power stations reflects population density: i.e. electricity is generated where people live. Places such as eastern China, the eastern United States and throughout western Europe are examples of this. However, this is not always the case. For example, major power stations are found in Wyoming (the least densely populated US state) and in some other sparsely populated western states where there are large deposits of coal. As it is easier and cheaper to transport electricity than coal, very large power stations have been built in these states and the electricity is then exported. California has for many years made it impossible to build new coal-fired power stations, due to state environmental regulations; consequently the power stations have been built in neighbouring states such as Arizona.

Conclusions

According to the IEA and the Organisation for Economic Cooperation and Development (OECD), given the growth of total energy consumption worldwide even with a projected rapid increase in the use of renewable energy (particularly wind in the short term), fossil fuels will continue to be the dominant source of electricity generation for many years to come (Figure 3.11). For the next 30–50 years, in the absence of global action, most major developing countries will expand their energy production with fossil fuels, until renewable energy becomes more available, more reliable and more cost effective. No country illustrates this more than China, which is currently installing more wind power than any other country but is also installing of the order of 1 GW of coal-fired power each week. Most of the new plants are more modern designs with a higher efficiency than power stations in many developed countries. These new power stations will be emitting CO_2 to the atmosphere for the next 40–50 years unless action is taken. Therefore, even though the long term goal must be the transformation of our energy systems to renewable energy, this will take time. Reducing the emissions from the expected continued use of fossil fuel is an urgent imperative. It is not a substitute for greater use of renewables, but an essential adjunct. We must have a portfolio of technological options for decreasing our emissions. The next chapter considers what some of these might be.

4 TECHNOLOGY OPTIONS FOR DECREASING CO_2 EMISSIONS

One of the simplest and most cost effective ways we can reduce greenhouse gas emissions is through improved energy efficiency and energy switching. Everyday changes such as more efficient light bulbs and appliances, improved insulation, passive heating and cooling in homes and walking or cycling to work, can all make a small but useful contribution. However, energy efficiency has to be linked to broader 'lifestyle' issues too: a study by the Australian Department of the Environment, Water, Heritage and the Arts in 2008 showed that efficiency gains from measures such as insulation, double glazing and so on, has largely been offset by a tripling in the average house size and the increased use of electrical appliances (such as large flat screen, energy hungry TVs). As a result, average per capita energy consumption in Australia has remained fairly constant.

Efficiencies can be achieved in the electricity generation process: in older coal fired electricity generation plants, as little as 30% of the energy is converted to electricity, whereas more modern ultra supercritical power stations can have efficiencies of 45% or more (Figure 4.1). The efficiency of China's generation fleet now surpasses that of the United States thanks to the addition of new equipment designed to expand China's electricity generation capacity. Energy efficiency is a critical part of any greenhouse strategy, but once the 'easy' improvements have been made, energy efficiency can become increasingly expensive.

We can reduce greenhouse gas emissions by transforming our energy systems through increasing the use of renewable energy systems, using nuclear power, switching to less carbon intensive fuels, and improving the way we use fossil fuels. Renewable energy options include solar, wind, hydroelectricity, wave, tidal, biomass and geothermal. Switching from coal to gas decreases greenhouse gas emissions significantly. In addition, it is possible to decarbonise fossil fuels such as coal and natural gas and use the produced hydrogen gas as the energy source. We can also avoid the emission of associated CO_2 through the process of carbon capture and storage (CCS). This chapter will briefly examine each of these options and the role they might play in decreasing global emissions.

Solar energy

Solar energy resources are vast, essentially inexhaustible, and there is no question that solar energy will play an increasingly important role in meeting energy needs globally. Recent decreases

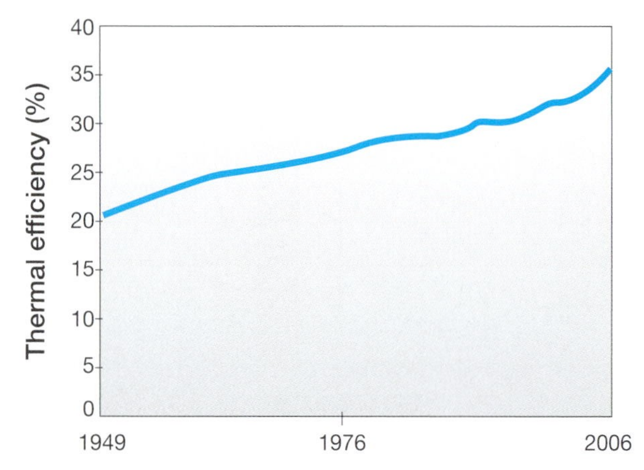

Figure 4.1 Improved energy efficiency of power stations is an important step being used by some countries to decrease their fuels and their CO_2 emissions. For example, China has increased the average efficiency of its large power stations by almost 50% over the last 50 years. (Data source: Seligsohn *et al.* 2009)

in the cost of solar panels, coupled with generous government incentives have made small scale solar electricity a feature of the suburban landscape of many developed countries. Despite this, fulfilling solar's potential to provide large scale electricity production appears to be some way off, not least because we have yet to develop cost effective energy storage at large scale. There are two types of solar power – solar thermal and solar photovoltaic.

Solar thermal

Most people are familiar with domestic solar hot water systems, where solar radiation heats water for use in the home (Figure 4.2). In industrial scale systems, the solar radiation is concentrated

Figure 4.2 Solar thermal provides the potential for producing near-base load power. At this demonstration plant in Newcastle, Australia, the solar power is concentrated onto the tower using a large number of mirrors. (Photograph courtesy of CSIRO)

through solar collectors or mirrors, which can be parabolic troughs or dishes that track the sun and focus the energy onto a working fluid, which is in turn used to drive a steam turbine. Temperatures can reach as high as 700°C depending on the nature of the collector. While other forms of solar energy create electricity directly (which is difficult to store), solar thermal creates heat, which is easier to store. This stored heat can then be used directly, or used later to create electricity on demand. While water can be used, large scale solar thermal plants use more complex fluids or chemical systems. Because solar thermal requires direct, undiffused sunlight, there are climatic constraints on the effective deployment of solar thermal, with a strong preference for hot desert locations. Such locations typically have low population densities, and long distance electricity transmission costs can be high (of the order of $1 million a kilometre for double circuit high voltage lines).

Solar thermal power generation is simple in principle and uses steam Rankine cycle turbines that are commonly deployed for the generation of electricity. But there is much work still to be done to maximise thermal storage, including optimisation of mirror collector configurations and heat exchange fluids and systems. Similarly, improvements are needed in the high energy receivers, and in the storage fluids and media to make them stable at high temperatures and able to handle rapid and marked variations in solar radiation. So-called power towers, with a raised, centralised focal point for the collectors are an example of a design providing improved efficiencies, better tracking systems and simpler mirror collectors.

Solar thermal systems are costly at the present time, but they can potentially be integrated with existing gas or coal-fired power stations to bring down costs. This concept is currently being tested, with the aim of fully integrating solar thermal in new-build stand-alone power stations rather than as an 'add-on' to existing stations. However, it could be decades before fully workable and cost effective systems become a significant part of the energy scene. The relative cost of solar thermal is discussed in Chapter 10.

Solar photovoltaic

Unlike solar thermal, solar photovoltaic (solar PV) panels produce electricity directly and can make use of reflected or diffuse solar radiation (Figure 4.3). Solar PV panels will work even under cloudy conditions, although obviously the systems are more efficient under direct sunlight. Electricity is generated as a result of the sun's energy hitting the solar panel, producing electricity by releasing electrons which travel from one side of a wafer to the other through a wire loop. The 'wafer' is most commonly made of super-pure silicon (99.9999% purity!), which is a costly and energy-intensive commodity to produce. Recent developments in wafer technology have focused on using less costly materials such as cadmium telluride or exotic compounds of copper indium and gallium. Whilst less costly than super-pure silicon, the geological distribution and resources of some of the other PV elements' are somewhat limited, which could mean that if they were to become favoured for solar PV panels, then their price could rise substantially. Added to that, there are concerns about the use of elements such as cadmium in PV wafers, because of its toxicity. However cheaper and more benign materials such as boron are also being investigated.

Historically, the cost of PV technologies has decreased by approximately 20% with each doubling of global module production. In the three years 2007–2009, PV module prices fell by approximately 50% and there is still scope to further decrease costs through new technologies such as:

- thin films and non-silicon materials
- improved industrial processes such as cell fabrication
- nano materials in hot carrier solar cells.

Figure 4.3 Solar photovoltaic cells produce electricity directly, but only operate for approximately 20% of the time. (Photograph courtesy of CSIRO)

The cost of small scale solar photovoltaic power is higher than almost any other option for electricity generation at the present time, but it has some significant advantages. First, it can be manufactured in modular form and fitted on unused roofs, so space for deployment is not a major impediment, though it may become an issue in the future with very large scale solar PV installations. Second, solar panels can be installed where the power is needed, meaning that solar PV can be a cost-effective option in remote locations or in developing countries that do not have an adequate electricity grid. Third, there can be a good match between the optimum period for solar PV electricity generation and the time of maximum demand for power. This is particularly the case in areas with hot summers which typically have high energy demand for air conditioning in the hottest (and sunniest) part of the day.

Despite all these potential advantages, at the present time solar PV is not cost competitive with most other energy options (see Chapters 10 and 11). To counter this, a number of governments (including Germany, Spain and Australia) have used feed-in tariffs (a form of subsidy) to encourage solar PV uptake. Under these schemes, governments mandate that a premium tariff be paid for solar PV–generated electricity that is fed into the grid. That premium tariff can be several times higher than the normal tariffs paid for conventionally generated electricity. In addition, mandatory renewable energy targets can be used to encourage the uptake of solar PV, so that the power company (and therefore the consumer) is required to accept this high cost (but low carbon) electricity. Solar power will meet an increasing proportion of our electricity needs in the future as the technology improves and costs come down.

Wind power

Wind power is currently one of the most mature and lowest cost renewable energy options (Figure 4.4). Wind turbines are a well established

Figure 4.4 Wind turbines provide the cheapest form of renewable energy, but only produce electricity for 30% of the time. (Photograph courtesy of CSIRO)

technology. Technology improvements over the past 30 years have tended to focus on increasing the size of the turbine blades from a diameter of 15 metres 30 years ago, to as much as 126 metres now, with larger ones planned. The height of the towers has increased to 140 metres (with higher towers planned), to take advantage of stronger and more consistent wind speeds at higher elevations. Construction materials are also being improved: carbon fibre and composite materials are increasingly used to bring down costs and weight. Because wind speeds are two to three times higher and more reliable offshore than in adjacent onshore areas, some countries such as Denmark and the United Kingdom have established large scale offshore wind farms. The disadvantage of offshore locations is that the cost of installation, maintenance and servicing can be very high.

Wind (like solar) faces the challenge of intermittency and the problem of producing reliable electrical energy. A range of options are being explored for storing wind energy for subsequent use when energy production drops off. Options include:

- electrochemical storage batteries
- thermal systems (based on high and low temperature media with an electrical output)
- chemical energy storage (such as hydrogen-based systems)
- compressed air energy storage
- pumped (hydro) storage systems
- high energy ultra capacitors and fly wheels for short term storage.

Along with stationary storage systems, a range of options are also under development for vehicular energy storage. Electricity can be sent to and taken from vehicle batteries while they are parked and plugged into the grid thereby providing electricity storage for at least part of their daily cycle of use.

Having a wide geographic spread of wind farms is a possible way to help smooth out some of the variability, based on the idea that 'the wind is always blowing somewhere'. However, in many parts of the world there may be insufficient land available to do this.

Wind is making a significant contribution to meeting our energy needs. For example, in March 2011 wind power accounted for 21% (4738 GWh) of Spain's electricity. This was the largest single component of Spain's energy mix that month, contributing more than nuclear (19%), hydro (17%), gas (17%), or coal (13%). The other extreme is illustrated by an Australian example; The Australian Energy Market reports that in South Australia (an Australian state which has a high proportion of wind power), total electricity demand in the state at 4:30 pm on a hot afternoon in February 2011 was 3399 MW, but only 19 MW was delivered by wind, despite more than 1000 MW of installed wind capacity. This 19 MW of windpower therefore provided only 0.5% of the total energy demand at that time.

Despite the fact that wind is partly predictable from wind forecasts and known weather patterns, a wind turbine will on average produce electricity for only 30% of the time. At present, other than pumped hydro, the only technically feasible large scale way of addressing the intermittency of wind power is by having other sources of back-up electricity, mostly provided by fossil fuel (usually gas) – based power sources. As mentioned previously, Denmark is used as an exemplar of the feasibility of a very high level of wind penetration in its electricity generation system, but it is able to do this by importing electricity generated from coal, nuclear or hydro in neighbouring countries to handle the intermittency. Most other parts of the world do not have such a diversity of power options and have to be more self-contained.

Many countries are seeking to increase their use of wind power, with targets of up to 20% wind-based electricity. While the capacity of grids to handle intermittency is improving, the configuration of the grid currently limits wind power to providing no more than approximately 20% of electricity needs. The cost of upgrading the electricity grid, coupled with the cost of providing back-up power from gas or other sources, has to be factored into the overall cost of wind power.

There are a range of other hurdles to the acceptance of wind power. Wind turbines can be visually intrusive and optimum sites from a wind speed perspective are often on aesthetically and commercially valuable landscapes such as coasts, headlands and hills. Noise can also be a problem. Typically, a few landowners gain direct financial benefit while others in the vicinity are not compensated for loss of amenity. There may also be some adverse environmental impacts such as disruption of the migratory path of birds. Research published by the National Academy of Sciences suggests that air turbulence created by turbines can warm ground surfaces at night by up to 1.5°C, increasing drying of the soil and potentially leading to the need for increased irrigation of crops. As a result of these issues, there is significant opposition to some wind farms.

Despite these challenges, wind power has a significant role to play in decreasing greenhouse gas emissions, and can potentially meet approximately 20% of our total electricity needs, but the full cost (including the cost of back-up power and modification of the electricity grid) is high. Also, community acceptance is, and will continue to be critical to whether a wind farm does, or does not go ahead.

Hydroelectric power

Hydroelectricity is a well established, cost effective and reliable energy option. There is nearly 800 GW of installed hydroelectric capacity worldwide, operating on average for 40% of the

Figure 4.5 Hydroelectric power is one of the most established large scale renewable energy sources. The Hoover Dam on the Colorado River in the United States is an outstanding example of modern hydro engineering but also serves to illustrate the very significant impact that hydro can have on natural river courses. (Photograph courtesy of United States Bureau of Reclamation)

time (Figure 4.5). The electricity is generated mainly through the water released from a dam passing through large turbines, but it is possible to use the natural downhill run of a river via a pipe into a power station and returned to the river.

A number of countries with high rates of precipitation and mountainous terrain, are major generators of hydroelectric power, including Norway, New Zealand, Brazil and Canada. These countries may still have some unexploited hydrocapacity. Other countries such as Australia (through its Snowy Mountain Scheme), the western United States and Western Europe have already developed much of the available hydro capacity. Future opportunities to develop new hydroelectric capacity are limited because of environmental concerns regarding the flooding of valleys and disruption of the ecology of rivers.

The disruption of communities can also be a major issue. The construction of the Three Gorges Dam in China required thousands of villages to be flooded, and disrupted the lives of millions of people. In most developed countries, community opposition makes it difficult to gain approval for new hydroelectric schemes, even at a very modest scale, despite the undoubted benefits in terms of providing reliable clean energy. Some developing countries, for example Brazil, are still pressing ahead with large scale hydroelectric schemes.

The greatest benefit of hydroelectric power in the future may lie in its capacity to store water (so-called pump storage), and release it when required to provide clean base load electricity. Realistically the availability of hydroelectricity is unlikely to increase significantly in the future and its relative proportion in terms of the total energy mix is likely to decrease.

Ocean energy

Tidal rise and fall can be used to generate electricity. Water is trapped at high tide in the mouth of a river by a barrier; when the tide falls, the water flows from the river into the sea via a turbine, which generates electricity. This type of electricity generation is not dependant on the weather or the seasons and is predictable, though not continuous. There are environmental considerations such as impact on estuarine ecosystems and changes to currents which can increase coastal erosion in some areas or produce siltation in others. The 240 MW Rance Tidal Power Plant in France (Figure 4.6) has been providing electricity to the French electricity grid since 1967. There are undoubtedly some opportunities for increased

Figure 4.6 Tidal power has significant potential for providing renewable energy, but there are few operational systems at the present time. A tidal power station has been operating on the Rance river in western France for almost 50 years, producing 240 MW of electricity on an ongoing basis. (Photograph courtesy of Électricité de France)

development of tidal energy, particularly in areas with large tidal ranges such as eastern Canada and parts of northern Australia, but community and environmental concerns may limit such developments.

Wave energy can be used to generate electricity using a range of configurations that convert the mechanical energy of the waves to electrical energy, usually by moving a magnetic shaft vertically through a coil. Alternatively air flow generated from the waves can be used to directly power a turbine. There is potential for using more wave energy and the resource is almost limitless, but the marine environment is very challenging – large waves can quickly destroy expensive equipment or break the moorings; transmission of electricity from offshore installations can be expensive and difficult, and seawater is a corrosive medium. At the present time there are a number of experimental wave energy systems installed around the world, but none are producing electricity on a commercial basis.

Because of the inherent technical difficulties in surface-positioned wave power systems, attention is now being given to anchored submarine systems such as the CETO System of Carnegie Wave Energy Ltd (Figure 4.7). The system uses the change in pressure resulting from the crest of a wave passing over the unit, which in turn delivers high pressure water onshore to a power unit. The concept which is currently being tested off the West Australian coast, does seem to offer a way around the problems of the extreme variability in wave pattern and sea states, and minimises the prospect of storm damage. Wave energy may have a future role to play in the

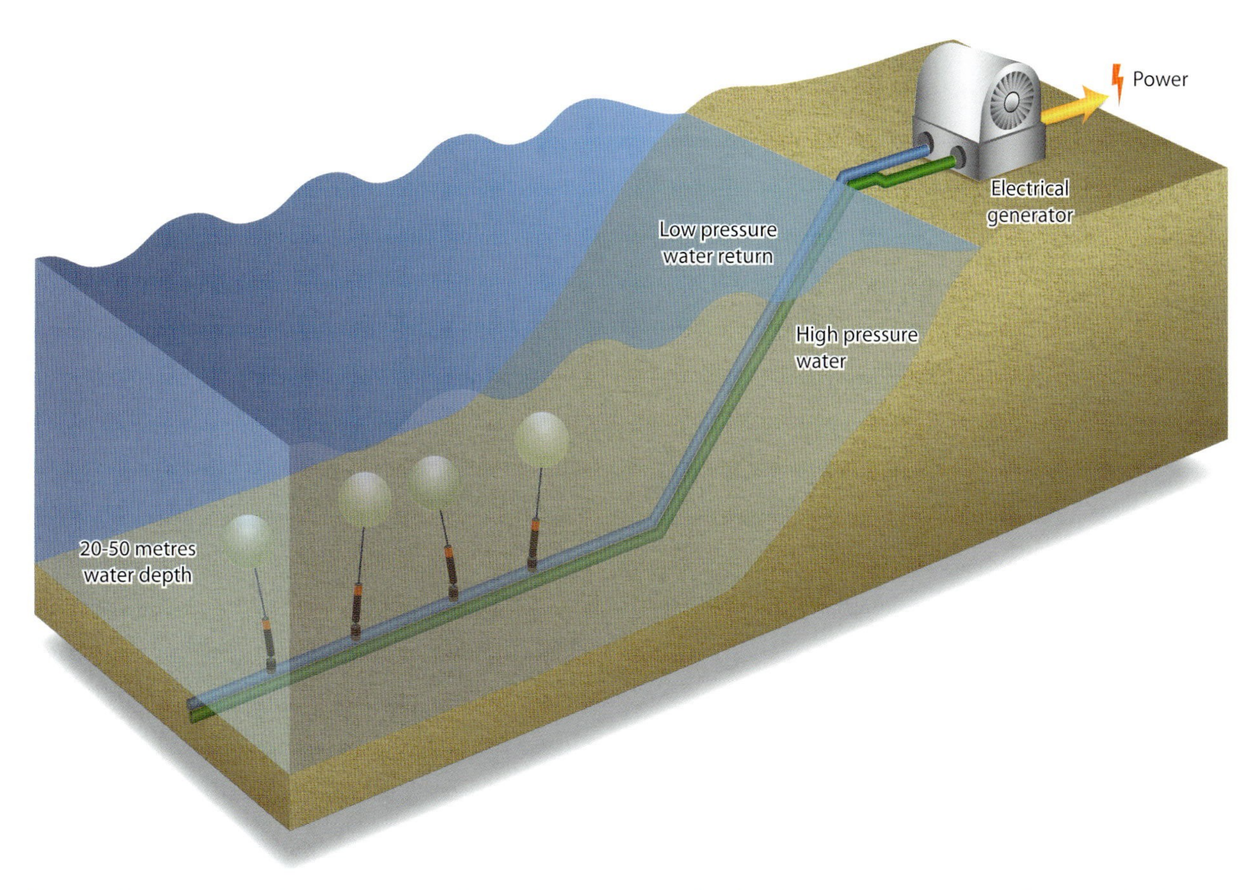

Figure 4.7 Many systems have been proposed to exploit wave power, but the destructive power of waves has proved to be a major problem. Carnegie Wave Energy Ltd's CETO system is fully submerged, helping to limit storm damage. Its wave-driven pumps deliver high pressure water onshore to drive an electric generator. (Image after Carnegie Wave Ltd)

energy mix but there appears to be no immediate prospect of it making a major contribution to global energy production.

Ocean Thermal Energy Conversion (OTEC) uses the temperature differential between shallow (warm) and deep (cold) ocean water to drive a heat engine. It is limited to areas where the maximum thermal gradient occurs in accessible locations. Because the temperature difference between warm and cold ocean water is quite small (20 degrees celcius or less), large amounts of water must be processed to extract useable amounts of energy.

The total energy resource available from OTEC is large, but the areas with the greatest potential for OTEC, in terms of maximum thermal gradient, are areas such as the central and south-west Pacific where there are few major population centres. The first test OTEC plant was built as a test facility in Cuba in 1930 and an OTEC plant was in operation for several years at the island of Nauru, sending 30 kW of power into the Nauru grid, demonstrating that it can be a useful clean energy option for oceanic islands. But OTEC has yet to be commercially exploited at scale and because of its very specific geographic distribution and massive water processing requirements, it is unlikely to play any significant role in meeting global energy needs in the foreseeable future. It may, however, be locally relevant.

In addition to OTEC, it has been proposed that deep ocean currents can also be used to turn submarine turbines. This option is being investigated by Taiwan with a view to using the Kuroshio current, a strong north-flowing ocean current on the west side of the North Pacific Ocean.

Biomass

More than a billion people depend on biomass, particularly wood or dung, for domestic heating and cooking. Biomass can also be used to produce electricity. Combustion of waste materials such as wood off-cuts and sawdust, harvestable stubble (wheat, rice, cotton), domestic and municipal garbage, waste from sugar cane (bagasse; Figure 4.8) and wood from dedicated forests, can all be used to generate electricity. Biomass can also be used in a power station in conjunction with coal (co-generation).

Biomass can provide electricity through:

- direct combustion, producing steam for a steam turbine
- direct gasification with a combined cycle turbine or through pyrolysis to produce a syngas fuel
- biochar.

Depending on the composition of the waste, it may produce complex mixed emissions, but the advantage of biomass is that it can potentially provide base load power that is carbon neutral, so long as replacement timber or other crops are grown. In most countries, there is only limited access to biomass, and at present it contributes only a very small percentage of total electricity generation.

A form of biomass energy production known as biogas involves the decomposition of waste from plants or animals in the absence of oxygen (anaerobic decomposition) to form methane gas which can then be collected and used as fuel to generate electricity. Biogas is now collected from landfill sites and sewage plants in many parts of the world. It is a relatively mature, although small scale technology. CO_2 is generated through the combustion of the methane, but this is a far

Figure 4.8 Sugar cane is one of the most effective crops for producing bioethanol and biomass.

better greenhouse outcome than letting the methane escape to the atmosphere from land fill sites or sewage plants.

Along with electricity generation there is extensive production of liquid transportation fuels from biomass, especially ethanol from grain or sugar cane. However, it is often unclear how much energy is used to produce the fuels. According to the United States Department of Agriculture, ethanol made from corn produces, on average, 34% more energy than that required to grow, transport and process the corn. Conversely David Pimentel at Cornell University concludes that ethanol from corn requires 29% more energy to produce than it can provide, and is therefore a net carbon emitter (a concept known as a negative Net Energy Value or NEV).

Large scale ethanol production may have the potential to decrease emissions, particularly if steps are also taken to sequester the CO_2 that is produced as part of fermentation during the production process. Given that these ethanol-related emissions are composed of pure CO_2, it could be relatively straight forward to capture the CO_2 and this is being done as part of the Decatur CCS Project in the United States.

An issue with the production of bioethanol is the amount of farming land it takes to grow energy crops. Some plants, such as the perennial highly productive sterile grass *Miscanthus*, are being considered as herbaceous feedstocks ('feedstock' is the term given to the raw material from which the biofuel is made) conversion to ethanol. *Miscanthus* is reported to be non-invasive, it stores carbon in the soil, has high water use efficiency and has low fertiliser requirement. However, further studies are required to ensure that there are no adverse environmental impacts. There is no question that this plant is capable of producing large quantities of biomass and some forms of *Miscanthus* have been used in power plants for 30–40 years.

In the United States, agricultural energy sources are projected to continue to be major providers of ethanol for transportation but scale is an issue. By way of example, to produce 130 billion litres of ethanol per annum (approximately 25% of US gasoline consumption) would require 33.7 million hectares of switch grass (20% of US cropland), 18.7 million hectares of corn (12% of US cropland) or 11.8 million hectares of *Miscanthus giganteous* (7% of US cropland). One option is to use the cellulose from residual field crops (wheat stubble etc) for the production of bioethanol, to avoid the food versus fuel problem. The problem with this approach is that crop residues play an important role in maintaining the soil organic carbon stock, although this can be influenced by factors such as crop rotation, tillage factors and the climate.

Combining bio energy with CCS may present an opportunity to actually decrease CO_2 in the atmosphere, by being 'carbon negative'. Some years ago, Australia's Cooperative Research Centre for Greenhouse Gas technologies (CO2CRC), a research organisation studying the entire CCS process, modelled the use of sugar cane trash for energy production coupled with CCS. Results suggested that it could be an effective mitigation option. More recently, the International Energy Agency (IEA) has suggested that by 2050, as much as 2.4 Gigatonnes of CO_2 per annum could be geologically stored by applying this concept (combining bio energy with CCS) to a range of biomass-related technologies producing heat, power, biogas refining or ethanol. To put this in perspective, approximately 300 million tonnes of CO_2 were emitted from biomass used for energy in 2009 and approximately 50 million tonnes of CO_2 from ethanol production. None of this CO_2 was captured. The Regional Partnership Decatur project in Illinois is testing the concept of geologically storing CO_2 emitted from an ethanol plant, and a number of other projects proposed for Europe and North America may also take this concept forward.

There is increasing interest in using biomass for electricity production either alone or with coal. Biomass undoubtedly has a role to play but the magnitude of that role is still unclear. A major constraint is the low energy density of biomass—the energy generated from a tonne of biomass is quite low compared to that available from a tonne of coal, for example. This in turn means that biomass cannot be cost effectively transported long distances because the CO_2 emitted during transportation could end up negating the carbon benefit arising from the use of the biofuels in electricity generation. But there is a possibility that CO_2 emitted from biomass could be mitigated with CCS to produce 'negative carbon emissions' and this is attracting increasing attention.

Geothermal energy

Geothermal power (Figure 4.9) is currently used in volcanic regions for domestic purposes, for industrial processes and for generating electricity, in more than 20 countries with a total of more than 10 GW of geothermal electricity generation. Its use is limited to areas with high near-surface geothermal gradients, typically volcanic areas on or near the margins of plate boundaries (Figure 4.10), such as the Philippines, California, Iceland, Indonesia and New Zealand. In these countries, geothermal heat provides reliable baseload power, but also has limitations and can be over

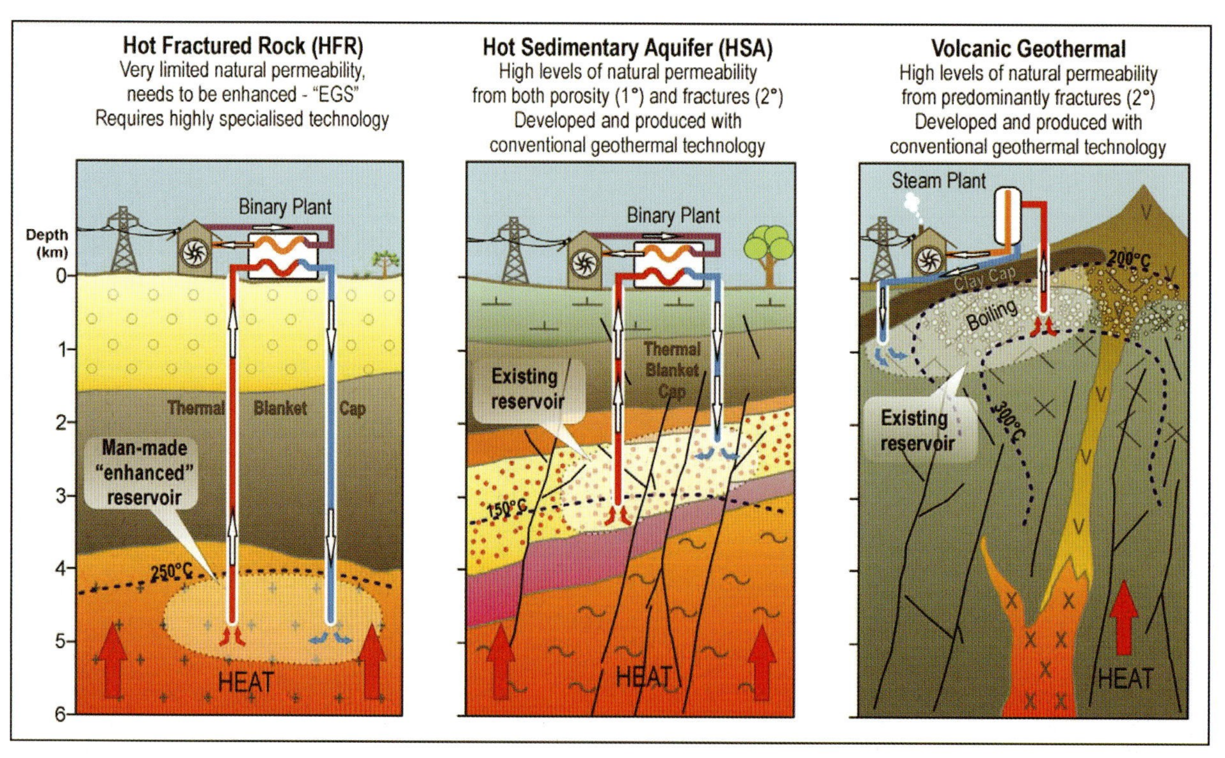

Figure 4.9 There are three types of geothermal systems that can be exploited for power generation: shallow volcanic geothermal; intermediate depth hot saline aquifers; and deep hot fractured rocks. (Image courtesy of Hot Rock Ltd)

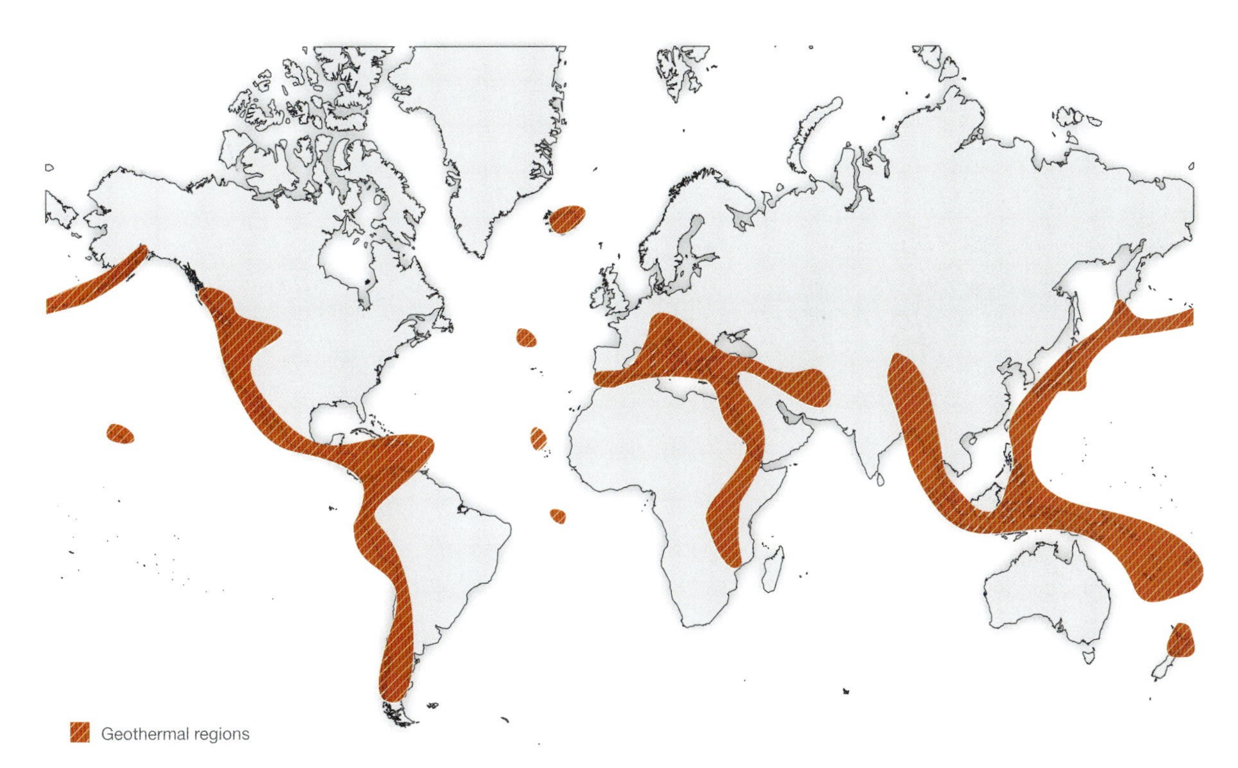

Figure 4.10 Volcanic geothermal regions are mainly found at active plate boundaries. (Adapted from Smithsonian Institution, Global Volcanism Program)

exploited. At the Wairakei power station in New Zealand, over-production lowered the pressure and temperature of the system, decreasing the amount of geothermal energy available for use, until remediation measures were taken. Nonetheless, geothermal energy can be used sustainably, provided careful consideration is given to the thermal and groundwater recharge of the system.

Hot sedimentary aquifers (HSA) are found in parts of the world where there are deep (usually saline) aquifer systems with higher than average geothermal gradients, and where there is a thermal blanket, such as shales or coals, overlying the HSA. In addition, the aquifer needs permeability, perhaps via faults or fractures, so that the hot water can be produced via a deep well.

The temperature of the water in a HSA is typically up to 100°C. It cools as it is transported up the well to the surface and consequently the useable heat is often insufficient for steam-based power plants. Binary power plants are an effective way of using lower grade heat, in that the working fluid (e.g. isobutane) boils at a relatively low temperature, producing vapour that drives the turbine generator and produces emission-free electricity. The cooled water is then re-injected either into the original HSA or into another suitable geological formation. The advantage that HSA systems enjoy over volcanic geothermal systems, is that they are more widespread. In addition HSA may offer opportunities for synergies with other energy technologies such as injection of CO_2 providing the opportunity to store the CO_2 while controlling pressure within the aquifer.

The third type of geothermal system – hot fractured rocks (HFR or engineered geothermal systems) have some similarities with the HSA and

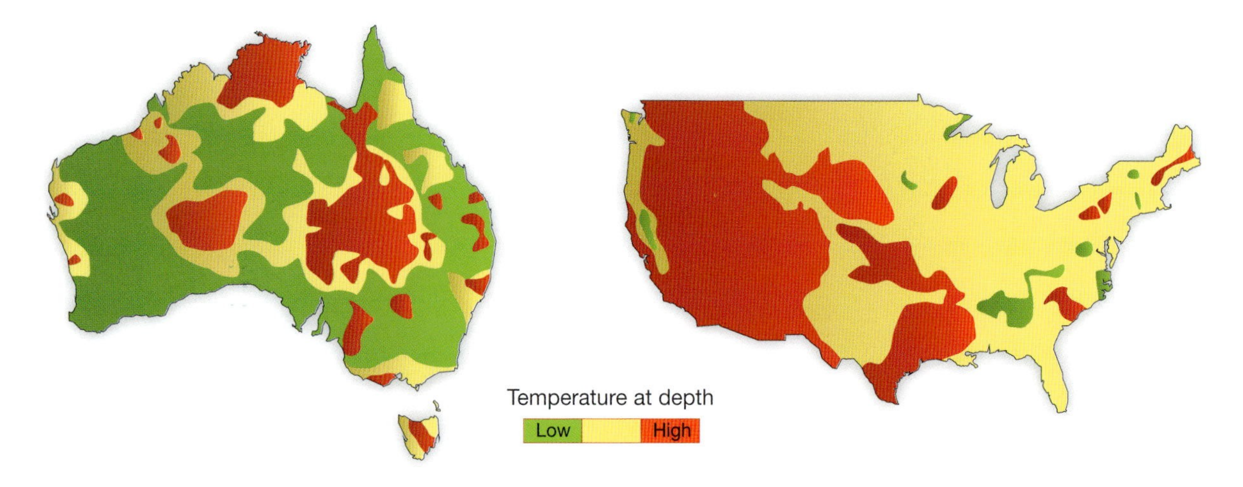

Figure 4.11 A number of regions in the world may be suitable for hot fractured rock systems. In Australia, there are three large regions with high temperature deep rocks. Much of the western United States is underlain by hot rocks, but some potentially suitable areas also occur in the mid-continent. (Data sources: Geoscience Australia; US Department of Energy/EERE)

volcanic geothermal systems but also several important differences. HFR systems occur at depths of 4–5 km (Figure 4.11), in granitic rocks containing small quantities of radioactive minerals which heat the rocks. For example, at a site in central Australia operated by the company Geodynamics, these temperatures reach up to 280°C. Because deep granite has little or no natural porosity or permeability, it is necessary to hydraulically fracture the deep hot rock in a preferred direction, so that water can be pumped into the granite where it is heated as it migrates along the fractures and is then circulated back to the surface in an essentially closed system. Steam is generated at the surface and used to drive a steam turbine to produce electricity.

Challenges with HFR systems include:

- the likelihood of losing the circulating water
- the extreme difficulties of drilling a well in very hot granite at a depth of 4–5 km
- finding suitable hot granites that can be preferentially fractured
- the fact that suitable hot granites do not necessarily occur where the electricity is needed, potentially resulting in high electricity transmission costs.

One way around this may be to take the electricity user to the HFR. Electricity-using industries suggested as suitable for location in areas where there are HFR opportunities include mineral processing and large data centres.

Nuclear power

At the present time, nuclear power produces 15% of the world's electricity, all of it zero emission electricity (Figure 4.12). There are more than 400 nuclear reactors operating in 31 countries. France is the country with the highest percentage of its electricity produced from nuclear power, at nearly 80%. Nuclear power is produced from fission – the nuclear reaction resulting from the splitting of uranium or plutonium atoms by free neutrons, which then release more free neutrons that, in turn, trigger more fission events. The rate of the fission reaction can be managed by controlling the number of free neutrons available for triggering fission events. The heat generated in the reactor heats a boiler which produces steam, which in turn drives steam turbines and produces electricity.

Figure 4.12 There are more than 400 nuclear power reactors operating around the world. The Davis-Besse nuclear power station in the United States, on Lake Erie, is a single pressurised light water reactor that has been operating since 1978, producing 889 MW (7706 GWhr of electricity). (Photograph: United States Nuclear Regulatory Commission)

There are abundant uranium resources, with identified high grade deposits sufficient to meet current world usage for 60–100 years (Figure 4.13). There are also large low grade uranium resources – for instance those associated with phosphate deposits, which could potentially meet all foreseeable demand well into the next century. In addition, with new fourth generation reactors requiring much smaller quantities of uranium, it is unlikely that availability of uranium will be a limitation on the future growth of nuclear power, although it may become more expensive as more marginal uranium resources are exploited.

Beyond cost, clearly a potential constraint on the uptake of nuclear power is the level of public opposition to the technology. Accidents such as those at Three Mile Island, Chernobyl and, most recently, Fukushima, have had a very negative impact on public perception, despite the overall very good safety record of the nuclear industry. The link with nuclear weapons (which existed in early installations, but has now been largely removed) in modern nuclear power stations is also still seen as a negative feature. Further, while handling and disposing of nuclear waste does not pose insurmountable scientific or technical barriers, the issue of nuclear waste is a very emotive one for many people and a key question for the future will be: how will people balance the benefits of nuclear power against the perceived risks arising from nuclear power and the associated nuclear waste?

Following the March 2011 earthquake and tsunami, and the resulting damage to the Fukushima Daiichi nuclear power plant in north-

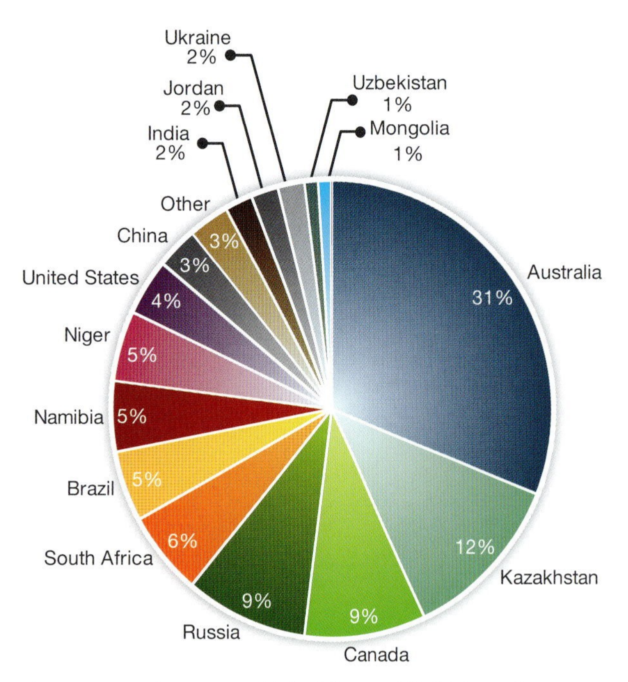

Figure 4.13 Almost one third of the world's high grade uranium resources occur in Australia, although Australia does not have any nuclear power stations of its own. (Data source: OECD/IAEA 2008)

eastern Japan, the country lost 10 GW of generating capacity. This resulted in major questions being asked about the future of nuclear power (which currently meets 30% of Japan's total electricity requirement) in a seismically active country such as Japan. However, it was not the earthquake but the tsunami that impacted catastrophically on the power plant. Clearly one of the lessons from Fukushima is that nuclear power stations must not only be designed to withstand major earthquakes but must also be sited to avoid the related consequences of an earthquake, notably a tsunami, but also including major landslides, breeching of large dams and redirection of major rivers.

If the Fukushima plant had been, say, 20 metres higher, it would have survived the earthquake and been out of reach of the tsunami. Clearly the siting of a number of nuclear power stations in Japan and other seismically active countries will now be reassessed. However, it is important to also remember that some non-seismic areas can be affected by tsunamis. For example, the coastal areas of eastern Scotland and north-eastern England are believed to have been inundated by a tsunami 8000 years ago as a result of the Storrega submarine slide off Norway where a vast 300km underwater landslide slid 800km into the Norwegian sea . The resulting tsunami affected more than 600 km of UK shoreline. There is also evidence the eastern United States coast may have been impacted by a tsunami resulting from a major landslide on the flanks of the Canary Islands.

None of this constitutes grounds for arguing that nuclear power has no future, but it does clearly indicate that more care must be taken not only in the operation of nuclear power stations (as indicated by the accidents at Three Mile Island and Chernobyl) but also in their siting, as indicated by the Fukashima accident. More rigorous siting requirements may add to the cost of nuclear but they are clearly necessary in order to give the community confidence in nuclear power. There may also be a need for more rigorous standards governing the holding of spent fuel rods on the same site as the power station and the siting of nuclear waste respositories.

An alternative fuel for nuclear fission reactors that is currently being considered is thorium. Thorium is three to four times more abundant than uranium but, unlike uranium, thorium atoms cannot produce a fission chain reaction. Before fission can take place, thorium atoms must first be converted to fissionable material by absorbing an additional neutron from a neutron source such as uranium. The need for this conversion and for the addition of a neutron source, leads to technical challenges in the fabrication and reprocessing of the fuel, making this approach difficult. There have been a few thorium reactors built in the past and several demonstration projects are currently under construction, but at this stage, thorium is unlikely to be a widely used nuclear fuel for many years.

Figure 4.14 A great deal of research has been undertaken into nuclear fusion as an energy source but for the present it is only at the experimental stage, such as illustrated in this fusion Tokamak reactor in the UK. (Photograph: European Fusion Development Agreement-Joint European Torus)

Nuclear fusion of lighter elements (such as the hydrogen isotopes deuterium and tritium) releases considerable amounts of energy and has the potential to provide almost limitless power (Figure 4.14), but after many years of research, fusion reactors are still in the experimental stage and are not expected to be a viable energy option for many decades to come.

Sequestering CO_2 through carbon capture and storage (CCS)

Storing (sequestering) CO_2 emissions is a technology that offers the opportunity to provide low emission power from fossil fuels and cut CO_2 emissions from some major industrial processes. Carbon capture and storage or sequestration (CCS) can be used where there is a major source of CO_2 such as a power station. Various systems can be deployed for capturing and separating the CO_2 from the flue gases and this is discussed in Chapter 6. The CO_2 is then transported (see Chapter 7) to a location where it can be stored (sequestered). The term 'storage' can be misunderstood in that it may be taken to imply that at some stage the stored CO_2 will be taken out of storage and used. In fact, the scope for extracting and using stored CO_2 is very limited. However, as the term 'storage' is in general use, it is used here to indicate the long-term removal of CO_2 from the atmosphere. There are currently three ways in which this can

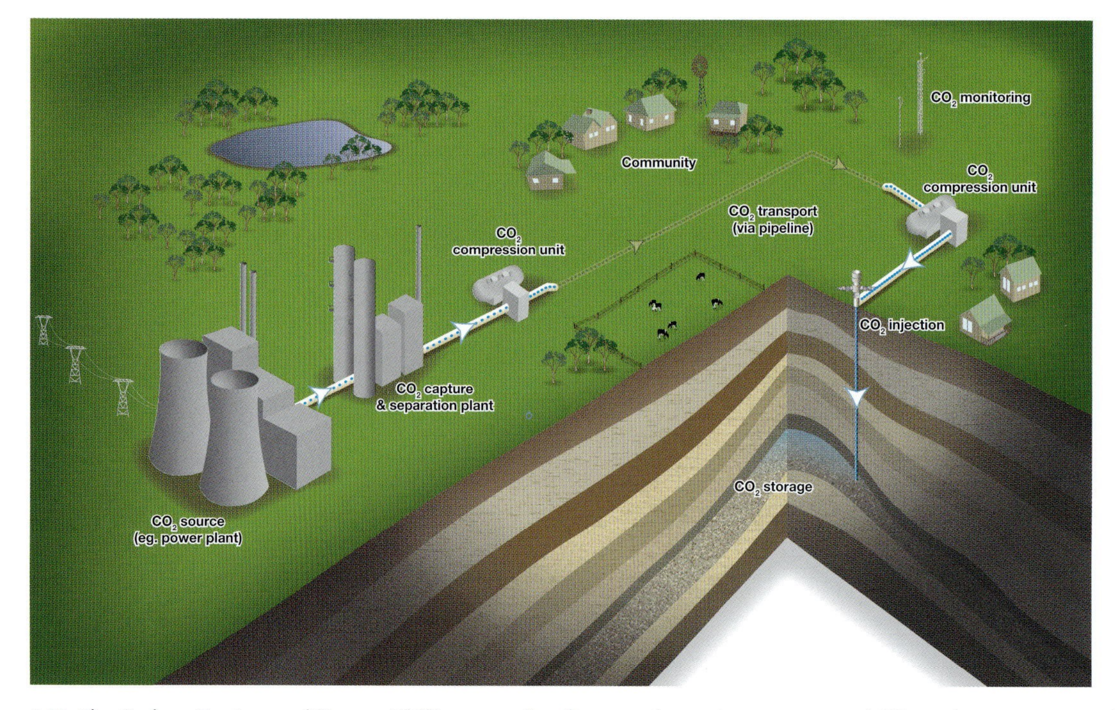

Figure 4.15 The Carbon Capture and Storage (CCS) process involves a major stationary source of CO_2 such as a power station or fertiliser plant, a plant to separate the CO_2 from the other emissions, a compression unit to compress the CO_2 so that it can be transported by pipeline, and a suitable storage site at which monitoring is undertaken. A critical element in any CCS project is successful engagement with the community.

be done – geological storage, ocean storage and mineral storage.

Geological storage (Figure 4.15) depends on the injection of CO2 into rocks that will permanently trap it at depths of around 1000 metres or more, where it will remain for geological time (millions of years). This technology is discussed in much more detail in Chapters 8 and 9.

Ocean storage (Figure 4.16) involves the injection of CO_2 into the deep ocean (at depths greater than 1000 metres), where it will remain isolated from the atmosphere for hundreds of years or longer. The concept of ocean storage was first suggested by Cesare Marchetti of the International Institute for Applied Systems Analysis in 1976. Since then, various methods have been proposed for injecting the CO_2 into the ocean at depth from a pipeline or from a ship, including injecting it as dry ice 'torpedoes' that would penetrate the sea floor.

The CO_2 dissolves into the water before it reaches the surface of the ocean and because it would be injected at depth, it would take hundreds of years for the dissolved CO_2 to circulate back into the surface waters. Alternatively, at greater depth, CO_2 hydrates are created (an ice-like form in which the CO_2 and water molecules form a net-like crystal structure) which would sink through the water column and collect on the sea floor. Also, at great depths, CO_2 is heavier than water and will sink into enclosed ocean basins to form a permanent 'CO_2 lake'. A further option may be the injection of CO_2 into unconsolidated sediments in the top tens of metres of the seafloor, which can be seen as intermediate between geological and ocean storage.

As a general rule, the deeper the CO_2 is injected in the ocean, the longer it will take to get back into the atmosphere, but ocean storage has not

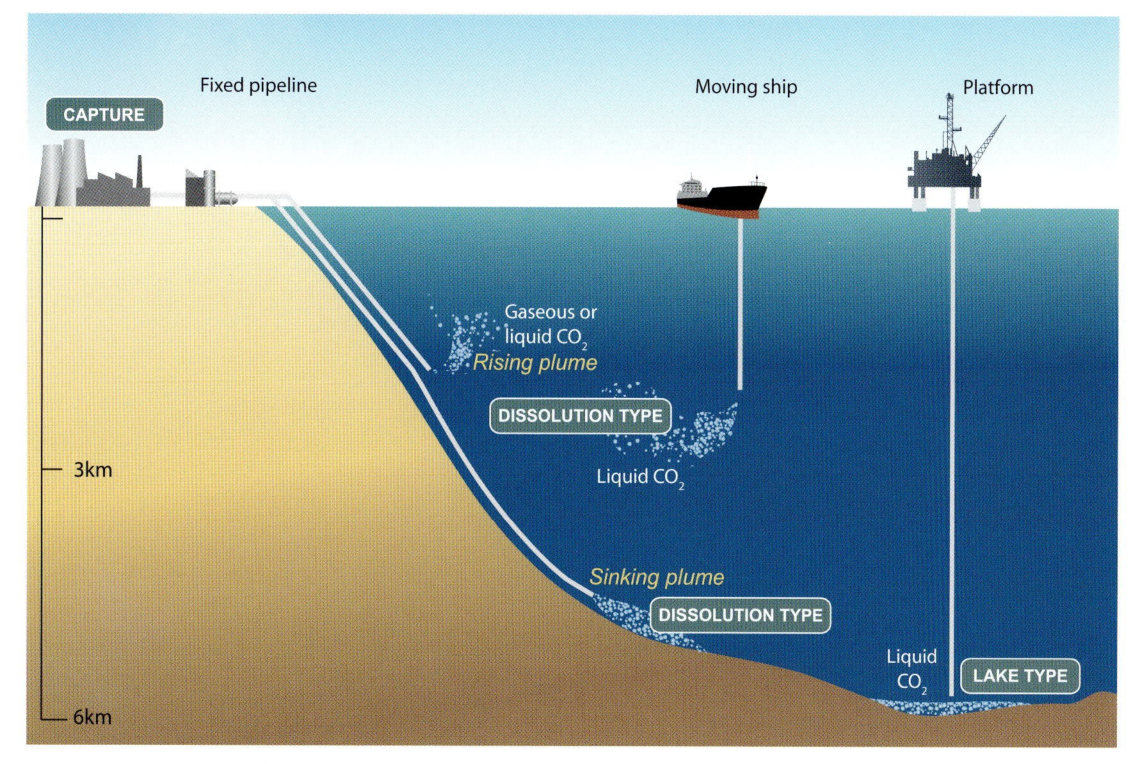

Figure 4.16 Various proposals have been put forward for injecting CO_2 in the ocean, but this storage option faces many challenges, not the least being significant community and government opposition to the concept.

been tested at a significant scale to date. Laboratory and field tests indicate that the impact of injected CO_2 on marine organisms in the immediate vicinity of the injection site is likely to be significant and the longer the injection goes on, the more widespread the impact.

Increasing atmospheric CO_2 over the past 200 years has resulted in an increase in the concentration of oceanic CO_2. This explains the rising acidity of the ocean surface waters where pH has decreased by 0.1 pH units over the past 200 years. Direct injection of CO_2 into the ocean would greatly accelerate the acidification process, potentially resulting in significant unintended consequences to marine ecosystems. Because of the likely effects of ocean storage of CO_2 on seawater, plus the attendant difficulties in reliably monitoring the CO_2 once it is injected into the water column, there is opposition to the concept of ocean storage amongst NGOs and Governments. The likelihood of ocean storage being used as a mitigation option is further diminished by the fact that both national and international laws and treaties banning the dumping of industrial waste in the ocean, are usually taken to include flue gases and anthropogenic CO_2. The sum of all these uncertainties is that there appears to be little likelihood of ocean storage being an acceptable mitigation option in the foreseeable future.

Mineral storage of CO_2 is also being investigated as an option. This involves the reaction of the CO_2 with magnesium and calcium to form new carbonate minerals, thereby locking up the CO_2. This option is considered in more detail in Chapter 8. Mineral sequestration is also used to treat highly alkaline bauxitic red muds from alumina plants (Figure 4.17), using waste CO_2 to lower the pH and form carbonate minerals. A further example of this approach is provided by

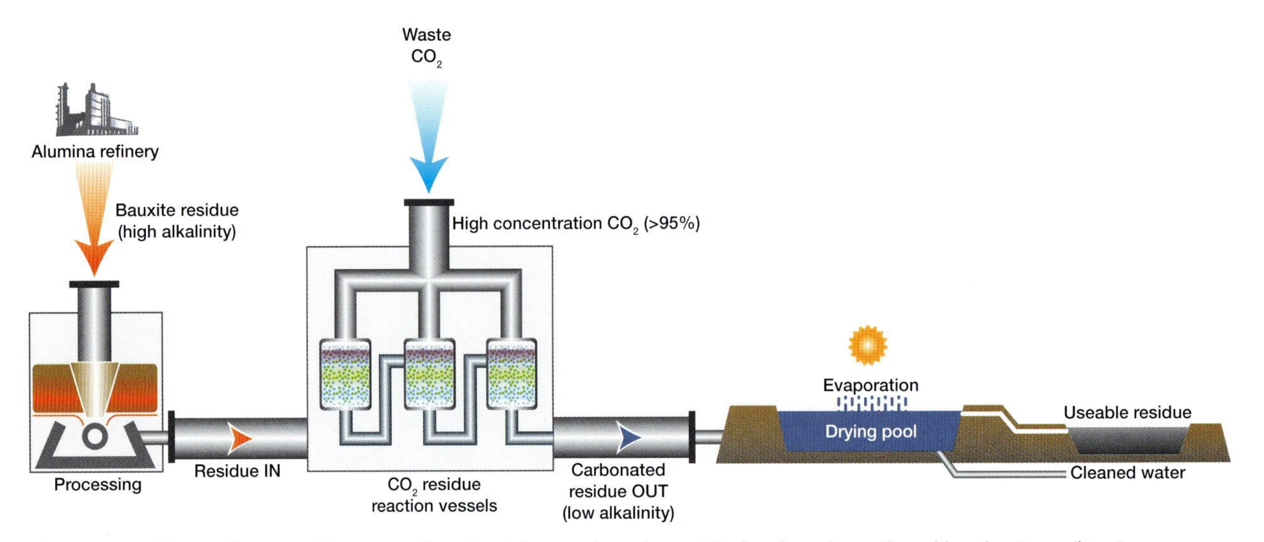

Figure 4.17 CO_2 can be stored in some mineral residues such as the residual red muds produced by alumina refineries. The process at the Alcoa alumina refinery, shown here schematically, has a number of environmental benefits and also stores the CO_2.

the Hazelwood capture project in Victoria, Australia, where captured CO_2 is used to treat alkaline ash slurry to lower the pH of the ash water and sequester the CO_2 through the formation of calcium carbonate.

A more challenging opportunity may be presented by asbestos, which constitutes an extremely hazardous waste product found in various parts of the world. Conversion of asbestos to a benign product such as limestone ($CaCO_3$) or magnesite (Mg CO_3) could represent a very beneficial way of storing CO_2 and addressing an environmental and health hazard, provided safe methods could be devised for handling asbestos.

A range of methods have been proposed for turning CO_2 into useful products such as building materials, but for the present the opportunities appear to be quite limited. The International Panel on Climate Change (IPCC) considered that no more than about 100 million tonnes of CO_2 is used annually at present (much of that in enhanced oil recovery) and that amount is unlikely to change to any degree in the foreseeable future. A useful step would be to substitute the use of geological CO_2 with captured industrial or power-related CO_2 for enhanced oil recovery. A study by Parsons Brinckerhoff concluded that there is a moderate opportunity for revenue from use of CO_2 and suggested that there may be a particular benefit for developing countries where there is a high demand for construction materials. However, the study also concludes that other than in very specific instances, use of captured industrial or power-related CO_2 is unlikely to make a significant contribution to global mitigation of CO_2 emission.

Algal sequestration is also proposed as a storage option and a number of projects are underway at the present time to test the proposal (Figure 4.18). In summary, the separated CO_2 (or the CO_2-rich flue gas) is pumped into vessels, or tanks, or ponds where the CO_2 is taken up by photosynthesizing algae. The algae can then be harvested and turned into a range of products such as transport fuels, pharmaceutical products or animal feed. The process of algal fixation may

Figure 4.18 Algae will take up CO_2. The algae can then be harvested and a variety of valuable products made such as biodiesel and animal feed. Algal sequestration can be a financially worthwhile activity but it is a niche commercial opportunity not a major storage option. (Image courtesy of AlgaePARC)

be commercially attractive in some circumstances, but it should not be seen as a significant storage opportunity, with the storage time measured in days or weeks rather than years.

Conclusion

In summary, there are a number of options for decreasing emissions associated with electricity production. As a mitigation approach, improved energy efficiency has the potential to be cost effective and sustainable in that it reduces emissions and saves money, whether through waste heat recovery, improved boilers in power stations, better insulation or more efficient appliances. Most national strategies have improvements in energy efficiency as their highest priority. However, developing an optimum mix of technologies to cost-effectively reduce CO_2 emissions is proving difficult. In no small measure this is because of the complex interplay of issues, including energy security, environmental impact, practicability, cost and community acceptance. In the next chapter, this interplay of issues is considered, along with the nature and complexity of the energy mix.

5 THE MITIGATION MIX

The previous chapter outlined some of the technologies that have the potential to decrease CO_2 emissions to the atmosphere, and in so doing pointed out that there is no such thing as a single technology solution but rather that a portfolio of technologies is required. The portfolio will need to vary to reflect local conditions, policy requirements and community expectations. It also needs to take into account the limitations of each of the technologies, including their impact on a range of external but related issues. So, before considering what a mitigation portfolio might look like, it is useful to consider the interplay of issues – energy–climate–water–population–food – and the impact of the key technologies.

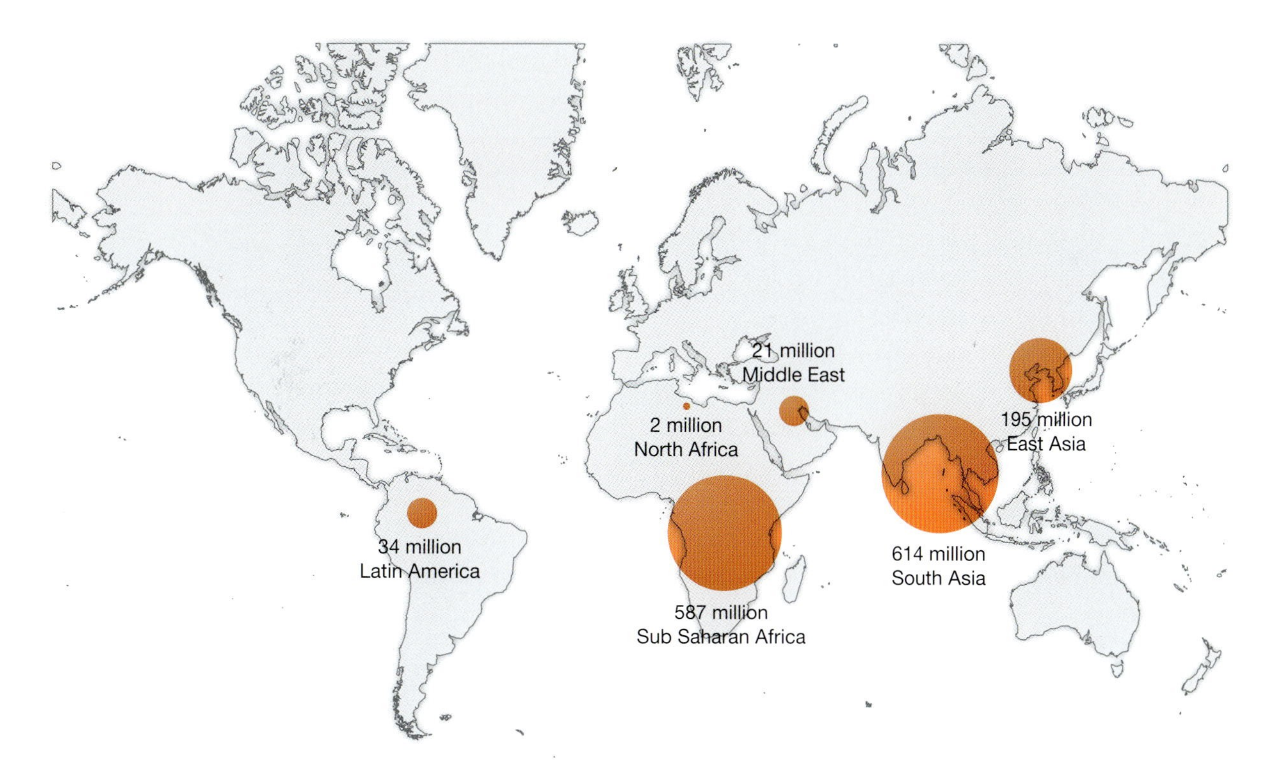

Figure 5.1 More than a billion people have no ready access to electricity, most of them in southern Asia and Africa. In these areas, the top priority is often to get affordable electricity to people who lack it, with greenhouse issues a much lower priority. (Data source: OECD/IEA 2010)

Population growth and the energy mix

There is an elephant in the room that has not received much attention so far in this book: population growth. This is perhaps the most intractable problem, because it goes to the core of what it means to be a human on this earth. At present there are 7 billion people on Earth, a billion of whom go to bed hungry most nights. Around a billion of them (often the same hungry people) also have little or no access to electricity and its accompanying benefits (Figure 5.1).

The world population will rise to 9 billion this century. It is argued by many that what is needed is population control. This is unlikely to happen by government edict. Harsh measures have been tried in China through its one-child policy, but the policy is no longer enforced with any great enthusiasm. In India, despite a more benign approach to population control on the part of the government, the population grows inexorably and will surpass that of China by mid-century or sooner.

Education and improved living standards offer perhaps the greatest potential to decrease population growth in the developing world (which often requires better access to electricity), but for the moment we are on a steep path of population growth and our strategy for tackling climate change must factor that in as 'a given'. Some might see that as defeatist, but 'realistic' is the more appropriate term and it is the assumption that is factored into all the energy projections of world bodies such as the International Energy Agency (IEA) and the Intergovernmental Panel on Climate Change (IPCC).

To feed those extra mouths we will need to increase land productivity, which is probably going to require new advances in biotechnology and the application of more fertilisers, particularly nitrogenous fertilisers and greater mechanisation of farms. Nitrogenous fertilisers are energy intensive to make and produce a lot of CO_2 as a by product. But it is not just the problem of feeding 2 billion more people at a fairly basic level, overwhelming though this is alone, it is also the added issue that those extra 2 billion and also several billion others in developing countries aspire to have access to the same material benefits as the developed world now enjoys.

Is there any way that these reasonable aspirations of the developing world can be met without a massive increase in the global carbon footprint? One way may be for the developed world to decrease its own expectations, and decrease its already excessive carbon footprint, thereby making available more 'carbon credits' for the developing world. Given that most people will not readily accept a decrease in their standard of living, a decrease in carbon intensity to counter the carbon impact of population growth will have to be achieved largely through technology improvements, just as improved biotechnologies will be needed to improve food production.

Biofuels in the mix

Accepting the inevitability of population growth serves to emphasise the importance of energy efficiency and new technologies generally, but also puts into question the place of some technologies in the mix. The development and use of biofuels is complex; in some circumstances they can be produced sustainably and can make a useful contribution to CO_2 mitigation but in other circumstances, the consequences in terms of loss of soil carbon, invasiveness, excess water or fertiliser use or impact on food production, may outweigh the benefits.

Growing feedstock for biofuels has increased competition for arable land, contributing to rising food prices and, in some instances, shortages and civil unrest. One possible way to avoid this is to grow biofuel fuel crops only on marginal, otherwise unproductive lands. This is being tested in many countries using the succulent plant *Jathropha curcas* (also known as physic nut) and in salt marshes where salicornia (also known as glasswort, pickleweed, and marsh samphire) is being grown as a biofuel option. Jatropha in particular has received a great deal of attention: one hectare of Jatropha can produce 400–600 litres of oil, which in turn can be modified to produce a high quality bio-diesel.

However, Jatropha can be poisonous and Western Australia has declared it to be a noxious weed, meaning that it cannot be grown as a commercial crop, or introduced into the state. Jatropha also requires more water than corn or sugarcane, which may limit its value as a biofuel crop to tropical areas.

For the moment, therefore, increasing production of biofuels will require more farm land at a time when more land will also be needed to grow more food. Given that the net energy benefit from corn-based ethanol (the dominant biofuel in the United States for example) is modest at best, it is questionable whether or not corn-based biofuels should be subsidised by the United States taxpayer, or any other taxpayer. The fate of existing subsidies is likely to depend more on the effectiveness of the farming lobby than the strength of the scientific or economic argument. Sugar-based ethanol production, which is more energy effective than other biofuels, including corn-based biofuel, is likely to continue. But all biofuels should be subjected to a full carbon and energy accounting, along with consideration of the water and fertiliser requirements, to ensure that

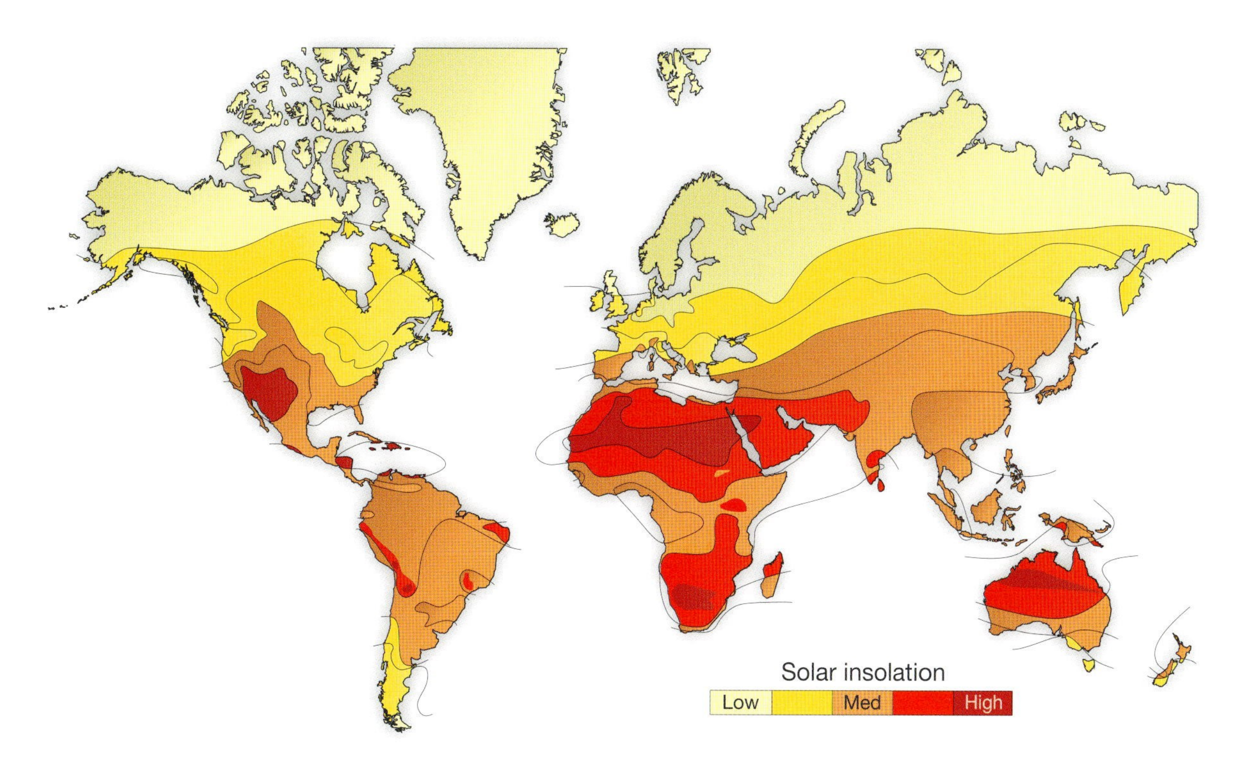

Figure 5.2 Desert and tropical areas in particular have high solar radiation resource. Some of these areas have quite low population densities, but there are parts of Africa and Asia where solar power could provide power now to millions of people who currently have no access to electricity. (Data source: NASA 2001; 3TIER 2011a)

a positive environmental, economic and carbon outcome is achieved.

Land requirements of different technologies

Will the need for living or growing space for an increasing population compete with the space required for the various energy technology sources, given that some of them can have a relatively large footprint? Taking 1000 MW of delivered electricity as our benchmark, what would be the land area occupied by the various options?

In the case of solar, it depends on the system used. Solar photovoltaics (PV) can of course be installed on 'unused' roofs, but large scale solar PV is likely to require a significant area of land, probably of the order of 25 km^2 to provide 1000 MW.

A wind farm delivering 1000 MW of electricity would require an installed capacity of around 3000 MW, which would have a total footprint of 50 km^2 assuming a grid pattern with separations of 3 diameters and 8 diameters between each wind turbine. However, food production could still be undertaken within say 90% of the wind farm footprint and therefore in terms of 'quarantined' land, the figure of several km^2 would be more representative of the footprint.

In the case of a coal-fired power station, taking into account the size of an open cut coal mine able to produce sufficient coal to provide 1000 MW of electricity for 40 years, plus the power station, plus a notional 100 km of railway line for transportation of the coal, the total footprint is of the order of 10 km^2. If the coal is imported to the power station then the footprint could be regarded as that of only the power station and the stockpile of coal, but of course the remainder of the footprint is elsewhere.

For relatively sparsely populated countries such as Canada, Russia or Australia, none of these land areas, when multiplied by total energy requirements, constitute a major barrier to deployment, but for densely populated countries such as those in Western Europe or Asia, the footprint of the various technology options could become a significant issue.

Depending on topography, hydro can have a very significant footprint and a major environmental impact. The land footprint of some other options, such as gas-based power, geothermal and nuclear, is likely to be quite small and the impact of tidal wave power on food production is negligible.

Overall, the footprint of energy systems is relatively modest, except biomass and bioethanol, which have a major impact on land that would otherwise be used for food production. Increasing urbanisation will also have an impact in terms of withdrawing land from active food production, unless we squeeze far more people into the existing urban areas, but this too has its impact on energy use.

Energy and water

How does energy impact on water and visa versa? Potable water is one of our most precious commodities and we must use it sustainably. In densely populated areas, it may need to be augmented by water recycling and/or desalination, but these processes are heavy users of energy. In a paper prepared for the World Policy Institute in 2011, Diana Glassman points out that electricity constitutes 75% of the cost of municipal water processing and distribution and 4% of all power used in the United States is for water supply and treatment. The major component of water use (70–80%) is agriculture, the majority of that in irrigation. Comparing water use for different methods of producing energy from oil and gas shows that water use is minor for conventional oil and natural gas, but much more significant for unconventional shale gas and coal seam gas. This is due to the gas extraction process for the latter. Per unit of energy produced, oil sands are very much higher in water use than either shale gas or coal seam gas.

However, according to Glassman, water use in shale gas, coal seam gas and oil sands pales in significance compared to the volumes of water used to irrigate energy biofuels such as soy or corn. Increased production of biofuels will therefore require more water and in some parts of the world, water availability will be a severe limiting factor.

The availability of water may also inhibit the growth of hydroelectricity, although environmental concerns are the major barrier in many regions. In some areas, climate change and decreased run off will limit the future growth of hydropower. On the other hand, pump storage – where water is pumped uphill when electricity is plentiful and cheap, and then allowed to run back down through turbines when extra power is needed – uses the same water over and over again and can be effectively linked with renewable energy. At the moment, pump storage tends to use off peak night time fossil-fuel based electricity (when the price is low) to pump water back into higher dams, making the 'green credentials' of some hydropower questionable.

Nuclear power requires large quantities of water for cooling. For this reason, nuclear power stations are sited on major rivers or on the coast. However, as discussed in Chapter 4, the impact of the 2011 tsunami on the Fukushima nuclear plant has highlighted the need to re-examine the siting of all coastal nuclear power stations.

Fossil-fuel based power systems usually require large amounts of water for cooling and steam generation and there is likely to be increasing use of air-cooled systems to mitigate this water use. Coal-fired power plants use on average twice as much water as a gas-fired plant. In contrast they use 30% less than a nuclear power station and, according to Dianna Glassman, only half the water used by a solar thermal system, athough this will depend on the fluids used in the solar system.

Stored CO_2 from CCS-equipped thermal power stations could potentially impact on deep groundwater resources. To avoid this, the geology, and the hydrogeology of a potential storage site, must be thoroughly investigated to minimise the potential of stored CO_2 leaking into groundwater or displacing saline groundwaters into potable groundwater resources. This is discussed in Chapter 9.

In addition CCS-equipped power station will use more water than one without CCS as a result of the extra power needs.

Geothermal power would impact on deep groundwaters though not necessarily on shallow groundwaters that contain potable water. Wind power impacts on surface water evaporation, as discussed in Chapter 4, and solar PV power affects surface water runoff. In desert areas, the need to wash solar panels on a regular basis may also contribute significantly to the water needs of solar. Solar thermal can have quite significant water needs, depending on fluids used. In summary, all energy impacts to varying degrees on water resources and this will need to be increasingly factored into decisions regarding future energy systems.

The strong links between water, population, food, and energy serve to emphasise the complexity of the issues we are dealing with and the fact that there is no one-dimensional answer to the energy-climate conundrum.

Renewable energy in the energy mix

So, what are the boundary conditions when considering our future options for decreasing CO_2 emissions to the atmosphere? First, the world's population will increase significantly in the coming decades. Up to 9 billion people will have expectations of higher standards of living, including ready access to water, more food, better health services, greater mobility, more material possessions and, along with all of this, far greater access to reliable electricity. A mismatch between these expectations and what is realistic begins to emerge when we consider technology options with the capacity to deliver 24/7 electricity (because that is what people expect) including:

- potential impact on water resources and food production
- potential to make deep cuts in emissions
- technological maturity
- current level of acceptability by the community.

Energy efficiency is a 'no brainer' in that it ticks all the boxes and should be pursued to the limits of its cost effectiveness. Solar PV has many positive features and has a high level of community acceptance, but it cannot deliver 24/7 electricity until we greatly improve our energy storage systems; there is an obvious need to invest far more in energy storage systems. Solar thermal power is still an immature technology and this is likely to be reflected in its high cost for many years to come, as discussed in Chapter 10. Wind is a mature technology but it is site specific and cannot provide 24/7 power (Figures 5.3 and 5.4). This again points to the need for greater investment in power storage systems as opposed to an approach that seeks to address intermittency by installing more wind turbines that only deliver power for one third of the time (and not necessarily at times when the power is

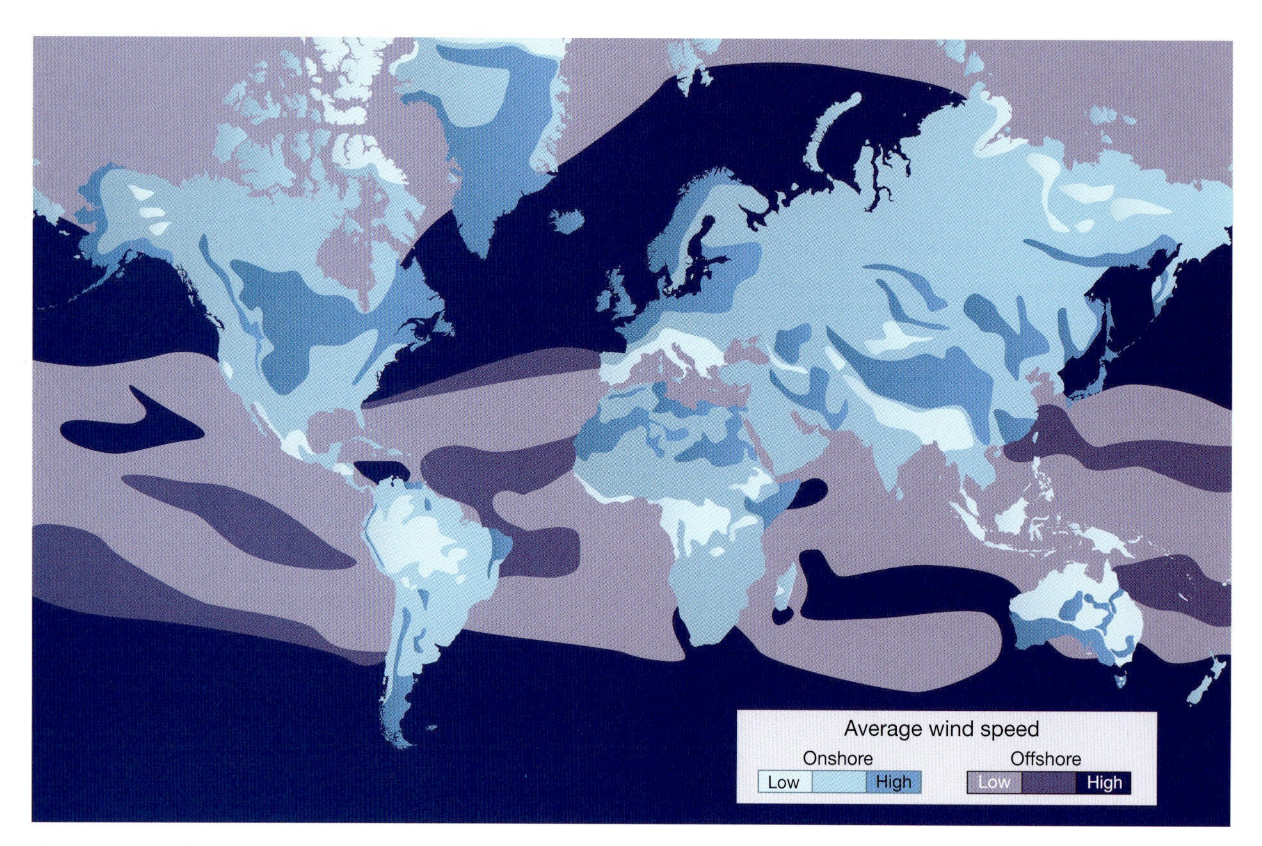

Figure 5.3 Wind power resources are site-specific. Offshore winds are generally much stronger than onshore sites, but offshore wind farms are expensive to build and maintain. (Data source: 3TIER 2011a)

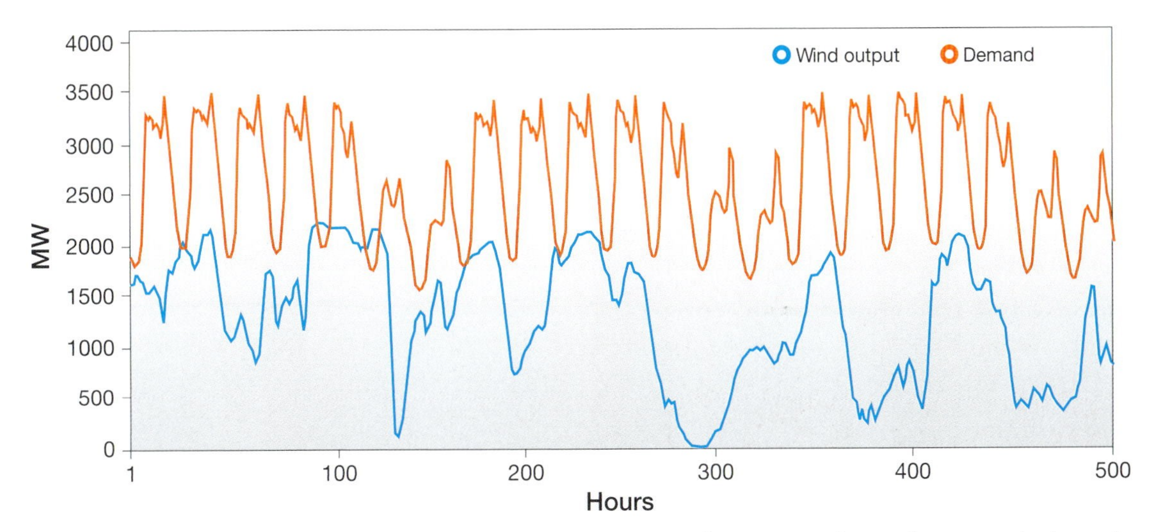

Figure 5.4 As illustrated in this plot showing wind output and power demand over a 3-week period in Denmark, demand and output seldom match. On occasions, wind is sufficient to meet lower demand levels, but there are also times when output falls to almost zero. Developing better power storage systems is critical to the future of wind power. (Data source: Soder *et al.* 2006)

needed). Concerns regarding visual impact, adverse impact on land values and potentially adverse impact on health from low frequency sound, may mean that wind farms will progressively be less acceptable to the community which could limit access to wind resources.

Hydropower has the potential to meet more of our electricity needs (as well as our water needs). However it will progressively decrease as a proportion of electricity supply because of the level of community opposition to the construction of new dams. The question of how to address this concern regarding dams is urgent. But perhaps new dams could be viewed more from the perspective of their value for pump storage as a way of addressing the intermittency of wind or solar.

Wave and tidal power are likely to be acceptable as an option by the community, in part perhaps because they are out of sight. But the reality is that the technology is still quite immature and will not deliver reliable large scale electricity or provide deep cuts in emissions for many years to come.

Geothermal power has the potential to provide 24/7 power, but it is only found in particular geological environments that do not necessarily coincide with the major population centres. While the notion of geothermal power is likely to be acceptable to the community, the reality is that other than in volcanic regions, geothermal is still an immature technology that is also unlikely to deliver deep cuts in emissions for many years to come. Nonetheless it will be part of the energy mix.

In Switzerland in 2007, small earthquakes were induced by a pilot hot fractured rock (HFR) project (see Chapter 4 for more detail on HFR) causing widespread community concern. Hot Sedimentary Aquifer technology is perhaps more promising in the shorter term as it is more widespread, and easier to exploit. However, it provides only low temperature heat.

Biomass is seen positively by the community and politicians alike, but this is likely to change as people consider the impact on food production (and food prices) and water resources. In addition, the massive land area that would be required for biomass to make a major contribution to CO_2 mitigation is likely to inhibit its uptake. Nonetheless biomass is likely to make a valuable contribution to decreasing CO_2.

Overall, renewable technologies enjoy a high degree of acceptability by the community. However, the reality of what the technologies can actually deliver, whether individually or collectively, and at what cost, is often at odds with people's hopes. In the short term, this is reflected in the high cost of renewable energy and the reluctance of people to pay the extra cost of that renewable energy. People like the idea of renewable energy – so long as somebody else pays for the extra cost! In addition while the community may like the idea of renewable energy in the abstract, this does not always translate into acceptance by a community that is directly affected by a new wind farm or a new dam.

These issues with renewable energy do not diminish its potential importance as a clean energy option, but it is important to be realistic about some of the potential hurdles and what renewable technologies can deliver and when. This must take into account the cost and recent experiences in a number of countries, where high subsidies (feed-in tariffs) have resulted in unsustainable costs on the public purse, highlighting the problem of cost, a topic which is considered in more detail later in this book.

There is a need to invest more, but also more wisely, in renewable technologies. We must recognise that most of them only provide intermittent power at the present time: rather than encouraging the installation of more intermittent power generation technology, we should put greater effort into improving power storage, whether through existing technologies such as pump storage or more innovative technologies such as fuel cells, compressed air storage, heat storage in graphite or chemical systems. Better energy storage technology is essential to enable renewable systems to reach their full potential and make a more effective contribution to decreasing greenhouse gas emission. This is preferable to proceeding on the assumption that renewable energy can continue to be backed up by fossil fuel (usually gas)-based systems, an approach that does not fully take into account the cost of that back-up. Renewable energy must be part of the mitigation mix but must also be assessed as to its full cost, its impact on land, water and food and its capacity to deliver deep cuts in emissions and within what timetable, taking into account the growth in population globally.

Non renewable energy in the energy mix

There is little doubt that nuclear power is perceived as the least acceptable non-renewable energy option by a large proportion of the community who regard it as 'too risky', despite its advantages as a clean energy source. Japan is re-examining its nuclear program; in Germany, the government has decided to close down its nuclear program; Italy has put its nuclear program on 'hold'. Conversely in Korea, the country closest to Japan, there has been no apparent increase in opposition to nuclear. Without in any way downplaying the Three Mile Island, Chernobyl and Fukashima accidents, the total number of direct and indirect fatalities and injuries from these accidents is low.

The perceived link with nuclear weapons, and the lack of transparency of many nuclear companies, have contributed to the unpopularity of nuclear power. The reality is that nuclear power has produced reliable zero carbon electricity for the past 50 years and long term safe waste disposal poses no insurmountable technical challenges. Nuclear power has the potential to make deep cuts in global emissions and, if other forms of energy prove to be more expensive or more problematic than people currently hope or expect, there could be a significant increase in nuclear power worldwide. But in the short term, uptake of nuclear power will be inhibited by vocal community opposition and unwilling governments. In this context, it is

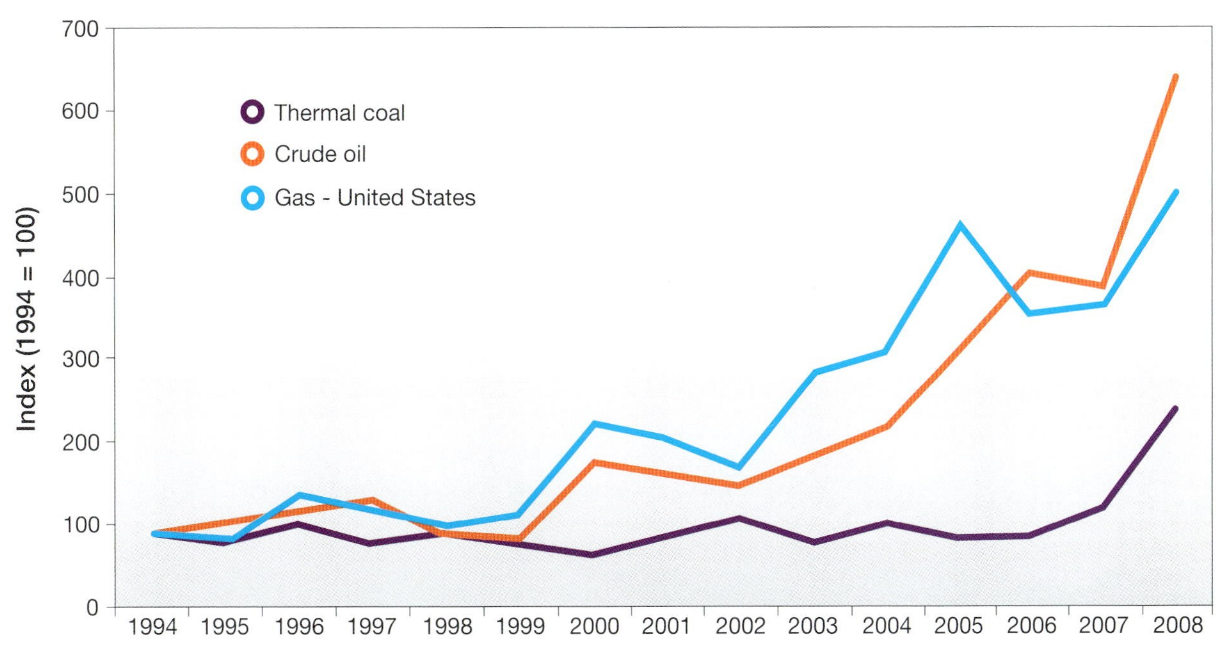

Figure 5.5 This graph showing how the price of coal, oil and gas has varied over time, illustrates that the price of thermal coal has been far less variable than the price of oil or gas, which in turn has a significant impact on future plans for fossil-fuel based power stations. (Data source: Garnaut 2008)

instructive to note that Germany's post-Fukashima declaration that they will close down their nuclear power stations, was made in the knowledge that any resulting short fall in Germany's electricity supply can be met by importing electricity from surrounding countries, including nuclear-generated electricity from France.

Prior to the Fukashima accident, nuclear power production in Japan was expected to double over the subsequent 30 years. Obviously the Fukushima accident has put a question mark over this. But Japan relies on nuclear power for 24% of its electricity and fossil fuels for 66% of its electricity generation (26% gas, 27% coal, 13% oil), so the reality is that Japan (and a number of other countries) have few short term options to achieve their emission reduction targets. Therefore, nuclear will continue to figure in the energy mix of many countries, including Japan. Globally, modelling by the IEA suggests that 6% of the world mitigation effort will come from deployment of nuclear power. If this target is not reached, then greater energy efficiency, more renewable energy and greater use of CCS will be required.

Switching from coal to lower carbon natural gas can significantly decrease CO_2 emissions per unit of energy and this will undoubtedly be an important part of the greenhouse strategy of many countries that are now heavily reliant on coal. But gas is more expensive than coal, its price has in the past been more variable and, while natural gas is very abundant, it is less abundant than coal (Figure 5.5).

Replacement of coal by natural gas is likely to have a high level of community acceptance, unless the substitution results in major increases in electricity costs. In the short to medium term, gas is undoubtedly part of the answer to greenhouse concerns, but in the longer term, gas becomes part of the problem in that it produces CO_2 emission – unless CCS is deployed. CCS applied to gas separation at the well head is likely

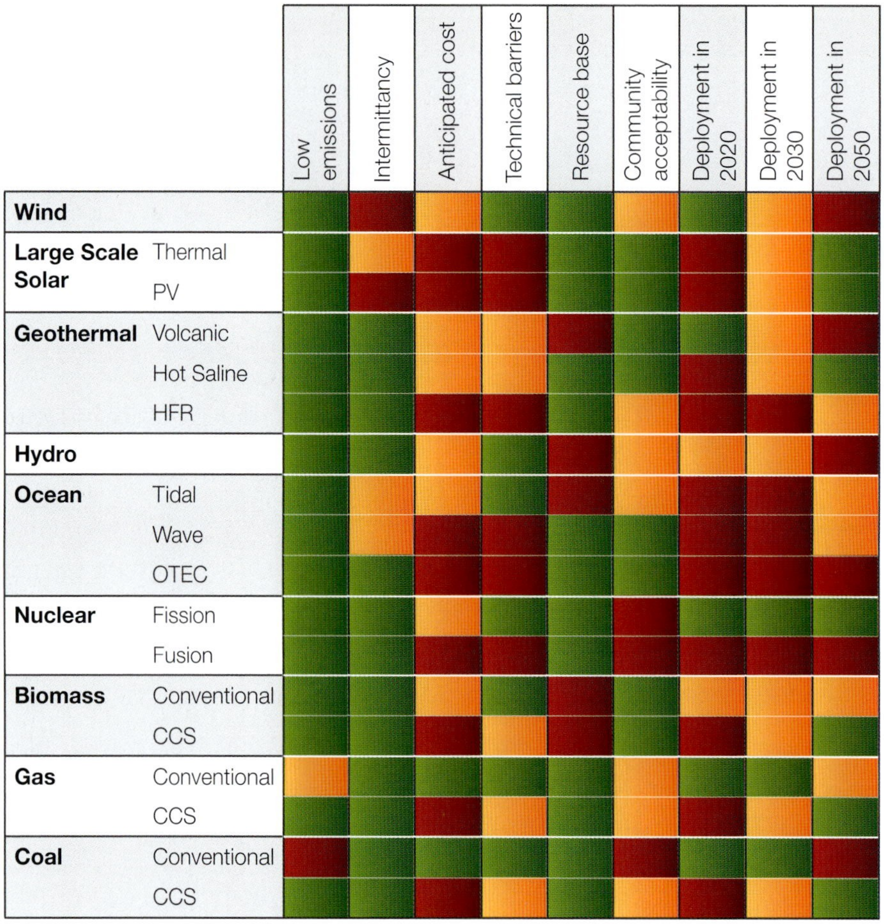

		Low emissions	Intermittancy	Anticipated cost	Technical barriers	Resource base	Community acceptability	Deployment in 2020	Deployment in 2030	Deployment in 2050
Wind										
Large Scale Solar	Thermal									
	PV									
Geothermal	Volcanic									
	Hot Saline									
	HFR									
Hydro										
Ocean	Tidal									
	Wave									
	OTEC									
Nuclear	Fission									
	Fusion									
Biomass	Conventional									
	CCS									
Gas	Conventional									
	CCS									
Coal	Conventional									
	CCS									

Figure 5.6 Large scale energy choices can be influenced by a range of issues including their carbon intensity, the reliability of supply (is it 24/7 electricity?) and so on. This matrix gives a view on these issues and also on the perceived readiness (or not) by 2020 and 2050. Of the clean energy choices, green represents a positive outlook for the technology, orange is neutral and red indicates a negative outlook.

to be favourably received by the community at large and projects such as the Gorgon Project in Western Australia will strengthen the overall level of community acceptance of gas and liquid natural gas (LNG). Application of CCS to gas-based electricity generation is as technologically and financially challenging as applying it to coal-based electricity.

There will be criticism of gas with CCS or coal with CCS because of the underlying wish of some environmental groups to cease all extraction and use of fossil fuels as soon as possible, despite the enormous difficulties in attempting to bring about such a profound change. Indeed to attempt such a change in just a few years would potentially destroy the economies of many countries and the world economic system as we know it. Profound changes in our energy system are more likely to occur over decades rather than a few years, and are likely to be implemented with a portfolio approach that includes CCS. According to the IEA, ongoing (and perhaps increasing) use of fossil fuels appears likely for many years to come. Coal is likely to be the fuel of choice in many developing countries and gas will increasingly be the fuel of choice in many developed countries, with CCS, providing a transitional technology to a lower carbon economy.

The energy mix in the medium to long term

Given that each of the clean energy technologies discussed so far has its positive features and its challenges (Figure 5.6), what is the energy mix going to look like in the medium and long term? Compared to the low cost of conventional coal-fired power generation, all other technologies are medium to high cost, and therefore moving away from existing conventional electricity generation will be expensive. Some of the renewable technologies face significant technical barriers before they will be in a position to offer a commercially viable option. For some renewable technologies the scope may be limited for expanding the resource base on which the technology depends, with examples of this including volcanic geothermal, tidal, hydro and biomass.

Based on an assessment of technical barriers, the need for energy security, the cost and the resource base, the IEA concludes that by 2020, unless there are some profound policy changes, conventional coal, gas and nuclear will continue to dominate the global electricity generation scene, with wind, hydro and perhaps solar thermal making only a modest contribution. This obviously is not good news for greenhouse gas emissions if it amounts to business as usual.

By 2030, according to the IEA, wind is likely to be a significant contributor to electricity generation (up to 20%) backed up by natural gas generation. HFR is likely to be still subject to technical difficulties, but HSA could be making a useful contribution by that time. Coal and gas with CCS will be a significant contributor, and biomass with CCS may also make a contribution. However, several of the technologies are seen by the IEA as not making a significant contribution even by 2030 – unless there are some profound policy changes, including:

- much higher prices for electricity
- a very high price on carbon, or
- a regulatory requirement to deploy particular technologies, or
- major R&D breakthroughs, or, more likely
- a combination of the above.

A number of these issues are considered in some detail in Chapters 10 and 11.

In an elegant approach to the topic of the future mitigation portfolio, Pacala and Sokolow of Princeton University have illustrated how the various technologies could contribute to greenhouse gas reduction, by representing them as a series of wedges that together have the potential to produce deep cuts in emissions over the 'business as usual' baseline emissions (Figure 5.7). This same approach was used by the International Energy Agency in 2008 to produce their so-called 'BLUE map' scenario, which models the magnitude of the problem and offers a view on the technology mix (Figure 5.8). End use efficiency is seen by the IEA as the largest single component of the mix, representing about one third of the total mitigation effort to 2050. The second largest component is increased use of renewable energy (a little over 20%), with CCS in power generation (and industry) at 19%. Fuel switching (primarily of coal to gas) constitutes

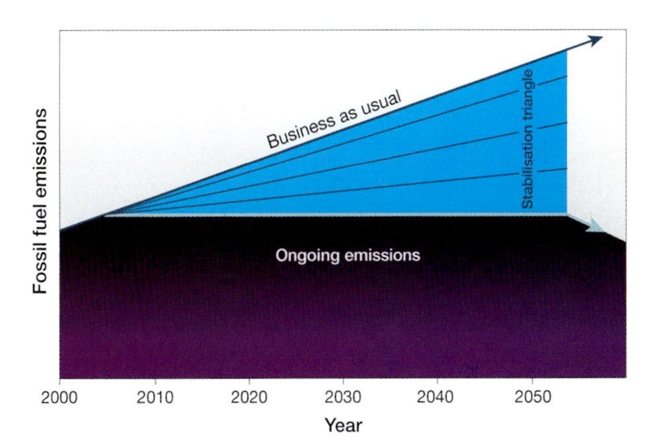

Figure 5.7 Sokolow and Pacala first used energy wedges of varying magnitudes to illustrate that a range of clean energy options will be needed to make the deep cuts in emissions required to meet emission targets. (Adapted from Pacala and Socolow 2004)

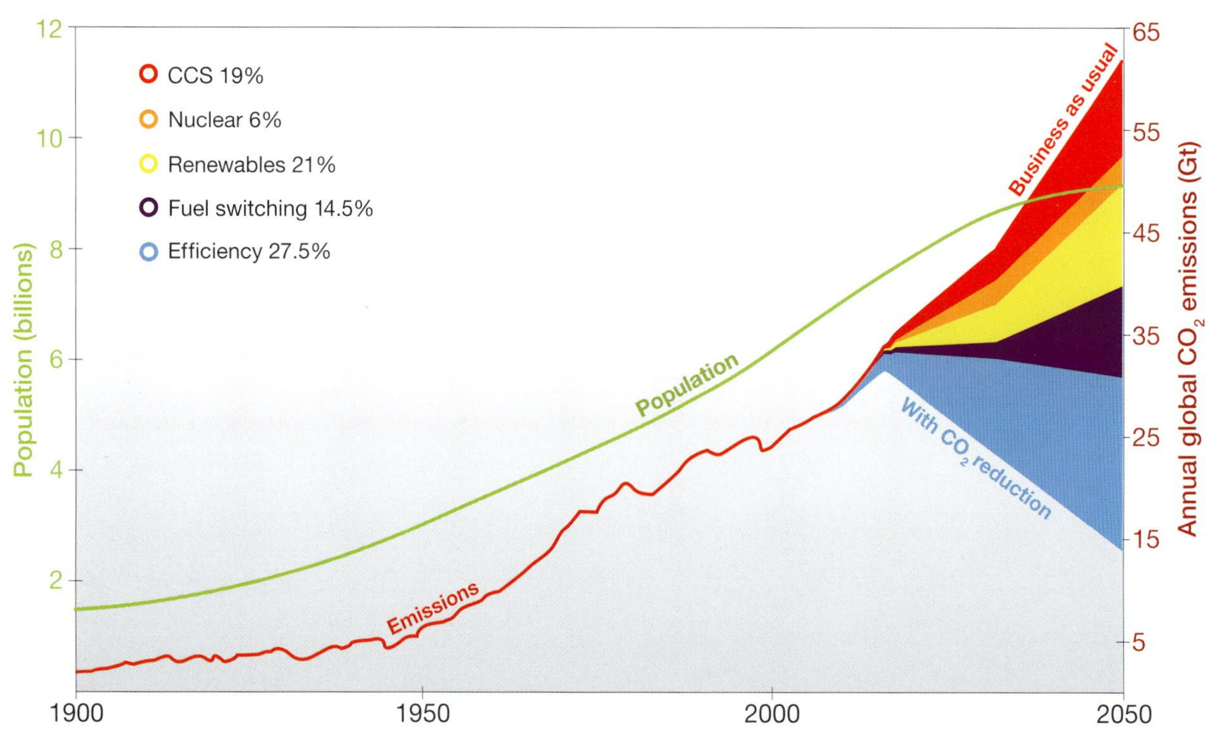

Figure 5.8 The wedge approach of Sokolow and Pacala has been adapted by a number of authors and organisations to indicate how an emission target might be achieved. This wedge diagram, adapted from the IEA, is underpinned by the 'blue map scenario', with the cuts achieved through increased energy efficiency (27.5%), increased renewable (21%) deployment of CCS (19%), fuel switching (14.5%) and increased nuclear (6%). (Data source: OECD/IEA,2008; United Nations 2004)

around 18% of the total reduction, while nuclear is about 6%.

Of course these outcomes depend on the assumptions in the model. But the overall trends are consistent with the view that there is no single solution to the greenhouse issue but rather that there has to be a portfolio of measures. The complexity of the solution is highlighted by the IEA projections for 2030 which shows that while renewable energy more than doubles, the total energy contribution from renewable energy is still relatively modest. The nuclear contribution remains constant, while gas shows significant growth. Alongside this, the IEA projection for the total increase in the use of coal is massive, dwarfing the increase in renewable.

A converse view is offered in a 2011 review of the potential of renewable energy by the IPCC Working Group 3, which modelled many different scenarios; one having up to 77% of the world's energy supply met by renewables in 2050. At the present time less than 13% of total primary energy supply is provided by renewable energy, so this would represent a massive increase over the next 40 years. In considering the individual components of its 2050 scenario, the IPCC suggests that a number of renewable energy sources will increase in absolute terms but decrease in terms of percentage of the total energy mix. Hydropower is projected to decrease as a percentage of total power production by 2050. Ocean energy is projected by the IPCC to be making only a very small contribution to the electricity supply by 2050, while wind, which meets only 2% of world electricity demand at present, is projected by the IPCC to be meeting more than 20% of world demand by 2050. The IPCC scenario considers that up to 30% of the world's electricity could be supplied by geothermal by 2050 compared to less than 1% at present time. Solar PV and concentrating solar thermal, which currently contributes less than 1% of the global energy supply is seen as likely to contribute less than 10% of electricity generation by 2050.

So in contrast to the IEA scenario, which indicates the need for action to address the massive increase in the use of fossil fuels, the IPCC review suggests that renewable energy can meet much of the challenge of decreasing greenhouse gas emissions and by implication that there is perhaps less need for a technology such as CCS. But is this really so?

First, it is important to emphasise that both the IEA and the IPCC numbers are projections based on scenarios and should therefore be treated with caution. Second, let us consider emissions outcomes using the extreme IPCC 77% renewable figure as the most optimistic outcome possible for renewable. According to this scenario, electricity production using fossil fuels would drop from around 13 600 TWh of electricity in 2010 to approximately 8400 TWh by 2050. If we take into account likely annual growth in electricity of 2.25%, even a 40% drop in fossil fuel use would produce sufficient CO_2 in the coming decades to add a further 100 ppm of CO_2 to the atmosphere, taking the atmospheric CO_2 concentration to around 500 ppm. In other words, even under the most optimistic models of the IPCC for the uptake of renewable energy, it would still be necessary to take action to mitigate emissions from fossil fuel-fired power stations. At the present time, the only technology option available to mitigate CO_2 emissions from large scale stationary sources using fossil fuels is carbon capture and storage (CCS).

The IEA suggests that there will be no uptake of renewable energy beyond 20% in 2050. Moreover, its projected 33% mitigation contribution from energy efficiency may be on the high side of what will actually be achieved. In other words, the 19% projected by the IEA as the mitigation contribution from CCS is realistic.

Indeed, there are grounds, based on the IEA figures, for suggesting that the mitigation contribution from CCS could be higher than 19% For example, if the projected contribution from nuclear is not achieved, then there may be greater use of fossil fuels with CCS. In addition, there is scope for applying CCS to the 18% mitigation contribution achieved mainly by switching from coal to gas.

Conclusions

Whether we take the 2011 IPCC view of what renewable energy might achieve by 2050, or the 2010 IEA view, it is quite clear that CCS must be an integral part of the future clean energy portfolio, globally and for many countries. There are also strong economic grounds for including CCS in the portfolio. The IPCC Special Report on CCS considered that inclusion of CCS in the range of measure would decrease the overall cost of mitigation by one-third. A recent study of Australia's Productivity Commission provided evidence of the staggeringly high cost of some existing renewable measures. The issue of cost is considered in Chapters 10 and 11.

Therefore for a range of reasons, it is necessary to see CCS as a key clean energy technology to meet 2050 emission targets. The following chapters consider CCS in detail, to enable the reader to understand the challenges of and the opportunities for CCS.

6 WHERE AND HOW CAN WE CAPTURE CO_2?

Directly removing CO_2 from the atmosphere

Can CO_2 be directly removed to decrease its atmospheric concentration? Plants can of course take CO_2 out of the atmosphere and biosequestration is part of most national mitigation strategies although there are difficulties in auditing the stored carbon. A number of techniques have been suggested for chemically removing CO_2 directly from the atmosphere.We have been chemically removing CO_2 from the atmosphere of submarines and space shuttles for many years, but it has never been attempted at a large scale, primarily because the technical and cost barriers appear to be insurmountable. Despite this, the option is receiving some attention.

One technology suggested to remove CO_2 from the atmosphere, is the 'soda-lime' process. This involves passing air through a sodium carbonate solution and then a calciner, to separate out the CO_2, which can then be stored. But the scale that would be required to decrease the concentration of CO_2 in the atmosphere by just a few parts per million is almost inconceivable because of the low levels of CO_2 and the immense volumes of air that would need to be processed. Another proposal is to enhance the rate of weathering: in the geological weathering cycle, CO_2 reacts with rocks rich in silicates to form carbonate minerals. The process is very slow, but could be sped up by grinding suitable rocks (such as aluminosilicates) to a very fine reactive powder. However, the cost of finely grinding large quantities of rocks would be enormous; there are potential health and environmental problems with finely ground aluminosilicates; and there would be major challenges in handling massive quantities of rocks and the resulting minerals. Enhancement of weathering rates to decrease atmospheric CO_2 is unlikely to work at a scale or at a cost that would make it practical. The process, known as mineral sequestration, is discussed in Chapter 8.

Enhanced fertilisation of the oceans by, for example, the addition of iron, will produce an increase in organic productivity and marine algal growth. Could this be used to decrease atmospheric CO_2? In some ways this could be seen as a marine version of enhanced terrestrial productivity, but the difference is that the marine ecosystem is less understood and less constrained compared to land systems, raising the possibility of unintended and unforeseen consequences. Enhancement of ocean productivity is likely to encounter a great deal of political and community opposition and is therefore unlikely to be able to play any significant role in decreasing atmospheric CO_2.

So for the most part, we must look to preventing man-made CO_2 from entering the atmosphere in the first place, by capturing it at the source, rather than trying to remove it from the air.

Capturing CO_2 emitted from various sources

The CO_2 emitted from mobile CO_2 sources such as vehicles, or from small-scale sources such as domestic gas heaters is too dispersed to be captured, which is why attention is focused on large scale stationary sources of CO_2 (Figures 6.1 and 6.2). Some examples of where we can potentially capture CO_2 include coal or gas or biomass- fired power stations, natural gas processing plants, iron and steel plants, oil refineries, cement plants, fertiliser plants and biomass generators.

The concentration of CO_2 in emissions varies with the fuel and operating conditions at the source. For example, in some existing coal-fired power plants, the flue gas concentration of CO_2 is in the range 12–14%, whereas natural gas fired boilers produce emissions with a CO_2 concentration of 7–10% (Table 6.1). Emissions from gas turbines have a CO_2 composition of 3–4%. The concentration of CO_2 in emissions from an iron blast furnace can be as high as 27%, while for a cement kiln they can be between 14% and 33%. At the other end of the spectrum, the emissions arising from the production of ethanol during the fermentation of sugar can be close to 100% CO_2. The concentration of CO_2 in emissions is a key factor (although not the only one) in determining how easy, or difficult it is to separate the CO_2 from the other gases: the higher the percentage of CO_2 in emissions, the lower the cost of capturing the CO_2 from the emissions.

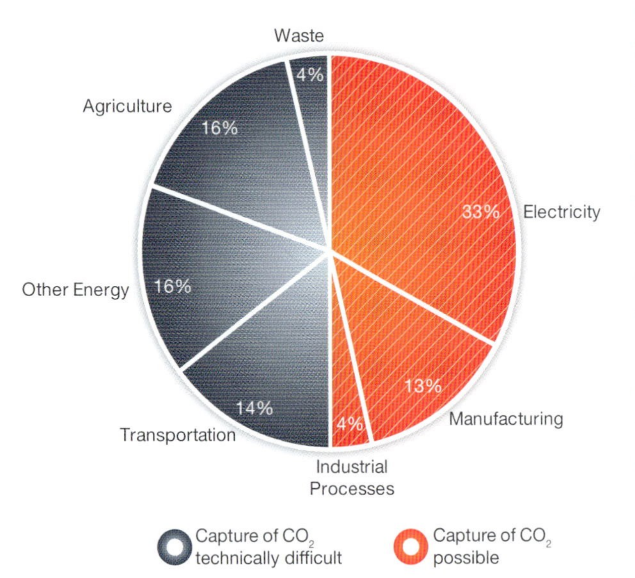

Figure 6.1 This example illustrates which Australian sources are potentially amenable to large scale capture of CO_2, with electricity production clearly representing the dominant opportunity for CCS. (Data source: Department of Climate Change 2009)

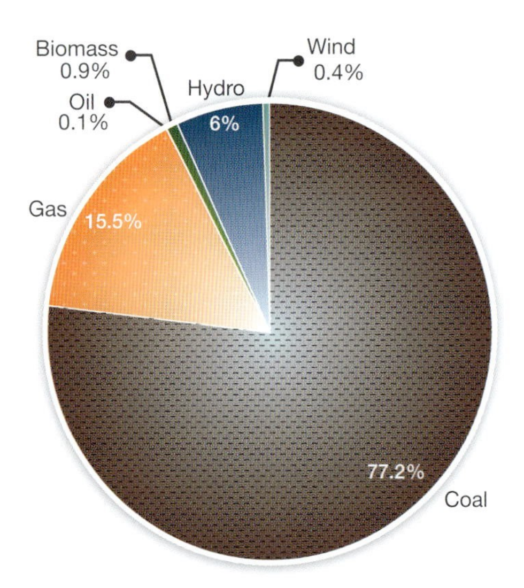

Figure 6.2 This example graphically illustrates the dominance of coal in electricity production, in Australia, with a minor role of gas and a very small component currently provided by renewable energy. A number of other countries have a similar energy profile (Data source: OECD/IEA 2005)

When combustion takes place in air (composed of 78% nitrogen), most of the nitrogen (which is unreactive) remains as nitrogen throughout the combustion process. Consequently there is a high concentration of nitrogen in the waste

Table 6.1 The concentration and pressure of CO_2 in emissions varies enormously depending on the source and the industrial or energy process. Although there are many variables that impact the implementation of CCS, generally, the higher the CO_2 concentration and the higher the pressure the easier it will be to capture the CO_2. For example it is easier (and cheaper) to apply CCS to an ammonia plant than to power station emissions. (Data source: IPCC 2005)

Source type	Carbon dioxide concentration in waste gas (%) of dry volume	Pressure of gas stream (kPa)
Coal-fired thermal	12–14	100
Natural gas thermal	7–10	100
Natural gas turbine	3–4	100
Iron blast furnace	Up to 27	200–300* 100**
Cement kilns	14–33	100
Sugar fermentation	100	100
Ammonia production	18	2800
Natural gas	2–65	900–8000

*before combustion **after combustion

gases. The CO_2 then has to be separated from the nitrogen (and other minor constituents) before it can be captured. However, if the fuel is combusted in an oxygen-rich gas (see the discussion on oxyfuel combustion later in this chapter), then the step of separation of CO_2 from nitrogen can be avoided or at least minimised. In other words each process has its particular set of emissions and its own challenges and opportunities when it comes to separation of CO_2 from emissions (see Box 6.1).

BOX 6.1: HOW IS CO_2 PRODUCED WHEN CARBON-BASED FUEL IS BURNT?

Burning any carbon-based fuel, such as wood or fossil fuels, converts the carbon content mostly, though not exclusively, into CO_2. Some nitrogen is converted into nitrogen oxides (NOx) and if there is sulphur present, this results in sulphur oxides (SO_x) in the emissions. Coal is quite variable in its chemical composition, but its combustion can be represented by the equation.

Coal (C_xH_y) + O_2 + other components → xCO_2+ yH_2O (water) + other gas flue components + heat.

If the combustion is incomplete, and the carbon is not fully oxidised, then there is also carbon monoxide (CO) in the gases produced.

Similarly, when methane (natural gas) is burnt, CO2 and water are the major products along with energy in the form of heat

CH_4 (methane) + $2O_2$ → CO_2 + $2H_2O$ + heat

CCS and gas production

Natural gas is largely composed of methane, with some heavier hydrocarbons, and varying quantities of impurities, notably CO_2, hydrogen sulphide, nitrogen and helium, which may need to be removed prior to use of the methane. While the natural gas from most gas producing wells has a CO_2 content of less than 5%, the CO_2 content of natural gas does vary enormously: some wells produce gas with as little as 1% CO_2; others produce gas composed mainly of CO_2. Carbon dioxide is non-combustible and reduces the heating value of the gas, to the extent that high concentrations of CO_2 make the gas unmarketable for domestic or industrial purposes. Therefore in most developed countries, the CO_2 content of natural gas is generally reduced to 2–4% before it is put into the pipeline. In some developing countries, high CO_2 gas is quite widely (and quite inefficiently) used for domestic and industrial purposes. However, if the natural gas is to be used for the production of liquid natural gas (LNG) then it is necessary to remove all the CO_2.

For these reasons, the removal of CO_2 from natural gas (along with components such as hydrogen sulfide), a process known as natural gas sweetening, has been practised by the gas industry for over 80 years (Figure 6.3). It is a simpler and cheaper process than removing CO_2 from the low pressure waste gas of a coal or gas-fired power plant, where the emissions also include water, oxygen, and other gases.

Some of the first commercial CO_2 capture operations have been established as part of the natural gas production process and include the Sleipner and Snohvit natural gas projects in the North Sea (see Sleipner project, Chapter 8; Snohvit Project, Chapter 7) and the In Salah natural gas project in Algeria (see In Salah project, this chapter). These projects each separate 1–2 Mt of CO_2 from natural gas each year prior to storage.

The $43 billion Gorgon LNG Project in Western Australia will separate out 3–4 million tonnes of CO_2 a year from offshore natural gas fields containing an average 14% CO_2, commencing in 2014–2015. The production of LNG involves the removal of any CO_2 from the original natural gas and then compression and cooling of the methane gas, which requires large amounts of energy, resulting in additional direct or indirect emissions of CO_2. LNG companies have already taken steps to decrease these fugitive emissions, through improved burners and more efficient cooling. However, to date none have announced plans to capture or store CO_2 from the actual LNG plant, which would be a much more complex and expensive process than the initial separation of CO_2 from natural gas. The capture of CO_2 from produced natural gas is likely to continue to be one of the early movers of CCS technology because it is relatively simple and cheap to carry out, compared to separation of CO_2 from power station emission, but it is likely that there will also be increased attention given to the application of CCS to LNG plants and to gas fired power stations in the future.

CCS and coal and gas-fired power generation

A very significant proportion of today's emissions of CO_2 come from coal (and to a much lesser

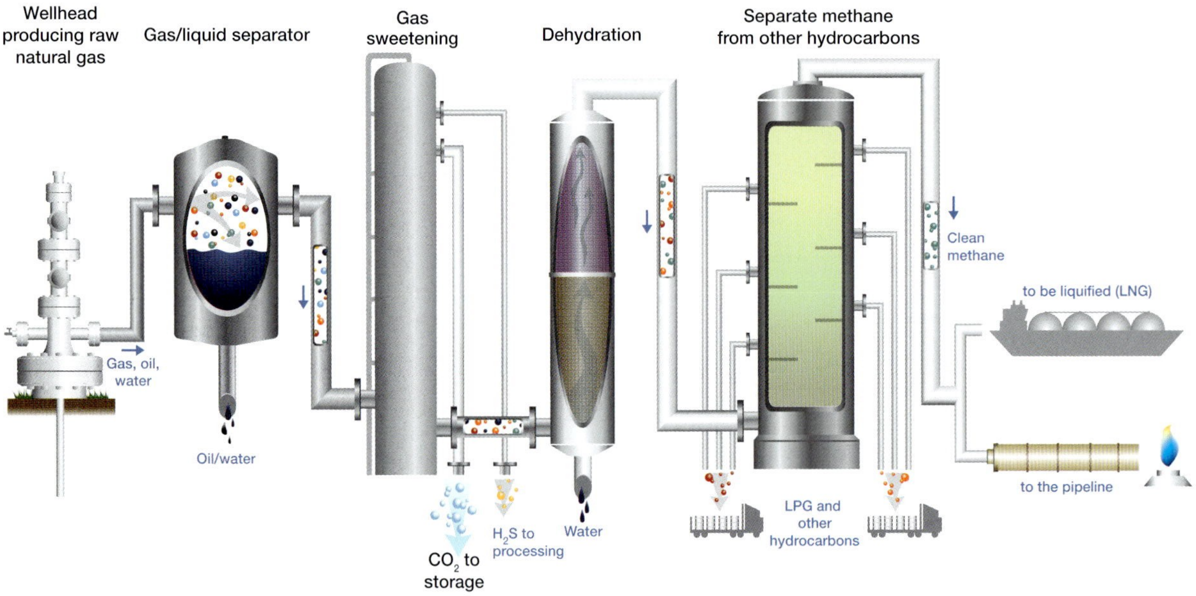

Figure 6.3 Most natural gas has some carbon dioxide in it when produced and often it is necessary to remove that CO_2 before the natural gas can be put into the gas network, or before it can be liquified to produce LNG.

extent gas) used for power generation. According to the IEA World Energy Outlook published in 2009, fossil fuel-based power plants around the world emitted 11.9 Gt of CO_2 in 2007, with coal-fired power plants accounting for 8.7 Gt.

Flue-gas concentrations of CO_2 are in the range 12 to 14% for coal fired power plants and somewhat less for gas. Increasing this to 90% or 100% CO_2 as part of the CCS process represents a significant technical and economic challenge. Nonetheless CO_2 is separated from waste gases at various power stations around the world, though mostly on a relatively small scale. The largest operational post combustion capture plant was, until recently, the Mountaineer power station in the eastern United States (see Alstom and American Electrical Power Mountaineer project, Chapter 9), but the challenge is to scale up capture to full plant size of, say, 500 MW or larger, as well as bring down costs and deploy the technology widely.

In a typical coal-fired power station (Figure 6.4), the coal is first crushed into a powder (pulverised fuel), which is then blown into a combustion chamber and burnt in air at temperatures of up to 1500°C. The burning gases heat water (in pipes) to form superheated steam. This superheated steam drives a series of turbines to generate electricity (Figure 6.5). When the steam leaves the last turbine, it is condensed so that it can be re-circulated to the boiler. A more efficient way of burning coal for production of electricity is achieved by using fluidised-bed combustion. In this process, a layer of fine material (such as sand), and finely ground coal, is blown into the combustion chamber using high pressure air. The fine 'sand' and finely ground coal floats (like a fluid); the coal burns and the sand transfers the heat to water pipes forming steam. Coal can also be gasified before combustion, to produce electricity (see under 'CCS and gasification' later in this chapter).

Natural gas can also be used directly in a gas turbine to generate electricity or can be combined with a steam turbine in natural gas combined cycle (NGCC) power generation (Figure 6.6). In this system, the gas is combusted and expands rapidly, which in turn drives a gas

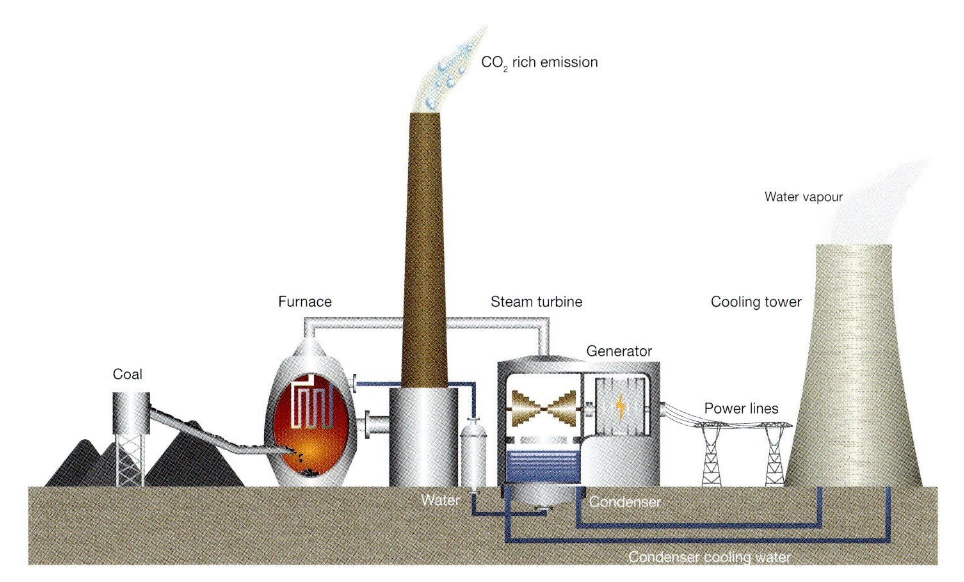

Figure 6.4 In this schematic diagram of a pulverised coal-fired power plant, crushed coal is injected into the furnace to produce superheated steam, which then drives the steam generator to produce electricity. The cooling tower emits water vapour and the CO_2 is emitted from the chimney stack.

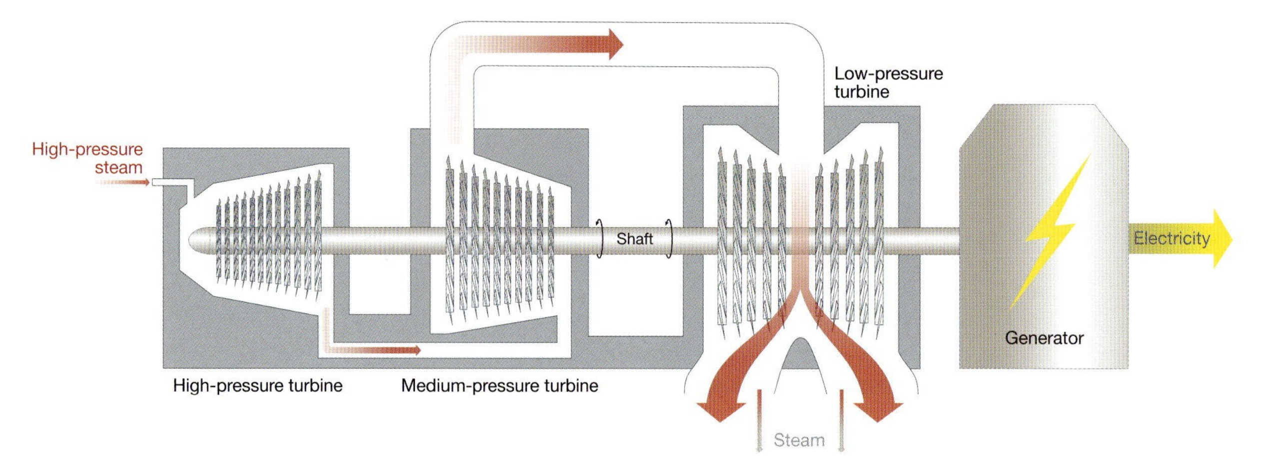

Figure 6.5 High pressure superheated steam from a coal-fired furnace drives the steam turbines which produce electricity.

turbine. The hot waste gas from the gas turbine is then used to heat water (in pipes) to drive a steam turbine. This combination of gas and steam turbines maximises the efficiency of the gas to electricity conversion (Figure 6.7).

Post combustion capture

Most of the world's existing fossil-fuel power plants operate by burning (combusting) the fuel in air and then using the energy created by that combustion to drive turbines. To capture the CO_2 from these power plants, it is necessary to separate the CO_2 from the exhaust gases after the combustion process, through a process called post-combustion capture or PCC. Post-combustion capture facilities can be retrofitted to existing power plants, and this is likely to be important if we are to reduce global CO_2 emissions significantly, given that the pulverised coal power plants now being built will still be operating in 30–40 years time.

In the past decade or so, many new power stations have been built in India and China. In China a massive electrification program is underway. Most of these power stations are coal-fired. Many of them incorporate new technologies and high levels of efficiency. It is unrealistic to expect that any of these power stations will be closed down before they have reached the end of their economic life and this is why technology for retrofitting of CO_2 capture plants needs to be

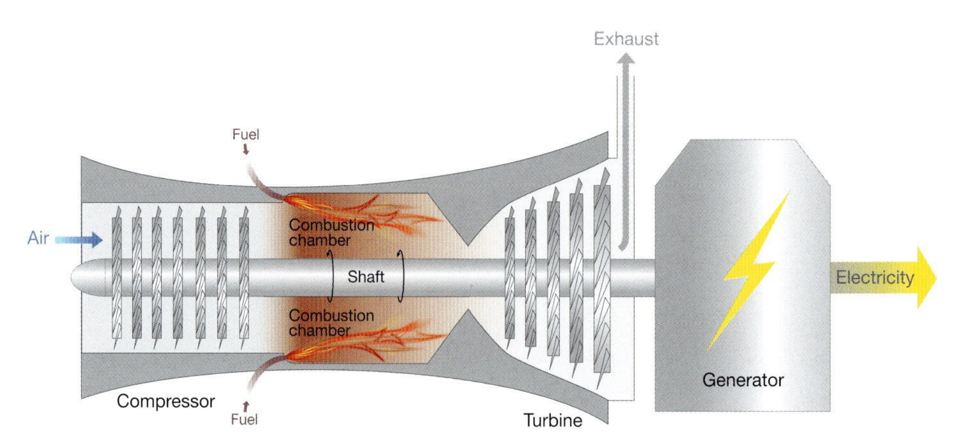

Figure 6.6 In a gas turbine, the expansion of gas in the combustion chamber drives the turbines to produce electricity.

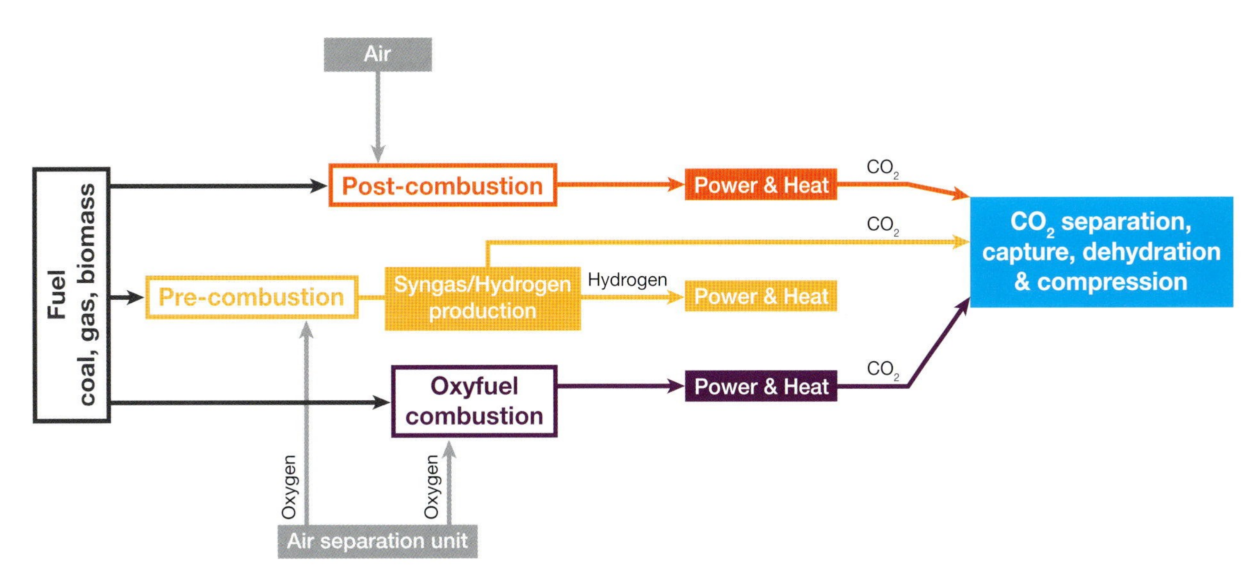

Figure 6.7 There are various combustion options for power generation and CO_2 capture. (Adapted from IEA 2009; IPCC 2005)

developed as a viable economic and technological option. Research efforts are currently focused on:

- reducing the cost of PCC
- increasing the efficiency of the capture process
- dealing with impurities such as sulphur oxides (SO_x) and nitrogen oxides (NO_x)
- upscaling current pilot plants and demonstration plants capturing tens of thousands of tonnes of CO_2 to full commercial scale requiring the capture of millions of tonnes of CO_2.

A variety of technologies can be used to capture the CO_2 after combustion and these are described in more detail later in this chapter (Figure 6.8).

Post combustion capture with oxygen-rich combustion

If fossil fuels are burned in an oxygen-rich atmosphere rather than in air (a process known as oxyfuel combustion or oxyfiring), a much purer stream of CO_2 is produced. To do this, nitrogen which makes up 78% of air, must first be removed using an air separation unit (ASU). In an ASU, compressed air is cooled to about –180°C and passed through a separation column. The oxygen condenses and collects at the bottom of the column, while the nitrogen is emitted from the top of the column. The oxygen is then used in the combustion process.

In the absence of nitrogen, the waste gas is composed mainly of CO_2, with some water vapour and trace amounts of other gases such as carbon monoxide, depending on the efficiency of the combustion process. Oxyfuel combustion has the advantage of making the CO_2 relatively easy to recover as it makes up most of the emissions. Burning coal in pure oxygen creates a very high temperature flame and therefore in order to retrofit a conventional pulverised coal power plant with oxyfiring technology, the combustion temperature needs to be reduced by recycling some of the flue gases back into the furnace. However, inert gases such as argon can build up in the flue gases and there is the potential for leakage of atmospheric nitrogen which in turn dilutes the emissions and increases the cost of CO_2 removal. New-build plants can be designed for higher operating temperatures, which should avoid the need to re-circulate gases and also minimise the potential for leakage of nitrogen into the system. Oxyfuel technology is

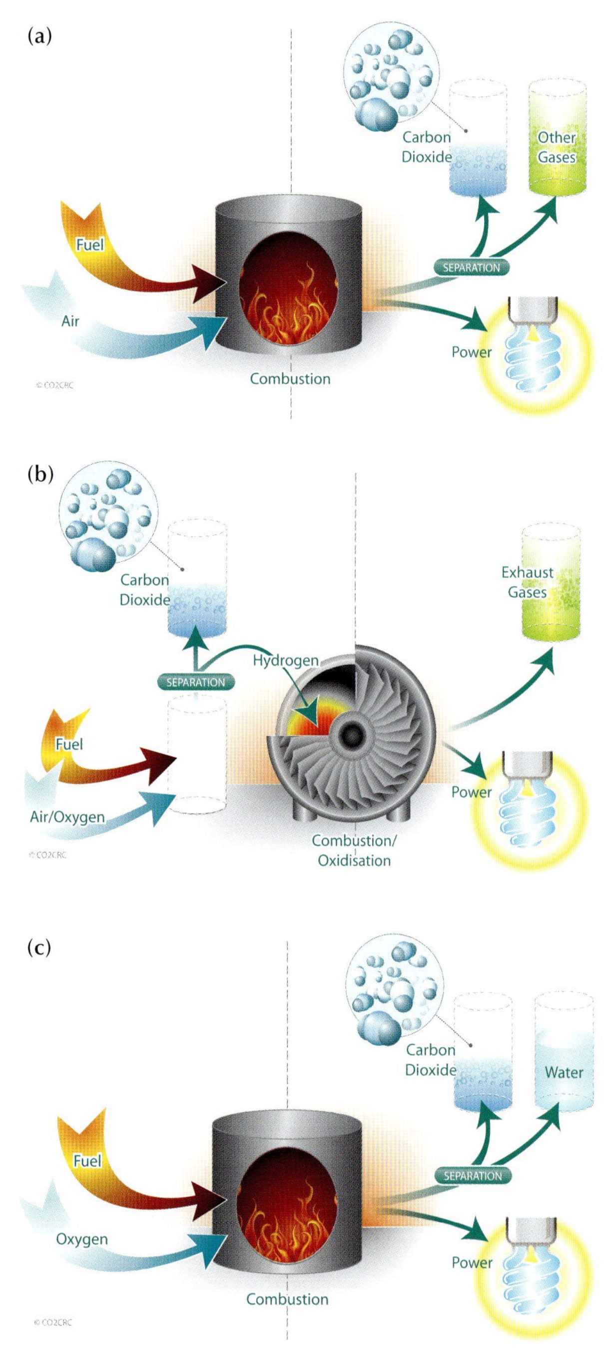

Figure 6.8 The three main combustion and separation options are shown here. Post-combustion (a) involves combustion in air which results in a low concentration of CO_2 in the emissions. Pre-combustion (b) involves gasification of coal to form hydrogen (and a concentrated CO_2 emission stream) before combustion of the hydrogen. Oxyfuel combustion (c) is much like conventional 'post combustion' except that combustion occurs in an oxygen-rich atmosphere, resulting in a high concentration of CO_2 in the emission stream.

currently deployed at pilot and demonstration scales, with the most advanced being the 30 MW purpose-built plant of Vattenfall at Schwarze Pumpe in Germany (see Vattenfall (Schwarze Pumpe) project and Ketzin, Chapter 9). There are several other projects involving retrofits to existing power plants that are under construction, such as at the Callide A power station in Australia. In the United States, the FutureGen Project in Illinois (see FutureGen project, Chapter 11) will be based on oxyfuel combustion and CCS.

A variation on oxygen-rich combustion is provided by the process known as chemical looping, in which a gaseous fuel such as coal or gas is reacted with a metal oxide at high temperature in the absence of air, forming a reduced metal oxide, water and CO_2 (Figure 6.9). The reduced metal oxide is then reacted in air and regenerated by oxidation. Any nitrogen or oxygen present is vented to the atmosphere and the metal oxide is returned to the fuel reactor. As in the case of oxyfuel combustion the process produces a concentrated CO_2 gas stream that can be fairly easily captured, but as yet chemical looping has not been tested at scale.

CCS and gasification

Gasification of coal involves reacting coal with water at high pressure and temperature to form a synthesis gas (syngas) containing carbon monoxide, CO_2 and hydrogen. The syngas can be used in an integrated gas turbine/steam turbine combined cycle (IGCC) plant to produce electricity, or it can be used to make methane gas, transportation fuels (synfuels) or other chemical products such as hydrogen. Using gasification to produce various products involves 'rearranging' the hydrogen and carbon atoms derived from the coal, to form long chain

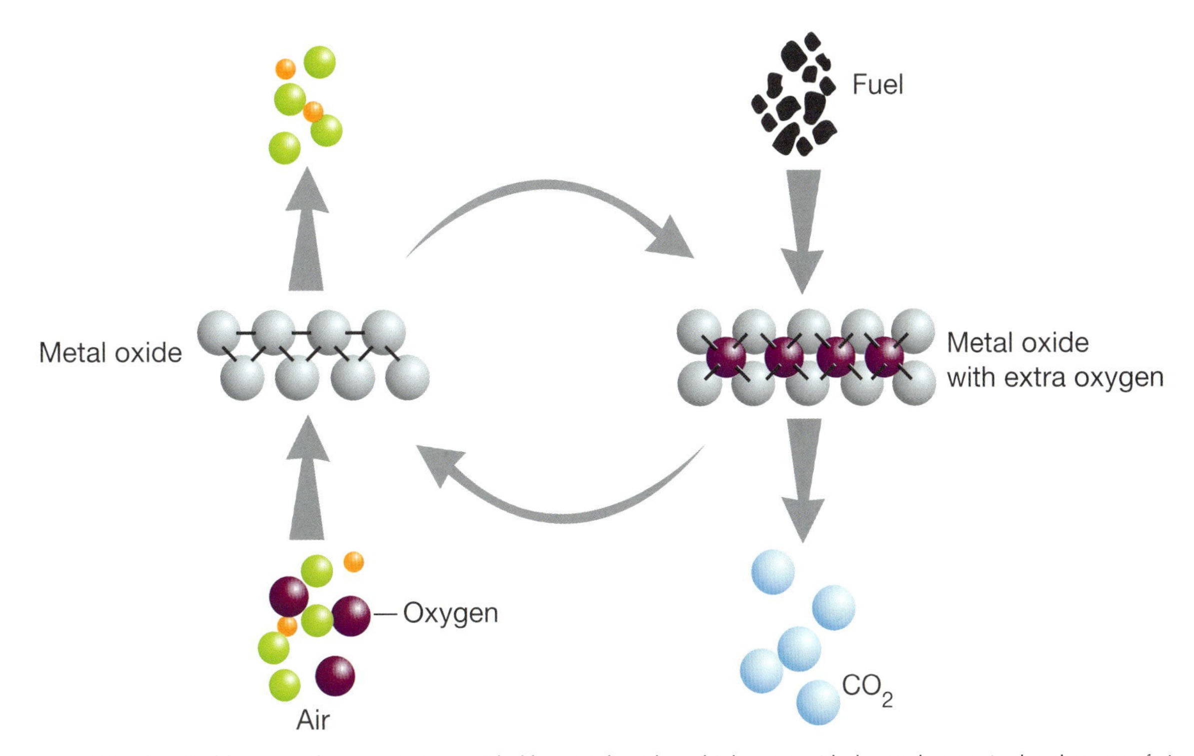

Figure 6.9 In chemical looping, the oxygen is provided by metal oxides which react with the coal or gas in the absence of air to produce heat and a relatively pure stream of CO_2. The metal oxide is then regenerated. At the present time, chemical looping is only at the experimental stage.

hydrocarbons (liquid and gaseous synfuels) and produces a highly concentrated CO_2 gas stream.

Gasification provides the opportunity to separate and capture the CO_2 before the fuel is burnt (combusted) in a turbine to create electricity. Once the syngas is produced, the CO_2 can be separated from any carbon monoxide and from hydrogen, a process known as pre-combustion capture. To maximise the opportunity to capture CO_2, the syngas can be reacted further with water, to convert the residual carbon monoxide to CO_2 which can then be captured. The hydrogen fuel can be burnt in a gas turbine, to produce clean electricity.

A different form of gasification, underground coal gasification, involves partial combustion of deep underground coal seams to produce syngas (Figure 6.10). The process does raise concerns regarding the potential for contamination of groundwater and clearly steps must be taken to ensure that this does not happen. However, the process has been practiced for a number of years in Russia, Central Asia and China and is being tested in a number of other countries. As part of the process, one borehole supplies oxygen and water/steam, a second borehole is used to bring the gas to the surface. As with other gasification processes, the emissions are rich in CO_2 and can potentially be captured although this is not currently being done at the present time

Gasification of coal can be used to convert coal to liquids (CTL) and as the demand for liquid fuels increases, there is increased interest in CTL as an alternative to conventional liquid petroleum, particularly by countries concerned about their dependency on imported liquid fuels. Synfuels can also be made through the steam reforming of methane in a gas to liquids (GTL) process. The problem CTL poses is that,

WEYBURN PROJECT

Oil production began at the Weyburn and Midale oil fields located in Saskatchewan, Canada, in 1954, peaked at more than 45 000 barrels per day in the mid 1960s and then steadily declined. This decline was counteracted in a number of ways including waterflooding, additional vertical production wells, additional horizontal wells, and, starting in 2000, CO_2 injection using anthropogenic carbon dioxide. CO_2 used at Weyburn is obtained from the Dakota gasification facility in Beulah, North Dakota (United States) and transported 320 km across the USA–Canada border. The CO_2-EOR (enhanced oil recovery) operation has resulted in substantial gains in oil production. By 2011 more than 3 Mtpa of CO_2 was being injected and permanently stored. Over the life of the operation it is projected that 26 Mt of CO_2 will be permanently stored at Weyburn.

Although the primary purpose of the injection is to increase oil recovery, the researchers at the University of Regina, together with collaborating organisations and the International Energy Agency for Greenhouse Gases, have conducted an outstanding 11-year monitoring program to comprehensively study the migration and effective containment of the injected CO_2. (Data source: PTRC)

compared to the production of conventional hydrocarbons, CTL is likely to significantly increase the amount of CO_2 emitted to the atmosphere as part of the liquid fuel production process, unless CCS is deployed. Gasification can take place using air or oxygen and there are examples of both types of operations.

Currently most CTL operations do not involve CO_2 capture although this can be done fairly readily because the emission stream has a high concentration of CO_2. The Great Plains Synfuels Plant in North Dakota gasifies coal to produce synfuels, other chemicals and 3 million tonnes of CO_2 a year which is then captured and transported by pipeline to Canada, where it is used for enhanced oil recovery in the Weyburn project (see box above), among other projects.

Synfuels plants can potentially be used to generate electricity, supplying it when demand and price is at a peak and producing synfuels in off-peak times. As previously pointed out, unless CCS is deployed, the production of liquids from coal (CTL) produces more CO_2 than the production of conventional oil which of course is in addition to the usual CO_2 emissions arising from their use in transport. The alternative is to use H_2 (generated through gasification) as a transport fuel.

The potential to turn coal into hydrogen has resulted in a number of energy companies investigating hydrogen as a fuel option for cars (in a hydrogen fuel cell, hydrogen and oxygen

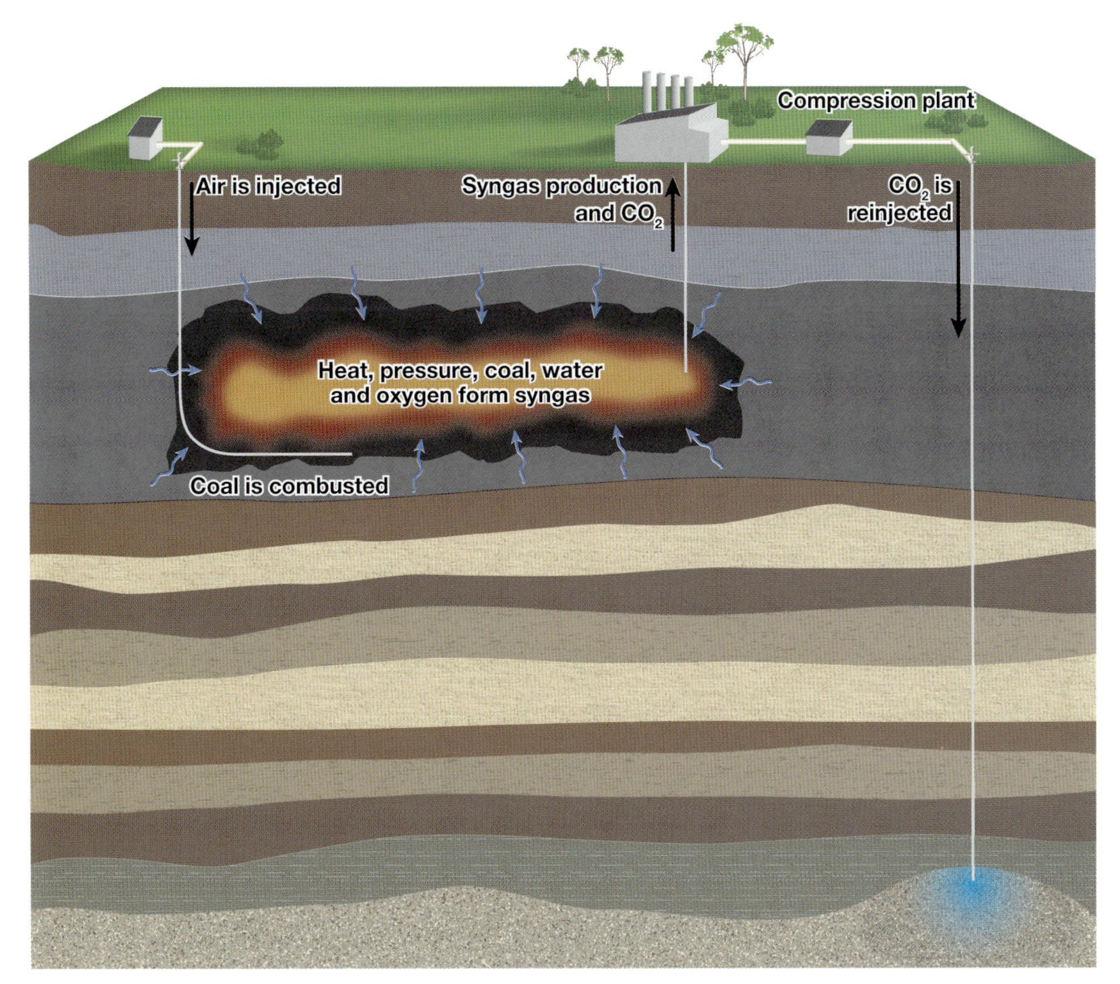

Figure 6.10 Underground gasification of coal involves partial combustion 1–200 metres (or deeper) below the earth's surface in coal-rich rocks to produce syngas, which can in turn be used for the production of chemicals or synfuels or combusted in a power station. Although not currently done, it is possible in theory to capture the CO_2 emissions associated with an underground gasification project.

combine to produce electricity and water) and the 'hydrogen economy' is often suggested as an 'emission-free' way forward. While hydrogen can be produced using renewable energy, given the volume of hydrogen required, any move to a hydrogen economy is likely to be based, at least initially, on hydrogen produced from fossil fuels or through high temperature electrolysis using nuclear power. The question then becomes whether it is more effective to use hydrogen or electricity in cars? There are also significant issues yet to be faced in developing the necessary hydrogen infrastructure as well as safety issues relating to the use of hydrogen as a transportation fuel. The advantage of hydrogen as a transport fuel is that it can be used for much greater distances than electric cars are capable of at the present time, but for short trips around town, electric cars are favoured.

CCS and industrial processes emitting CO_2

Some industrial processes emit fairly pure CO_2 which has the potential to be captured and separated relatively cheaply. Such processes may include the manufacture of fertilisers, cement and iron and steel.

CCS and cement manufacture

Cement plants operate at a high temperature (>1200°C) in order to decompose the limestone (calcium carbonate) and produce cement (Figure 6.11). For each tonne of cement produced, half a tonne of CO_2 is emitted, both from the fossil fuels used to create the high temperatures in the calcination process, and from the actual decomposition reaction. The concentration of CO_2 in the cement-related emissions can be as high as 33%, which is significantly higher than the CO_2 concentration in the emissions from coal or gas-fired power stations. Therefore it is potentially easier and cheaper to capture emissions from a cement plant than from a power plant.

In developed countries, cement manufacture is usually a modest contributor to total CO_2 emissions, but in many developing countries cement manufacture is a major source of CO_2 emissions. For example, although China is increasing the efficiency of its cement production, the Chinese cement industry produces 700 million tonnes of CO_2 emissions a year, which is higher than the total CO_2 emission from most developed countries. One of the challenges of the industry is that most cement plants are quite small and therefore there are limited opportunities for large scale capture. Whilst various polymers are being developed as an alternative to conventional cement, it is likely that limestone-based cements will continue to dominate the construction industry for many years to come. IEA mitigation modelling assumes the capture of 1.4 Gt of CO_2 from the cement industry by 2050 in order to meet proposed emission targets. If this contribution is to be made, there is a need for more work to be undertaken into the application of CCS to the cement industry.

CCS and iron and steel production

Coking coal, an essential ingredient in blast furnaces, reduces iron oxide to iron, with CO_2 as a by-product of this process (Figure 6.12).

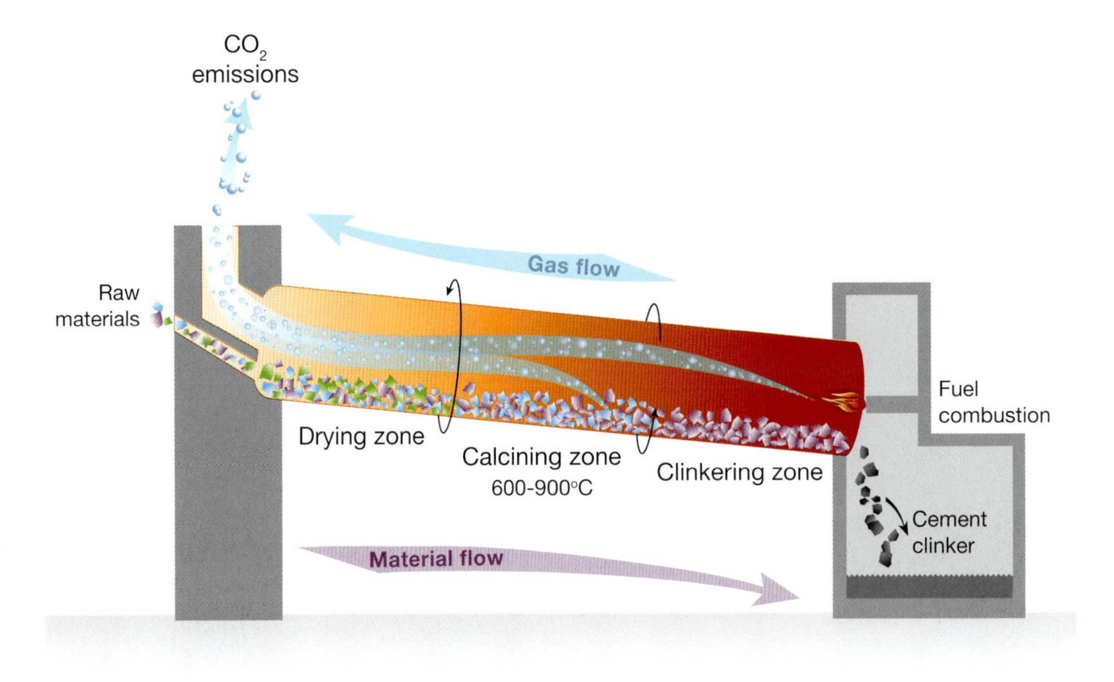

Figure 6.11 Cement manufacture is a major source of CO_2 globally. The CO_2 is derived from the limestone fed into the kiln and the fuel used to heat the kiln.

The smaller scale alternative is to use an electric arc furnace for steel-making, but this process is a heavy user of electricity which at the present time is obtained almost entirely from fossil fuel-based power stations. Whilst other reduction processes are under development (using hydrogen for example), for the present there is no real alternative to using coking coal in the reduction process.

Also as part of the process, limestone is decomposed to form quicklime (calcium oxide), which removes impurities as slag and produces CO_2 as a by-product (see Box 6.2). In steel making, oxygen is used to remove excess carbon from the iron, creating more CO_2. Coal will continue to be an essential component of iron and steel-making for the foreseeable future, although co-firing with waste plastics and biomass is likely to increase. There is a need for more work into the opportunities for the application of CCS to iron and steel production, but there are no insurmountable technical barriers.

BOX 6.2: PRODUCTION OF CO_2 DURING STEEL MAKING

$2C + O_2 \rightarrow 2CO$ - carbon monoxide is produced from coal

The iron oxide in reduced to metallic iron

$$9CO + 3Fe_2O_3 \rightarrow 6Fe + 9CO_2$$

CCS and ammonia production

Ammonia (NH_3) is an important component of chemical fertilisers (Figure 6.13). It also has many industrial uses, including as a solvent in carbon capture. The manufacture of ammonia requires hydrogen, currently produced mainly from steam reforming of methane or light hydrocarbons, using catalysts in the presence of air to form ammonia. Ammonia can also be produced via a coal gasification process carried out in conjunction with an air separation unit, which provides a source of pure nitrogen to react

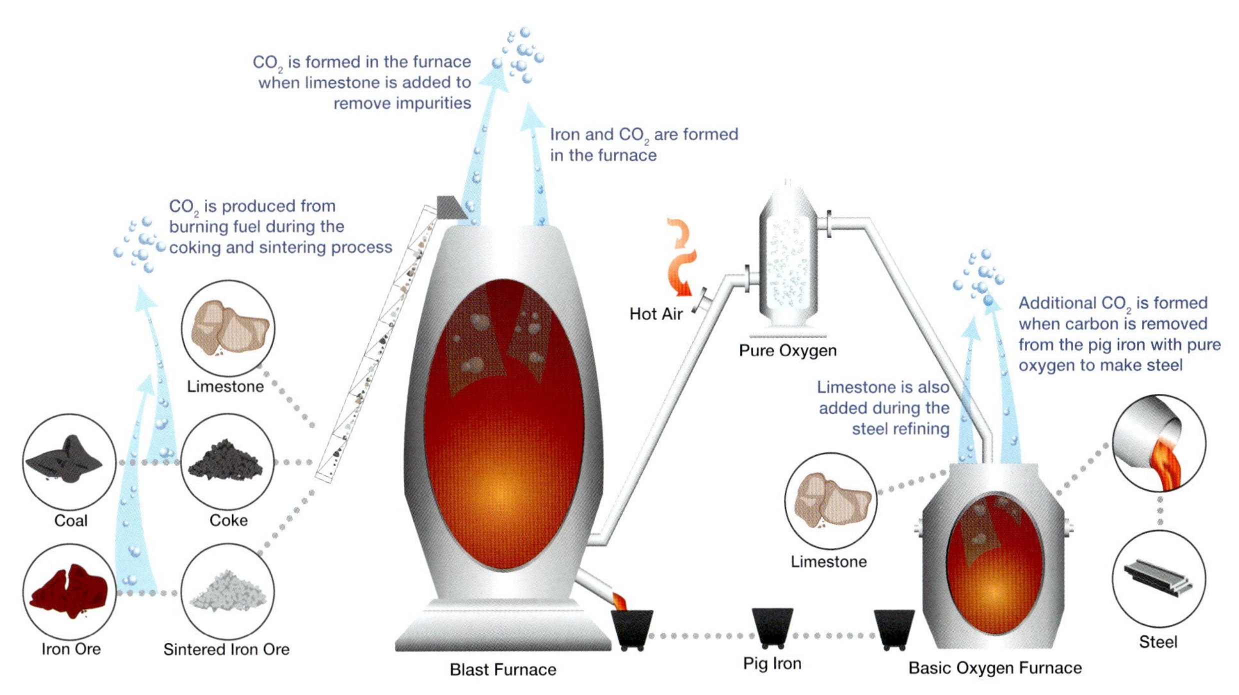

Figure 6.12 Coal (or coke) is an essential component in the production of most iron and steel and is important both as a fuel and for reducing the iron ore to iron. The emission stream has a high concentration of CO_2, which makes iron and steel production an early target for CCS.

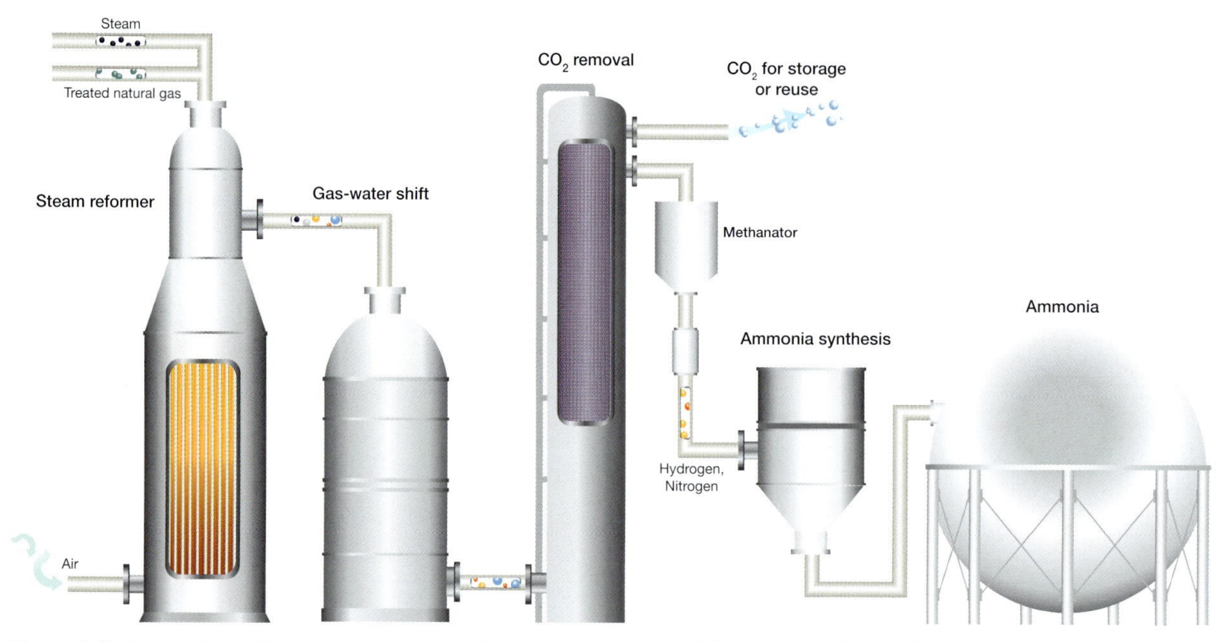

Figure 6.13 Ammonia, an important component of various chemicals and fertilisers, can be made using natural gas or coal to produce hydrogen then ammonia. A concentrated stream of CO_2 is usually emitted, but this has the potential to be captured.

with the hydrogen to form ammonia. There is no technical impediment to the application of CCS to a fertiliser plant, as the emission stream is CO_2-rich.

CCS and oil sands-derived oil

Heavy oil can be extracted from oil sands using a variety of methods, the most common being to inject steam into the sand (Figure 6.14). The process is very energy intensive because of the need to heat the steam (usually using natural gas), which produces large quantities of CO_2. Alternatively some of the bitumen can be gasified to produce syngas to supply the energy to heat the steam. A relatively pure stream of CO_2 is produced as part of this process and this can potentially be captured. The Province of Alberta in Canada has a major initiative underway to decrease emissions of CO_2 from its oil sand industry, including a proposal to transport the CO_2 by pipeline for use in enhanced oil recovery.

Technologies for separating CO_2 from emissions

There are four main technologies for separating CO_2 from a gas stream, that are presently operational or being developed:

- absorption in a solution
- separation using membranes
- adsorption onto the surface of a solid
- low temperature (cryogenic) techniques.

Liquid solvents

Liquid solvents are the most common method currently used to capture CO_2 (Figure 6.15). They have been employed commercially for many years, to capture CO_2 from a variety of sources. The solvent absorbs the CO_2 by a process in which the CO_2 undergoes a reversible chemical reaction with the solvent, or by a physical process in which the CO_2 is 'physically' bound to the solvent without any chemical reaction occurring. Solvents are already used on a large scale to separate CO_2 from natural gas in the 'gas sweetening' process and are also used to capture CO_2 emitted by coal gasification plants such as at the Great Plains Synfuels Plant in North Dakota.

Despite these commercial examples, it is a challenge to economically deploy large scale post-combustion capture at coal-fired power stations. The flue gas from a typical coal-fired power plant is largely nitrogen, with approximately 12–14% CO_2. A number of steps are needed to concentrate and capture this CO_2 from flue gases (Figure 6.16). First, the flue gas needs to be cooled and impurities removed. Second, flue gases containing CO_2 and nitrogen are fed into the bottom of an absorber tower, while solvent is fed in at the top. As the flue gas bubbles up through the tower, it makes contact with the solvent as the gas moves up and the solvent moves down. The gas that comes out of the top of the tower contains mainly nitrogen with the CO_2 retained within the solvent. The dissolved CO_2 is removed from the solvent in a regenerator or stripper tower, where the separation of CO_2 from the solvent is achieved by heating the solvent to above 100°C. Finally, the solvent (now minus the CO_2) is returned to the top of the absorber tower to be used to absorb more CO_2.

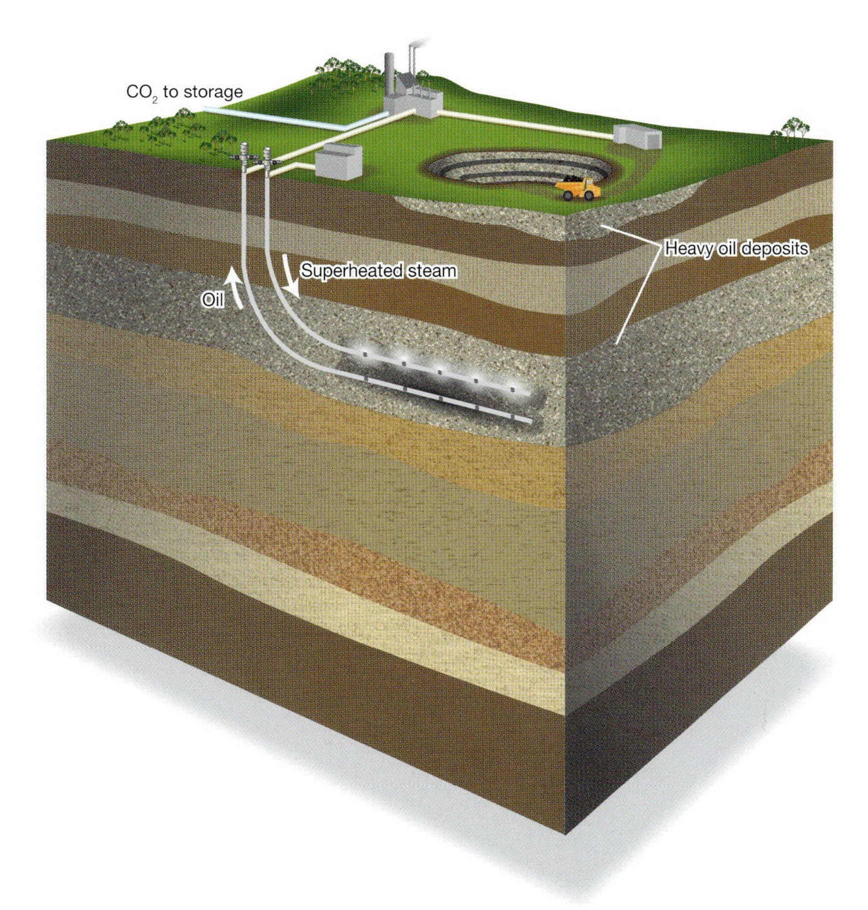

Figure 6.14 Production of oil from oil sands is becoming an increasingly significant source of CO_2. Oil shales and coals are also likely to become an important source of liquid fuels in the future. All these processes are energy intensive and produce large quantities of CO_2. While not currently applied, CCS is essential if these emissions are to be avoided. A project in Alberta, Canada, shown schematically here, aims to be the first major oil sands project to apply CCS.

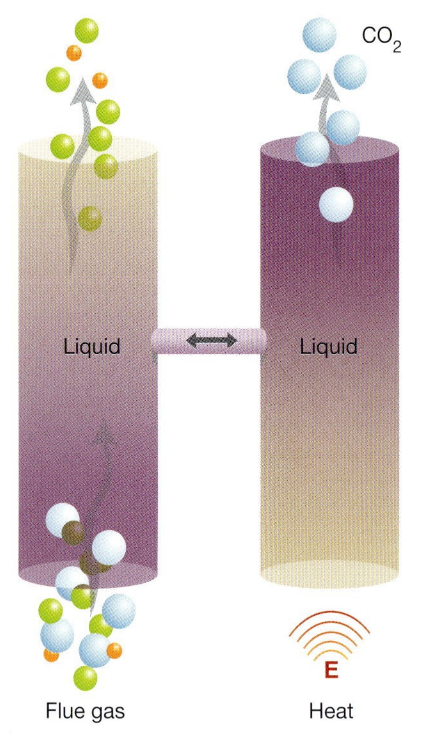

Figure 6.15 In this schematic representation of the absorption process, flue gases are passed through a solvent which preferentially absorbs the CO_2 which is then captured.

A significant amount of energy, commonly of the order of 20–30% of the total energy produced at the power station (referred to as the energy penalty), is used to heat the solvent and operate pumps and fans. It is therefore very important to design systems which will re-use heat efficiently and reduce the energy penalty. Other performance improvements can be achieved by modifying packing material in the columns to improve the absorption of CO_2, or by using better solvents. Together, these can potentially halve the energy penalty.

There are a variety of solvent types, with amine-based solvents the most common. However, amines do undergo solvent degradation over time. Concerns have also been raised about the environmental and health impact of amines, particularly if they are to be used on a very large scale for post-combustion capture. Therefore other solvents such as chilled ammonia are being trialled, as are solvents such

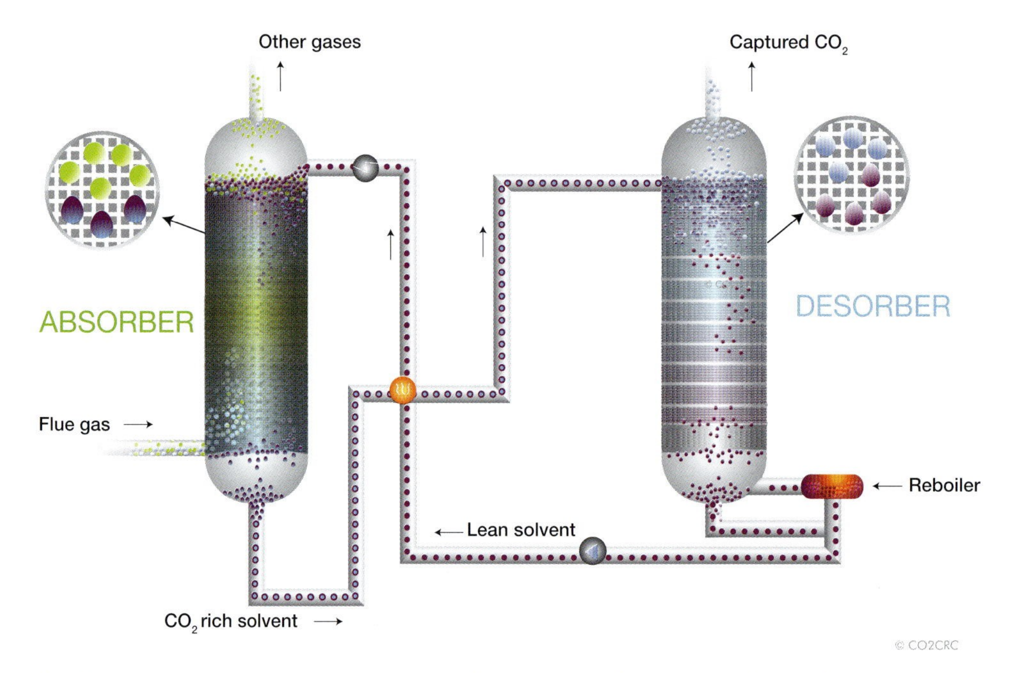

Figure 6.16 The absorption process shown in the previous diagram is actually more complicated than depicted and involves circulation, heating and recirculation of the solvent in order to separate and capture the CO_2 and regenerate the solvent. There are significant energy losses during this process and one of the challenges is to decrease those energy requirements. Amine absorption is currently the most widely used technology for CO_2 capture.

as potassium carbonate which can operate at high temperatures. Whilst these other solvents have a slower rate of reaction than amines, they do have the potential to be more environmentally friendly.

At full scale, the volume of solvent required for a power station would be very large. For example, based on current formulations, a 500 MW coal-fired power station would need on the order of several million litres of amine solvent in the capture system, with about half in the absorption tower and the other half in the desorption tower. Upscaling the solvent process is one of the key challenges and creative approaches being considered include such measures as replacing the steel vessels with cheaper concrete vessels that can be built at a much larger scale. These sorts of developments, together with improved process integration (such as the use of waste heat) have the potential to significantly reduce energy costs, driving down the overall cost of post combustion capture.

Membranes

Membranes, acting like a very fine sieve, can preferentially remove CO_2 from a stream of mixed gases. They are usually made up of polymers or ceramics and can also be used in conjunction with liquids. Membranes are used in many industries, including for natural gas separation, for the production of ammonia and hydrogen and for the separation of nitrogen from air. However, at present they are not used on a large scale for separation of CO_2 from flue gases. There are two main types of membrane systems that can be used to capture CO_2. One separates out the CO_2 from other gases (gas separation membranes); the other allows CO_2 to be absorbed from a gas stream into a solvent via a porous membrane barrier (membrane gas absorption; Figure 6.17).

Gas separation membrane systems have a key advantage over solvent systems: the equipment is

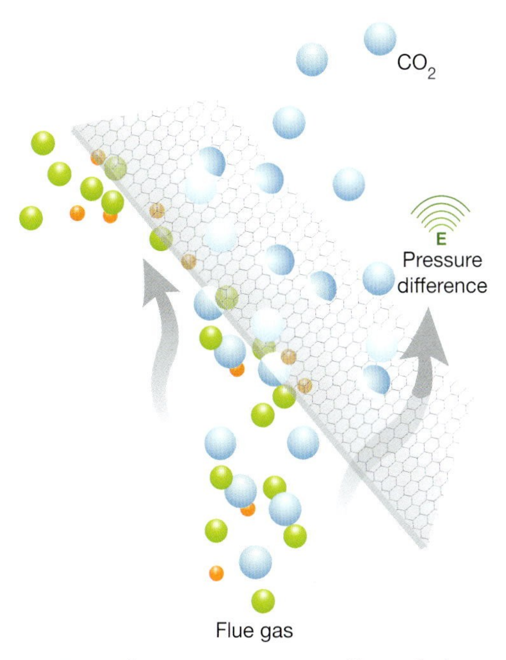

Figure 6.17 Membranes act as sieves or filters which can preferentially allow some gases (such as CO_2) to flow through the membrane while other gases are retained. It can also work in reverse with the CO_2 trapped and the other gases allowed to pass through. The pure CO_2 is then captured.

smaller, more versatile and requires no solvent. This cuts material costs and reduces environmental concerns. The membrane acts as a semi-permeable barrier through which CO_2 passes through more easily than do other gases. The CO_2 is driven across the membrane by a pressure difference. The energy required to maintain this pressure difference represents a large proportion of the operating cost of the technology at the present time.

A range of materials and configurations is used to make membranes (Figure 6.18). Early membranes for CO_2 separation were made from cellulose acetate, a natural plastic made from wood or cotton. Generally 'glassy' polymers are good at separating CO_2 from other gases, but over time, they tend to become less effective, which is a challenge when using membranes with wet flue gases. Composite membranes, with a combination of glassy and rubbery polymer segments, combine the structure of the glassy

IN SALAH PROJECT

The In Salah Project in central Algeria, a joint venture between Sonatrach, BP and Statoil, has been producing natural gas since 2004. Because the natural gas contains approximately 5.5% CO_2, the joint venture decided that rather than vent the CO_2 to atmosphere, the project would be used as a proving ground to demonstrate industrial scale geological storage of CO_2 as a mitigation option. The CO_2 is separated from the natural gas stream at the Krechba central processing facility using solvent technology, before being transported 14 km to the injection site. There it is injected into a highly saline downdip portion of the natural gas formation at a depth of approximately 2 km, via three horizontal wells. Approximately 1 Mtpa CO_2 are stored at In Salah.

The Project is undertaking extensive assurance monitoring to confirm that the CO_2 will be permanently stored. The monitoring includes integrity analysis, tracers, downhole and 3D seismic monitoring, as well as satellite based InSAR (interferometric synthetic aperture radar) studies that have detected the very slight arcing of the surface above where the CO_2 is being injected. Together these techniques have enabled the Project to assess the overall plume migration, well integrity, caprock integrity and formation pressure and confirm that the stored carbon dioxide is behaving as expected. (Image: courtesy BP)

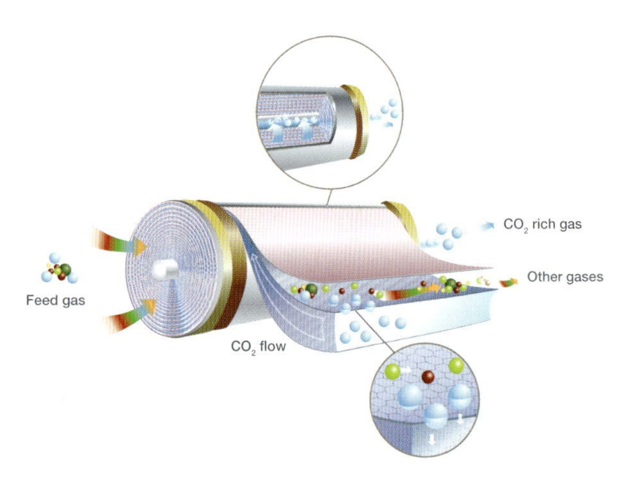

Figure 6.18 A range of configurations is used to optimise the effectiveness of membranes and minimise their size. In this example, membrane sheets are wound into a spiral structure to optimise the CO_2 separation process.

polymer (giving good selectivity for CO_2) with the higher permeability of a rubbery polymer (allowing more gas to flow through). But like all polymeric membranes, they do not operate well at high temperatures; ceramic membranes, which are able to handle higher temperatures, represent an alternative approach.

Membranes can be used in conjunction with solvents, with the membrane separating the flue gas from the liquid solvent, thereby reducing flow problems that occur when liquids and gases meet (Figure 6.19). The CO_2 passes through the membrane and is then absorbed by the solvent. However, this may lead to a smaller contact

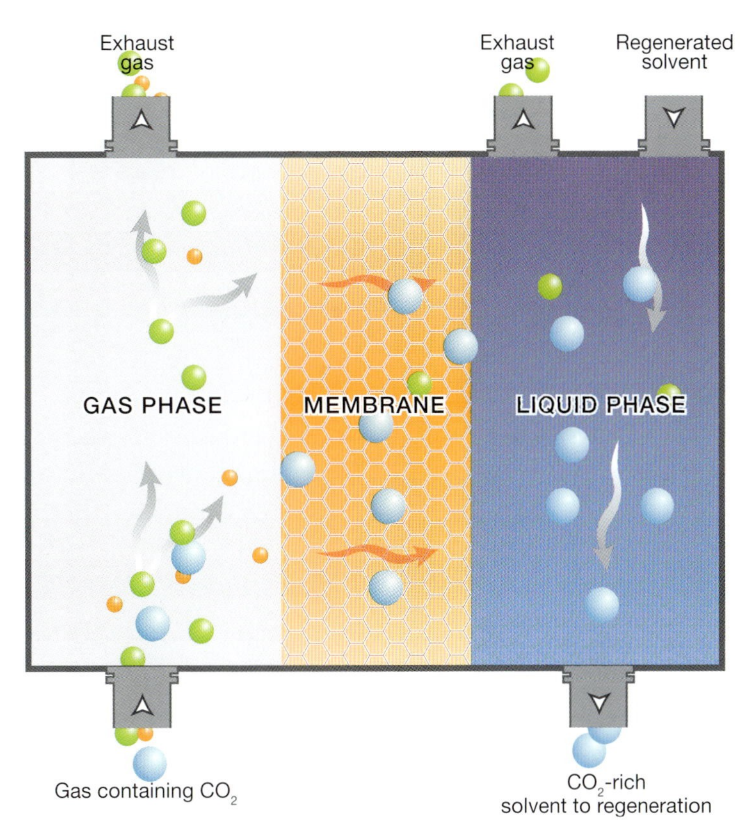

Figure 6.19 It is possible to use a combination of solvents and membranes to separate out CO_2 from flue gases and this is shown schematically here. The membrane separates out some, but not all of the gases, which then pass through a solvent that removes the remaining gas, producing a pure CO_2 stream.

surface area between the liquid and the gas, limiting the absorption efficiency compared to other solvent absorption techniques. The CO_2 is then removed from the solvent by heating. Using membranes in conjunction with solvents can reduce the size of the equipment required to absorb the CO_2.

Depending on the purity required, membrane separation at a coal-fired power station may need to be a multi-stage process, with different membranes needed to separate different gases. The separation process needs a pressure or concentration difference to be maintained between the two sides of the membrane, which requires energy. More research and development is required to enable membranes to be used at higher temperatures, with lower energy requirements and to enable them to be upscaled to the necessarily massive scale that will be required. Nonetheless, membranes are a promising technology for CO_2 capture.

Adsorption

Adsorption of CO_2 from other gases (Figure 6.20) involves several steps. First the CO_2 in a gas stream is preferentially attracted to the surface of a solid material (the adsorbent) and becomes bound to that material by physical forces or chemical bonds. The physical conditions are then changed to release the CO_2 from the adsorbent and there are a number of ways of doing this. In thermal swing adsorption (TSA), the CO_2 removal is triggered by an increase in temperature, but this is energy intensive and slow, since the entire mass of the adsorbent must be heated. In vacuum swing

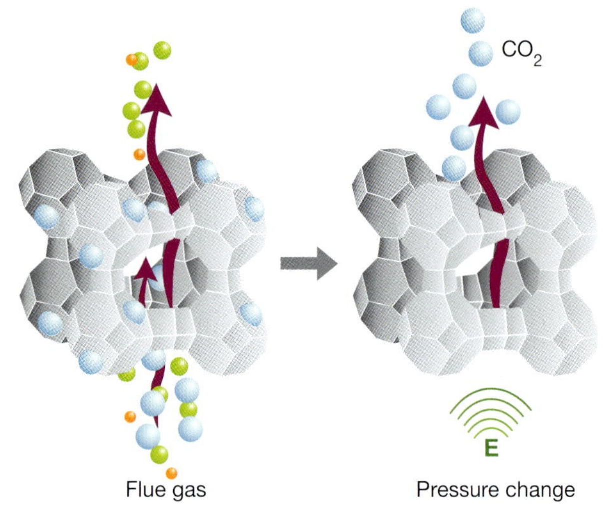

Figure 6.20 Adsorbents are able to preferentially attract (adsorb) molecules of CO_2. When the temperature or pressure or electrical field is changed, the material then releases (desorbs) the pure CO_2. This method is receiving a lot of attention from researchers, but is not presently used in large scale commercial activities.

adsorption (VSA), the CO_2 removal is triggered by reducing the pressure to near-vacuum conditions, a process that requires less energy than TSA. Pressure swing adsorption (PSA) is a process similar to VSA except that the decrease in pressure is usually close to atmospheric pressure. Finally electrical swing adsorption (ESA) involves applying a voltage to heat the adsorbent to release the CO_2.

There are various types of adsorbents. One of the most common, zeolite, is a mineral that both occurs naturally and can be synthesised. It has a porous structure and can be used to separate out molecules on the basis of size. Carbon structures (carbon nanocages) containing nanopores ranging in size from 2–50 nanometres, are prepared by decomposing organic material such as coal, wood or coconut shells at a temperature of about 500°C in the absence of air. Another type of molecular structure, composed of metal ions linked by organic bridging and with a highly porous structure, can also act as a molecular sieve or attract gases to its surface.

New adsorbents continue to be developed with higher surface areas and greater capacity to attract carbon dioxide at the expense of other gases in the flue stream. They all face the challenge of minimising the energy used in the adsorbtion process to change the pressure, the temperature or the electric field. Developing a commercially viable adsorbent-based system of capture requires a material that is cheap, environmentally friendly, water tolerant, impurity tolerant and works at high temperatures. An adsorbent with all these properties has yet to be identified, which is why there is a need for more research.

Low temperature separation

Low temperature separation is the fourth approach that can be taken to CO_2 separation (Figure 6.21). If a gas stream is cooled to sufficiently low temperatures, the CO_2 will condense and the other gases can be released to the atmosphere. Depending on the pressure at which the process is carried out, the cold CO_2 will be a solid or a liquid. The cooling process uses a significant amount of energy, but that energy requirement can be reduced if the separated CO_2 is then warmed against the hot incoming flue gas. This warming also results in CO_2 at a higher pressure, reducing the energy requirement for compression.

An alternative approach to low temperature separation is through the formation of CO_2 hydrates (an ice-like form of CO_2 molecules surrounded by a cage of water molecules). These can form when cold water is passed through flue gases, physically trapping the CO_2 in the hydrate cage and releasing it when the hydrates are heated. Energy is required to cool the water and to reheat the hydrates. The commercial

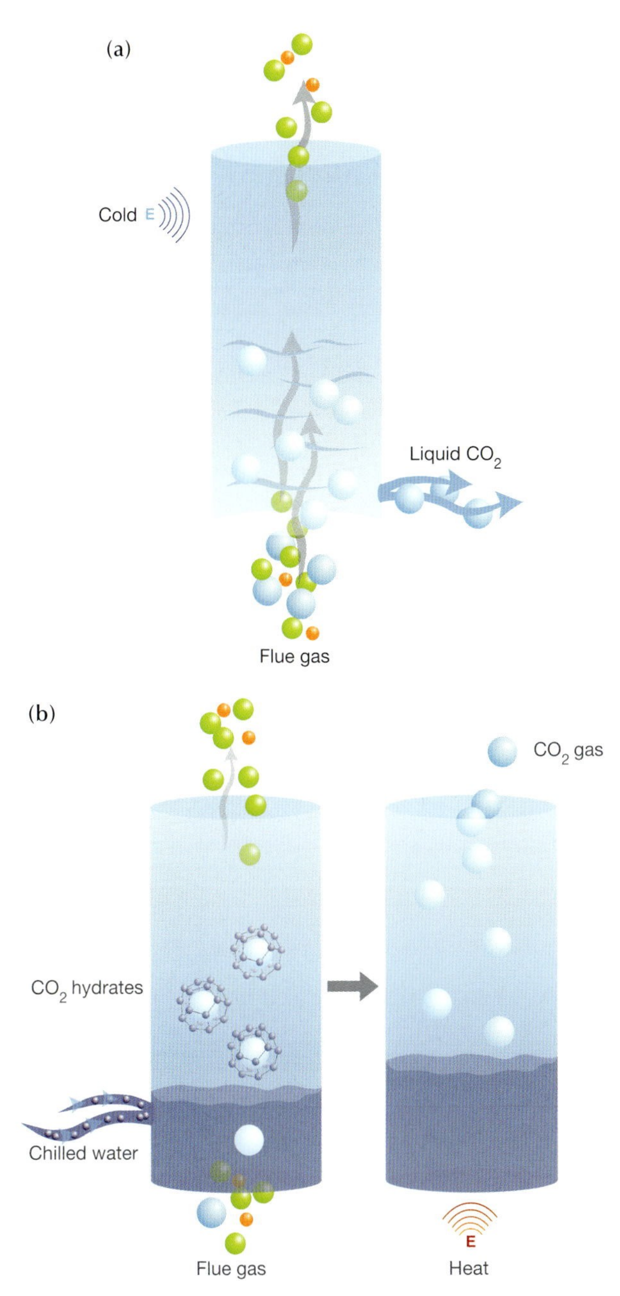

Figure 6.21 Low temperatures are being used on an experimental basis to separate CO_2 from flue gases. At low temperatures, the CO_2 condenses (a) or form as hydrates (b). The large amount of energy required to chill the systems is currently an impediment to deploying the method commercially, other than for natural gas.

application of low temperature techniques has, to date, largely been to streams of CO_2 at concentrations higher than 90% and has tended to be implemented in natural gas processing, but there may be scope for future application of the technology to pre-combustion and oxycombustion processes.

Conclusions

In summary, there is a range of large scale sources of CO_2, the most important of which (in terms of the amount of CO_2 emitted) is conventional coal fired power plants. In the future, oxycombustion and gasification will become increasingly important, but that is likely to be some years off. Natural gas will become increasingly important in power generation and although the amount of CO_2 emitted is half that of an equivalent coal-fired power plant, we should plan for the future need to separate and capture CO_2 from gas-based power systems.

There are a number of sources that produce relatively small amounts of CO_2, but where the concentration of CO_2 in emissions is high, notably gas separation facilities and some industrial processes. These represent 'low hanging fruit' where capture can be applied now at a relatively low cost compared to capture at coal or gas-fired power stations.

There are a number of techniques for separating and capturing CO_2; each with their advantages and disadvantages. Some work well on specific emission streams but not on others. Some may face environmental challenges in handling the material, while others can be quickly 'poisoned' by small quantities of impurities. All, to varying degrees, require additional energy, whether to remove the CO_2 from a solvent, drive gases through a membrane, change the pressure field to desorb the CO_2 or freeze the CO_2 from a gas stream. For this reason, a great deal of research is being directed at decreasing the energy use associated with capture by integrating the whole process – power generation, capture and separation – to more effectively use waste heat and minimise the energy penalty.

Does the need to optimise the entire system mean that it will not be feasible to retrofit CO_2 capture to existing power plants? Not at all, and there are a number of systems around the world where post combustion capture (with or without oxyfuel combustion) has been retrofitted, such as Vattenfall in Germany, Mountaineer in the United States and Callide A and Hazelwood in Australia. There is no doubt that 'new build' provides the best opportunities for process integration and for bringing down capture costs. However a whole range of other issues enter into consideration when comparing the cost of 'clean' electricity from a retrofitted capture system with a new power and capture system, including the cost of land, the residual value in the plant, and so on. The work of the International Energy Agency Greenhouse Gas Research and Development Programme (IEAGHG), the Electric Power Research Institute (EPRI) and, most recently, the work by Minh Ho and Dianne Wiley for CO2CRC, suggests that in some circumstances, the retrofit of capture systems can be significantly less costly than new build. It is important to stress that this needs to be carefully considered on a case-by-case basis, but it is a very important conclusion, which gives some hope for tackling emissions from some of the many existing coal-fired power stations around the world, which are likely to continue to release CO_2 to the atmosphere for many years to come, unless action is taken to limit those emissions through retrofitting of CCS.

7 HOW CAN WE TRANSPORT CO_2?

Major existing sources of CO_2, such as power stations and industrial plants, were located on the basis of their proximity to raw materials (such as coal or limestone), or to customers in towns and cities or industrial estates, or the presence of pre-existing rail, road or harbour infrastructure. The need to mitigate the CO_2 that they emit was not a consideration in siting any of these facilities. Therefore it is reasonable to assume that once CO_2 is separated from other emissions, it will need to be transported to a suitable storage site. Occasionally this may fortuitously be just a short distance away, but in some cases it could be a distance of tens to hundreds of kilometres. Fortunately, CO_2 can be transported as a solid, a liquid or a gas and can be carried via pipeline, ship, train or truck. For the most part it is likely to be transported by pipeline. Globally there are over 6000 km of CO_2 pipelines currently in operation, with the majority being in the United States (Figure 7.1). In other words, the transportation of CO_2 is a mature technology.

The properties of CO_2 obviously have a major influence on the transport method used. Carbon dioxide normally needs to be dried before it can be transported. This is because CO_2 containing free water can form carbonic acid, which is very corrosive, and also because CO_2-hydrates can form, which might block a pipe or cause other operational difficulties.

Carbon dioxide will cause asphyxiation at high concentrations. A typical standard for maximum exposure to CO_2 is 3% over 15–30 minutes (short term exposure), and 0.5% CO_2 over 8 hours (as a time-weighted average). If there is a small leak of CO_2 it may disperse into the atmosphere, but because CO_2 is denser than air, it can also accumulate in low-lying pockets of land and therefore transporting CO_2 through a variety of landscapes will require careful planning and perhaps the placement of CO_2 sensors in critical areas, such as those used in cellars in Italy, where natural volcanic CO_2 can accumulate and represent a health hazard.

One option to minimise risk may be to include a compound in the CO_2 that has an unpleasant smell so that it can be easily detected. Odourisation, as it is called, is commonly used for domestic natural gas and liquefied petroleum gas (LPG). However, the cost of odourising millions of tonnes of CO_2 could be very high and there may be less costly ways of addressing any potential risks.

Other safety issues relate to the risk of cold burns or frostbite, should high pressure CO_2 escape from a pipeline. The issue of risk of leakage is

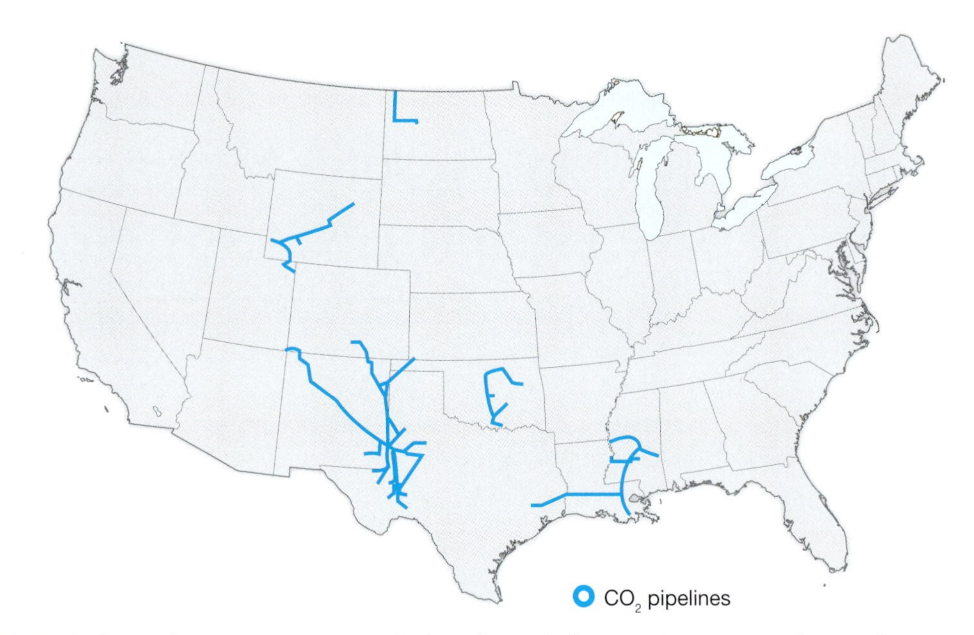

Figure 7.1 The United States has an extensive network of pipelines which transport approximately 50 million tonnes of CO_2 per annum. These mainly carry CO_2 from geological sources to enhanced oil recovery (EOR) projects. However, there is increasing interest in using anthropogenic sources of CO_2, which could potentially result in a significant reduction in US CO_2 emissions. (Adapted from Ciferno)

considered later, but in general, CO_2 which emerges from a pressurised condition will rapidly expand, forming dry ice and cold CO_2 gas. The dry ice then sublimates slowly into the atmosphere and the gas disperses. This phenomenon both reduces the amount of gaseous (and mobile) CO_2 released and provides an early warning sign.

Key issues in transportation of CO_2 via pipelines

There is a great deal of experience in the use of pipelines for transporting gases, liquids and slurries. For example there are millions of kilometres of oil and natural gas pipelines networked around the world (Figure 7.2). Over the past five decades more than 6000 km of CO_2 pipelines have been constructed and there is a great deal of experience and knowledge of operating dedicated CO_2 pipelines. Their safety record has been excellent with only a handful of reported incidents. CO_2 pipelines are still considered an unfamiliar technology in some countries but the main issues relating to their use are well known, including:

- phase changes (CO_2 changing from liquid to gas or vice versa) that may prevent the compressors and pumps from moving the CO_2 along
- CO_2 forming carbonic acid in the presence of water leading to corrosion of the pipeline
- water in the CO_2 stream freezing and forming solid hydrates that can block equipment and cause a dangerous 'sandblasting effect'
- the need for increased compression due to the presence of impurities.

Liquids and gases can flow (which is why they are suitable for pipeline transport) depending on their viscosity; the lower the viscosity the more easily they will flow. Gaseous CO_2 has a very low viscosity and therefore will flow very easily. However, viscosity is not the only issue. While gaseous CO_2 may flow very easily, its density is

SNOHVIT PROJECT

Following the success of the Sleipner Project, Statoil initiated a second industrial scale capture and storage project in 2008, the Snohvit LNG Project. Located off the far northern coast of Norway, the Snohvit gas field contains 5–8% CO_2. The produced gas is transported by pipeline to land-based processing facilities for the CO_2 separation process using solvent based technology, with 700 000 tonnes of CO_2 per year separated from the natural gas stream. The CO_2 is then piped 160 km through a dedicated pipeline back to the production platform in the Barents Sea for reinjection into a deep saline aquifer below the natural gas reservoir, at a depth of 2600 metres.

very low (2 kg/m^3) and to store a million tonnes of CO_2 as a gas, would require the transport of many cubic kilometres of gas, which is clearly not practical. On the other hand, liquids have a high density compared to gas and at high pressure CO_2 will become a dense liquid (Figure 7.3). There is an additional phase known as 'supercritical' which occurs under particular temperature and pressure conditions (operating pressure of 73.9 bar and temperature of 31.1°C for pure CO_2) when CO_2 has the density of a liquid and the viscosity of a gas.

So while it is possible to transport CO_2 as a gas, it is more cost effective to compress it to a dense liquid-like supercritical state. If CO_2 is transported in a pipeline, lower temperatures can be tolerated as long as the pressure is increased, but it is important to avoid multiple-phase CO_2 in the pipeline, as this can cause flow instability.

The pressure in the pipeline is dependent on the hydraulics of the pipeline, which are determined by flow rate, composition, distance, diameter and configuration. In order to maintain the fluid in a dense state there may be a need for additional repressurisation stations. Compression requirements depend on the composition of the gas mix in the pipeline, and while the gas may be mainly CO_2, the presence of small amounts of impurities can significantly change the repressurisation requirements. If the pressure and temperature conditions of the pipeline are such that the CO_2 were to change from a liquid to a gas, or vice versa, it could be detrimental to the pumps, compressors and other components. Additionally, rapid pressure swings along with the unstable flows they create, can decrease efficiency and under extreme conditions lead to pipeline integrity issues. However, the critical

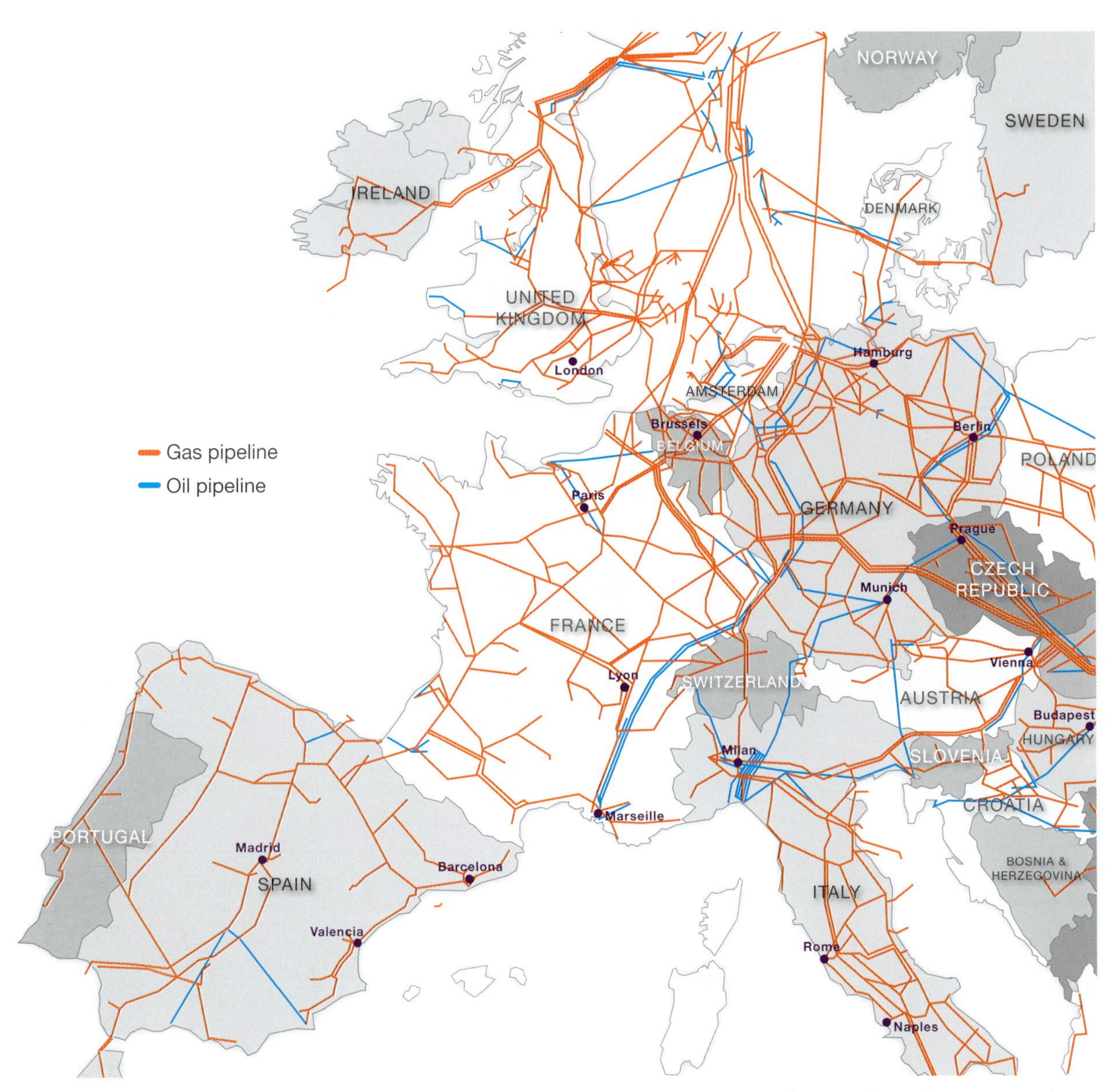

Figure 7.2 Many parts of the world have very extensive pipeline networks mainly for natural gas (methane) but some are also used for transport of crude oil. The natural gas network in Europe is particularly extensive, with many pipelines passing through heavily populated areas. (Data source: International Gas Union)

parameters are well known and assuming that proper design and safety considerations are followed, phase changes in the pipeline can be controlled and avoided. Existing pipelines that were designed for transport of other gases or liquids may be unsuitable for transport of CO_2.

One of the key issues in pipeline design is that in the presence of water, CO_2 reacts to form a weak acid, carbonic acid, which can corrode steel pipelines by as much as 10 mm/yr according to a report by Seiersten and Kongshaug. Consequently steps must be taken to prevent the acid forming, or the pipeline must be constructed out of non-corrosive materials such as stainless steel. As both options are costly, optimisation between the two is necessary. For example, in the case of a relatively small amount

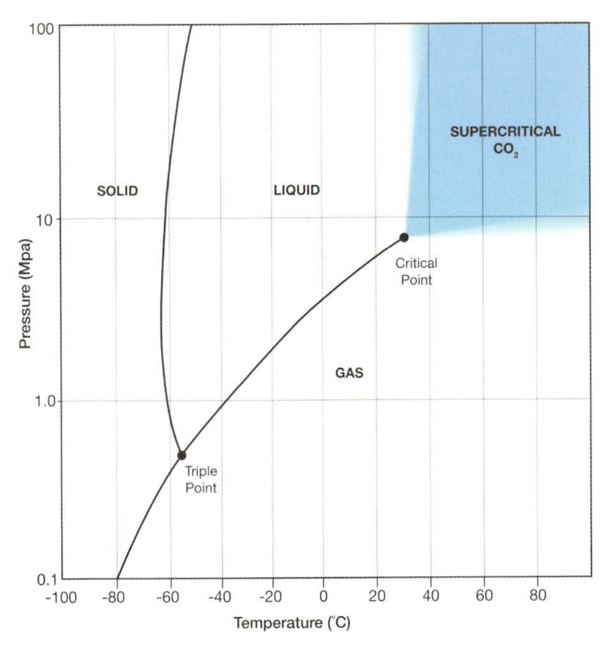

Figure 7.3 The behaviour of carbon dioxide is dependent on temperature and pressure. At very low temperatures CO_2 is a solid; at higher temperatures and pressures CO_2 is a liquid; at normal temperatures and pressures CO_2 is a gas. Beyond what is termed the critical point, CO_2 is in a supercritical state and has the density of a liquid, but other properties (such as viscosity) are more like those of a gas. (Adapted from IPCC 2005)

of CO_2 transported a long distance, it may be more cost effective to remove any water and build the pipeline out of the cheaper steel. If a large amount of CO_2 is to be transported a short distance, then it may be more cost effective to not remove the water and instead use a more expensive corrosion-resistant stainless steel.

The presence of water can pose another hazard in a CO_2 pipeline through the formation of hydrates, which are 'ice-like crystals' composed of a CO_2 molecule surrounded by a 'cage' of water molecules. These ice crystals can block valves and pumps and clog pipelines. Hydrates can also form in natural gas pipelines but there are established procedures to prevent them from forming. A leak or a controlled discharge of hydrates from the pipeline can have a sand blasting effect on equipment and can be dangerous to people in the immediate vicinity.

Understanding the nature of CO_2 escape from pipes or wells

In order to better understand the behaviour of CO_2 when escaping from a pipe or well, it is helpful to describe a demonstration project where several tonnes of high pressure CO_2 were released from a narrow pipe, simulating an escape to the atmosphere from a pipeline or a well (Figure 7.4). First, because the gas was at high pressure, releasing it through a narrow opening produced rapid expansion and cooling of the gas. It also chilled the air, so that there was a visible white mist, composed of CO_2 and water vapour, with a diameter of up to 5 metres or so rising about 3 metres above the ground. The CO_2 then fell to the ground as tiny particles of 'dry ice', forming a large pile of very cold particles. That night there was a heavy dew and water condensed as ice on the surface of the cold dry ice. This in turn provided very effective insulation and the pile of cold CO_2 particles remained there for several days, until finally it was necessary to bring in trucks and remove it.

This is not to trivialise the undoubted risks of cold burns and frost bite from handling solid CO_2 or dealing with a fractured pipeline, but it does demonstrate that CO_2 is not necessarily the 'deadly' 'poisonous', or 'toxic' gas that it is

Figure 7.4 Release of high pressure CO_2 into the air produces a visible plume of very cold CO_2 and water vapour.

sometimes described as. It all depends on the circumstances, the concentrations and the way an incident is handled. The risks involved in transporting CO_2 are known and can be managed using standard industry procedures, underpinned by an effective regulatory regime.

Issues posed by other substances in the CO_2 stream

While the discussion has focused on CO_2, it is unlikely that CCS will involve the transport (or storage) of absolutely pure CO_2; there will always be trace quantities of other substances such as methane (in the case of CO_2 from natural gas or liquefied natural gas projects) or hydrogen sulphide or nitrous oxide from coal-fired power stations. Small quantities of some of these impurities can have a significant impact on the pressure in a pipeline and may require additional compression along the pipeline leading to extra costs.

Some of the impurities can also be chemically active and can result in corrosion of the pipeline. This can require the use of stainless steel for the pipeline at significant extra cost. The alternative is to remove the impurities but this too can add significantly to the cost. By way of example, purchasing slightly impure CO_2 from a natural geological source may cost a few tens of dollars a tonne, whereas the cost of pure, food grade CO_2 (where essentially all the impurities have been stripped out) can be hundreds of dollars a tonne.

Siting, construction and monitoring of CO_2 pipelines

CO_2 pipelines are built to high engineering design standards, in the same way that all high pressure cross country pipelines (such as those for natural gas) are specified to ensure that all safety and regulatory standards are met. It has been noted that the risks associated with CO_2 pipelines are likely to be similar to those of natural gas pipelines but with the notable difference that unlike methane, CO_2 is not explosive or flammable and is therefore an inherently less risky gas to transport than methane.

One of the earliest considerations in establishing CO_2 pipelines is access to the land which will carry the pipeline. A pipeline will require a legal right of way and a regulatory framework within which the appropriate safety and access requirements are defined. Depending on the route the pipeline takes, re-compression stations using pumps or compressors may be needed. Designing and building the pipeline, will also be influenced greatly by such issues as population density and topography. Topography in particular will affect the overall cost of the pipeline, impacting on the number of compressors required as well as the difficulty of construction.

Pipelines are usually buried on land or in the seabed in shallow water, but in deep offshore operations, they are left exposed. Pipes are usually welded together before being placed in position. Despite their apparent rigidity, it is possible to wind a pipeline onto a large reel, which is then used to place an offshore pipeline in position. Pipelines can be coated and have cathodic protection (to protect from corrosion). A pipeline design normally builds in the capacity to block off sections for maintenance and safety inspections (for example regular pipeline isolation valves every 30 km).

As planning progresses for CCS projects, it is anticipated that pipelines for CCS projects may need to be built large enough to take the flow from additional sources in the future. This means that they will need to have access points along the way for extra CO_2 entering the main pipeline. This 'common access' framework will facilitate the future development of collection hubs.

Gas pipelines have a very good safety record. They are regularly inspected, sometimes via remote instruments at intervals along the pipeline; visually on the ground or from the air, or by remotely operated undersea vehicles. Internal inspection and cleaning is carried out by a pipeline inspection gauges (PIG) which are put into the pipeline and propelled along by the gas pressure, not only cleaning the line, but also detecting deformation and corrosion.

CO_2 transportation by road, rail and sea

Whilst the focus so far has been on transport by pipeline, CO_2 is commonly transported by road in purpose-built tankers. First, any water is removed and the CO_2 is then compressed ready for transportation as a liquid or dense gas. Small amounts are delivered in cylinders as compressed CO_2 gas, but bulk CO_2 is delivered by cryogenic tankers holding 10–20 tonnes of CO_2. This method of transport is satisfactory for relatively small quantities of industrial or food industry CO_2 or for demonstrating the concept of CCS, but is not a viable option for large scale CO_2 transport to a storage site, particularly as large amounts of CO_2 would be emitted during the transportation process. Transportation of CO_2 by rail tanker to a storage site may be more feasible than road transport, to the extent that it would have a lower carbon footprint. However, it is unlikely to be able to compete with a pipeline in terms of cost or the volume of material that can be transported.

Shipping large amounts of CO_2 may be an option in the future. Currently there is one purpose-built CO_2 transport ship, the 'Coral Carbonic', which has been transporting CO_2 since 1999. It can carry up to 1382 tonnes of CO_2. The ship transports food-grade CO_2 in the Baltic and North West Europe region. Transport by ship may become feasible in conjunction with LPG activities, due to the fact that the transport of both LPG and CO_2 requires ships which operate with cryogenic chambers at low operating temperatures and high pressures (a temperature of minus 48°C and pressure of 12 bar in the case of LPG compared to minus 50°C and 7 bar for CO_2).

However, an even more significant opportunity may arise from LNG. LNG is normally transported as a cryogenic liquid at a pressure of 1 bar and a temperature of minus 162°C. Carbon dioxide is transported at around 7 bar and minus 50°C, so a ship for CO_2 would require far greater pressurisation and far less refrigeration than would be required for LNG. In addition, the time taken to purge a ship in the changeover from LNG to CO_2 (and vice versa) could be significant. Nonetheless it is worth posing the question: would it be feasible to transport LNG by ship in one direction, and transport CO_2 in the other direction for injection into a depleted gas reservoir or other suitable geological site, assuming that engineering could overcome the hurdle of differing temperature and pressure requirements?

There are some obvious potential opportunities to take the concept forward. For example, Australia, which has a very large offshore storage potential, each year sends many millions of tonnes of LNG to north-east Asia, a region which may have little in the way of assured storage capacity. Could north-east Asian CO_2 be 'back-loaded' to offshore Australian storage sites? The same concept might work for Qatar, the world's largest LNG exporter, and a country also likely to have very significant storage capacity. There are profound technical, financial and political hurdles to be overcome before such a concept could become a reality, but given the global nature of the greenhouse gas problem, perhaps it is time to think 'outside the box' and be less concerned about the barriers and more concerned about making deep cuts in global emissions?

Reducing transportation costs: CO_2 hubs

Transporting CO_2 long distances will add significant costs to a CCS project (Figure 7.5). Long distance pipelines are expensive and if they are only used to transport a small amount of CO_2 then the cost per tonne of CO_2 will be prohibitive. There is clearly a benefit in seeking economies of scale: for example by building larger diameter pipelines to take far greater quantities of CO_2 , thereby bringing down the cost per tonne of CO_2 transported. If a number of sources of CO_2 can be combined to create a critical mass, then this will spread the transport cost and may also mean that a storage site can be used more cost effectively.

The general concept of a CO_2 hub, was first introduced by Statoil in the late 1990s and then applied by CO2CRC in 2000 as part of a proposed 'Gladrock' project, sited on the central Queensland coast in the vicinity of the towns Gladstone and Rockhampton, where there are a number of large scale CO_2 sources. The concept was then extended to the state of Victoria with the CarbonNet project as well as to projects in Alberta, Rotterdam, Western Australia, the United Kingdom and other locations. The concept is likely to be important in taking CCS forward by driving down costs and optimising storage sites.

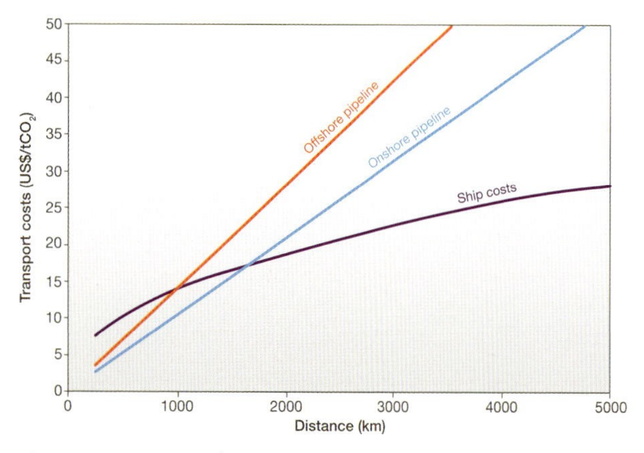

Figure 7.5 CO_2 can be transported by onshore or offshore pipelines or by ship. For distances beyond 1000 km ship transport may be cheaper than pipeline transport. However, this will be highly dependent on the local condition including the port facilities, water depths and so on. (Data source: IPCC 2005)

A number of issues would need to be resolved on a hub-by-hub basis and each hub will have its unique problems. However, the starting point is the same, namely several major CO_2 sources – perhaps including: power stations, cement, iron and steel and fertiliser plants, gas separation facilities and so on – with cumulative emission of perhaps tens of millions of tonnes of CO_2 per annum (Figure 7.6). With such a diversity of sources, standards would need to be set for the composition of the emissions entering a common user pipeline, otherwise trace contaminants from one emitter could adversely impact upon another user.

The 'common user' approach, has proved beneficial to natural gas pipelines, and in the case of CO_2 benefits could include a single approvals process, simplified regulation, storage optimisation, a smaller footprint compared to that arising from multiple pipelines and most importantly, minimisation of transport costs. The hub concept for CO_2 is already a reality in the United States to the extent that the pipeline network already exists for transporting CO_2 mainly, though not exclusively, from natural (geological) sources to sites for enhanced oil recovery. The Beulah-Weyburn pipeline transports anthropogenic CO_2 several hundred kilometres for example, as does the La Barge project in the western United States. It is likely that this network will grow to handle greater demand for enhanced oil recovery and will also offer ever more opportunities for tapping into anthropogenic sources of CO_2.

CO_2 hubs proposed for Australia

An ambitious proposal for CO_2 hubs has been developed for Australia, as part of the deliberations of the Carbon Storage Mapping Task Force. This study has identified seven hubs

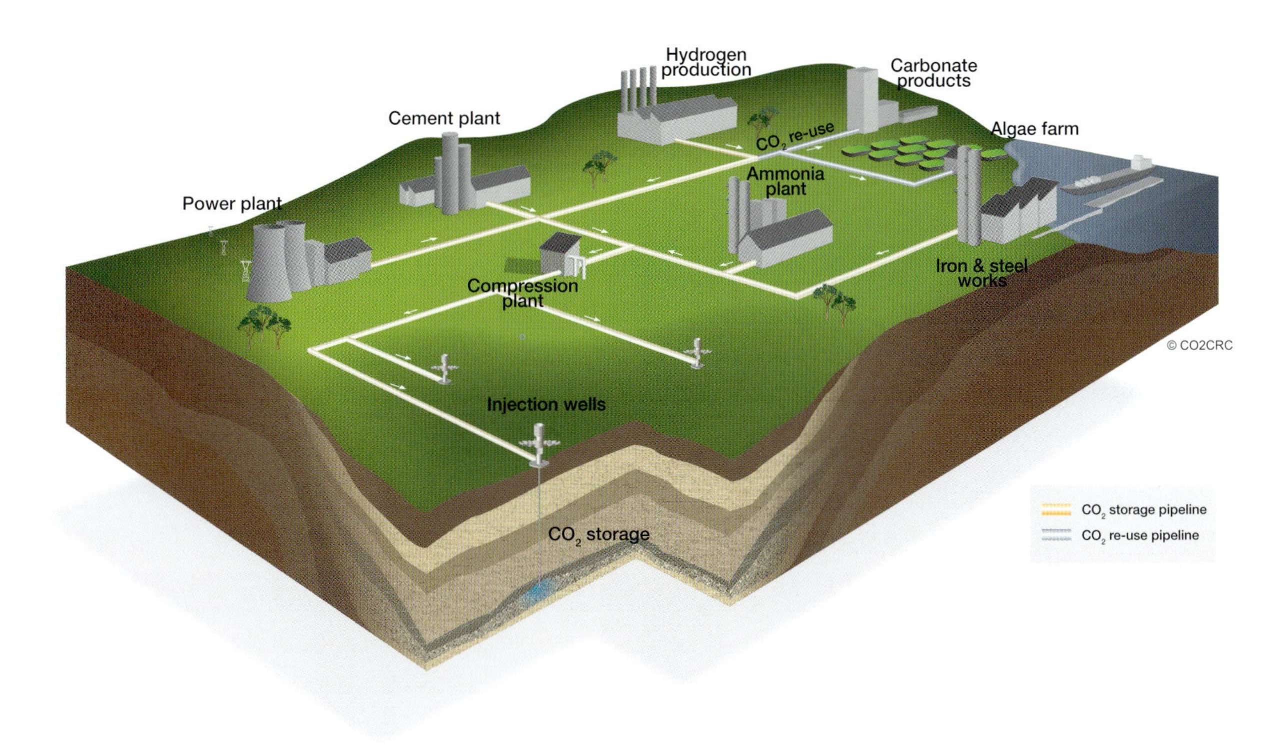

Figure 7.6 The development of a CO_2 hub linking together a range of CO_2 sources, opportunities for reuse of CO_2, and a common user pipeline and storage facilities, has the potential to significantly decrease the cost of CCS.

where already there are major emission sources, plus three additional locations that are likely to be sites of major CO_2 emissions in the future. Several, though not all of these, have identified storage prospects.

The most promising of these is a hub located in south-eastern Victoria, the area with Australia's largest single 'agglomeration' of emissions (more than 60 million tonnes CO_2 per annum) and potentially an excellent opportunity for geological storage of CO_2 in the offshore Gippsland Basin, with a likely storage capacity of many gigatonnes of CO_2. This has been proposed as the CarbonNet Project, which is intended to be a commercial-scale CCS project owned by the Victorian Government and potentially involving power generation, coal to liquid projects, and equipment and technology companies.

The CarbonNet Project incorporates the staged development of a multi-user foundation CCS network over the next 10 years and beyond, and the development of a CO_2 transportation infrastructure to link high CO_2 emission sources to proven geological storage sites. Underpinning this high level concept are a series of technology solutions dealing with capture and storage. Critical to the proposal is the identification and development of safe and secure commercial-scale storage capacity in the adjacent Gippsland Basin. This will be done through a multi-user hub-and-spoke CO_2 storage and transport system, with development of a large scale storage site offshore. Beginning in perhaps 2015–2020, regional overland pipelines (spokes) with a total transport capacity of 20 million tonnes of CO_2 per annum will connect stationary CO_2 sources to a larger CO_2 pipeline (hub) that will run to an offshore storage site.

The precise form of CarbonNet and its governance structure are by no means certain at this stage. For example, CarbonNet could be a

AUSTRALIAN CCS FLAGSHIPS PROJECTS (CARBONNET, QUEENSLAND, COLLIE SOUTHWEST)

Victoria has the world's second largest reserves of brown coal and the state government is eager to explore opportunities for using the coal. With this in mind, the Victorian government has proposed the CarbonNet Project with the aim of bringing together a number of major existing and future carbon dioxide emitters using brown coal in a single project that would result in deep cuts in their emissions. The project hopes to build a 20 Mt capacity trunk line with smaller capacity branch lines running to individual emitters from pre-combustion, post-combustion, and industry sources. The CO_2 would then be transported offshore for storage in the Gippsland Basin, which is known to have excellent storage potential. As the basin is also an important oil and gas producer, any storage activities will need to be planned to ensure no adverse impact on hydrocarbons and other natural resources.

Industry Partners Stanwell and Xstrata Coal along with other groups are currently developing the Queensland Flagship Project for a demonstration capture and storage project with the current focus being the identification of suitable storage sites in the Surat Basin of Queensland.

Initially aiming to store carbon dioxide from the proposed Perdaman fertilizer facility near Perth, the Collie Southwest Project is planned as a regional CCS infrastructure project which will eventually include coal fired power generators and various industry emitters. The project has been funded to undertake an exploration program to identify potential storage sites in the south Perth Basin.

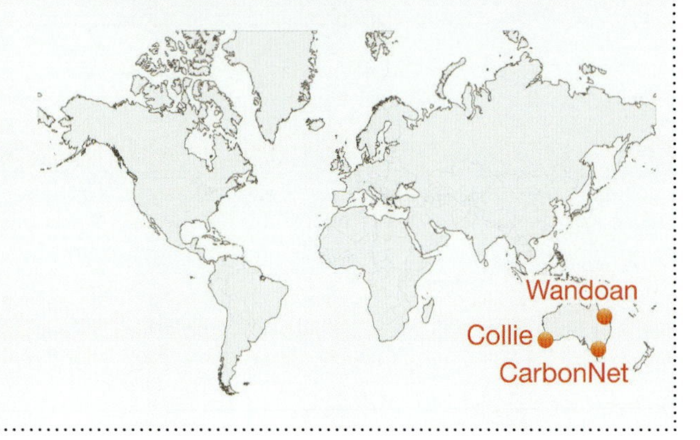

single all-encompassing company (or government entity) that oversees everything (funding, operations, and so on). Alternatively, it could be a government entity that exists primarily to cover residual risk, while a series of other companies (whether private or private-public partnerships) handle specific components of the CCS supply chain (transport, storage, and so on). It may proceed on the basis of a regulated private entity for example, or perhaps through a government 'clearing house'?

Obviously at this stage there are a number of uncertainties, but there are several drivers that are likely to take the proposal forward:

- there is a very large existing regional source of CO_2 emissions
- there are many billions of tonnes of brown coal in the region and the State Government is keen to see the resource exploited, and recognises that this will only happen if the CO_2 emission issue is addressed
- by bringing together a number of sources, CCS costs can be brought down quite significantly.

Nonetheless, it is likely to be some years before CarbonNet becomes an operational reality.

The proposed Rotterdam Climate Initiative CO_2 hub

Similar concepts are being developed elsewhere, with the Rotterdam Climate Initiative being one of the more advanced (Figure 7.7). Rotterdam is a

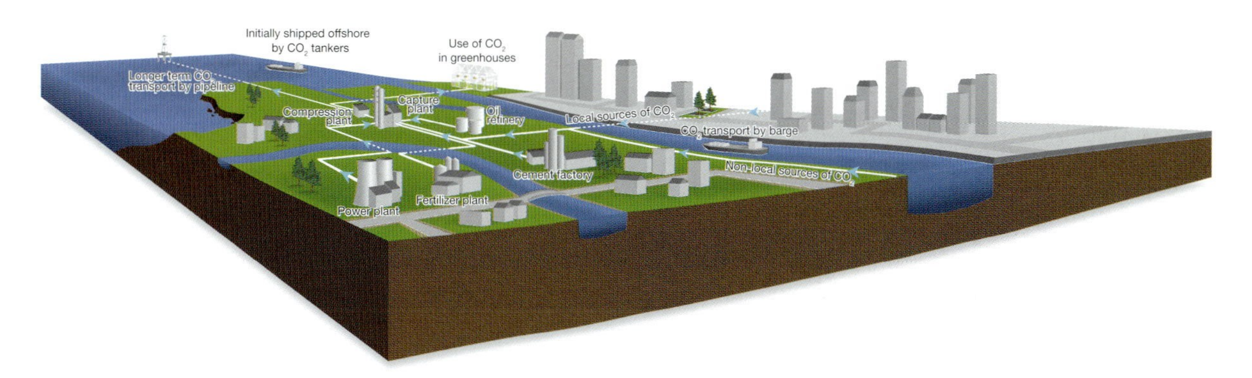

Figure 7.7 The Rotterdam Climate Initiative has developed a comprehensive proposal for a CO_2 hub, based on a range of potential emitters, with CO_2 transport by pipeline, river barge and ocean tankers. Beneficial use of CO_2 in greenhouses is also factored into the plan.

major international port in the Netherlands, with a large CO_2-intensive industrial sector that includes refineries, coal and gas fired power stations and chemical plants. A large number of companies are involved in taking the project, which is a partnership with the Clinton Climate Initiative. The proposal has Rotterdam at the hub of a large scale CO_2 network that would extend not only throughout the Netherlands but possibly also as far as the Ruhr area of Germany. CO_2 would be transported by pipeline, or by river barges down the Rhine to Rotterdam and then onward to potential storage sites in the North Sea.

The concept is that initially (by 2015) the network would transport and store at least 5 million tonnes of CO_2 a year, increasing to around 20 million tonnes a year by 2020–2025. At first, storage would be in depleted gas fields located up to 100 km offshore, but once these are filled it would be necessary to move storage operations further offshore. Initially transport would be by ship as shipping CO_2 in tankers will require less capital expenditure than pipelines and offers flexibility if multiple depleted gas fields are to be used for storage. But as the quantity of CO_2 being transported increases and the storage locations move further offshore, it is anticipated that a CO_2 pipeline would be the most cost effective transport option after 2025.

As in the case of CarbonNet, there are a number of uncertainties regarding the Rotterdam Climate Initiative, including the nature of the commercial venture to take it forward and of course the source of funding. But again, the key point is that one of the best approaches to bringing down costs is to develop CO_2 hubs, where a large group of major CO_2 emitters in close proximity can collectively transport and jointly use major storage opportunities in the most cost effective manner.

Conclusion

There are no 'show stoppers' in transporting CO_2. It has been undertaken safely and cost effectively for many years. Transport of CO_2 is not an inherently 'risky' activity but there is a need for an underpinning regulatory regime that will ensure all risks are appropriately managed. While most transport will be by pipeline, the option of long distance ship transport should not be discounted and may offer innovative international opportunities for co-ordinating CO_2 transport with other forms of liquid transport in the future. Finally, there is scope for bringing down costs through multi user hubs for transport of CO_2.

8 STORING CO_2

Once CO_2 is captured and transported from a major stationary source, we can store that CO_2 in a number of ways. It can be 'stored' in useful products. For example, CO_2 can be taken up by algae and then we can use the algae to provide valuable products such as animal feed or liquid fuels which can be sold. This can be useful as a way of offsetting capture costs, but such use does not result in long term storage of carbon. There are also some technological challenges to be overcome, notably how to harvest the algae at scale and how to minimise the massive area of land required for algal ponds. Algal fixation of CO_2 can be a valuable niche opportunity and may provide cost offsets, but with no expectation that major decreases in global CO_2 emissions will result.

A 'niche opportunity' also exists in the production of some useful minerals from CO_2. The most useful 'sequestration product' is likely to be in building materials such as concrete, because of the massive quantities used worldwide and because it can store CO_2 for tens to hundreds of years or more. There are barriers to this use of CO_2, not the least being the reluctance of the construction industry, to move to new and unfamiliar bulk building materials. Nonetheless, there is scope for incorporation of 'high CO_2' materials in construction, though the CO_2 stored will only ever be a fraction of total anthropogenic emissions. Sequestration in mineral waste generated during the production of alumina, is a good example (discussed in Chapter 4) of a useful 'by product' – in this case, the outcome of storage is an environmental improvement achieved by decreasing the alkalinity of the red muds as well as improving the sand by product, a building material that is used in construction and landscaping.

Storage within the ocean water column (discussed earlier) is not likely to be a storage option that will be acceptable to the community. This leaves geological storage of CO_2, as the most viable large scale option and it is this option which is the focus of the remainder of this chapter.

Why geological storage over other forms of storage?

What are the grounds for giving so much emphasis to the use of geological storage of CO_2 as a mitigation option, given the inevitable uncertainties about the nature of the deep subsurface and the many years CO_2 would have to remain there to make a real difference to atmospheric concentrations? The answer is in part an outcome of the oil and gas industry, where through more than a century of exploration for, and production of, petroleum and the trillions of dollars invested in order to understand the subsurface, we now know a great deal about the rocks and fluids deep below the surface in many parts of the world. Applying the lessons learned from the oil and gas industry in such areas of research as geology, geophysics rock mechanics, fluid flow, drilling, computer modelling and simulation, enables us to predict what will happen to CO_2 in the subsurface. In addition there are many parts of the world where CO_2 has been naturally stored in the subsurface for thousands to millions of years, which also provides us with important insights into the behaviour of CO_2 underground. Using this knowledge, how do we choose a suitable storage site for CO_2 and how do we ensure the long-term storage of that CO_2?

Identifying suitable geological CO_2 storage sites: sedimentary basins

The starting point of any search for a storage site is to identify suitable rocks. These are found mainly in sedimentary basins. The various geological options include storage in depleted oil and gas fields, deep saline aquifers (DSA), unmineable coals, shales and possibly basalts and other volcanic or igneous rocks (Figure 8.1). Whilst this full range of opportunities will be considered, the main focus here is on rocks found in sedimentary basins, because of their massive storage potential.

Sedimentary basins result from ongoing tectonic processes in the earth's crust. The pulling apart of fault blocks can produce a rift valley; the overthrusting of one part of the earth's crust onto another can produce a downwarp of the earth's surface. Changes within the deep crust or mantle can produce upwarping and a mountain range in one area, and downwarping and a basin in an adjacent area. In short, the key elements in producing a sedimentary basin are:

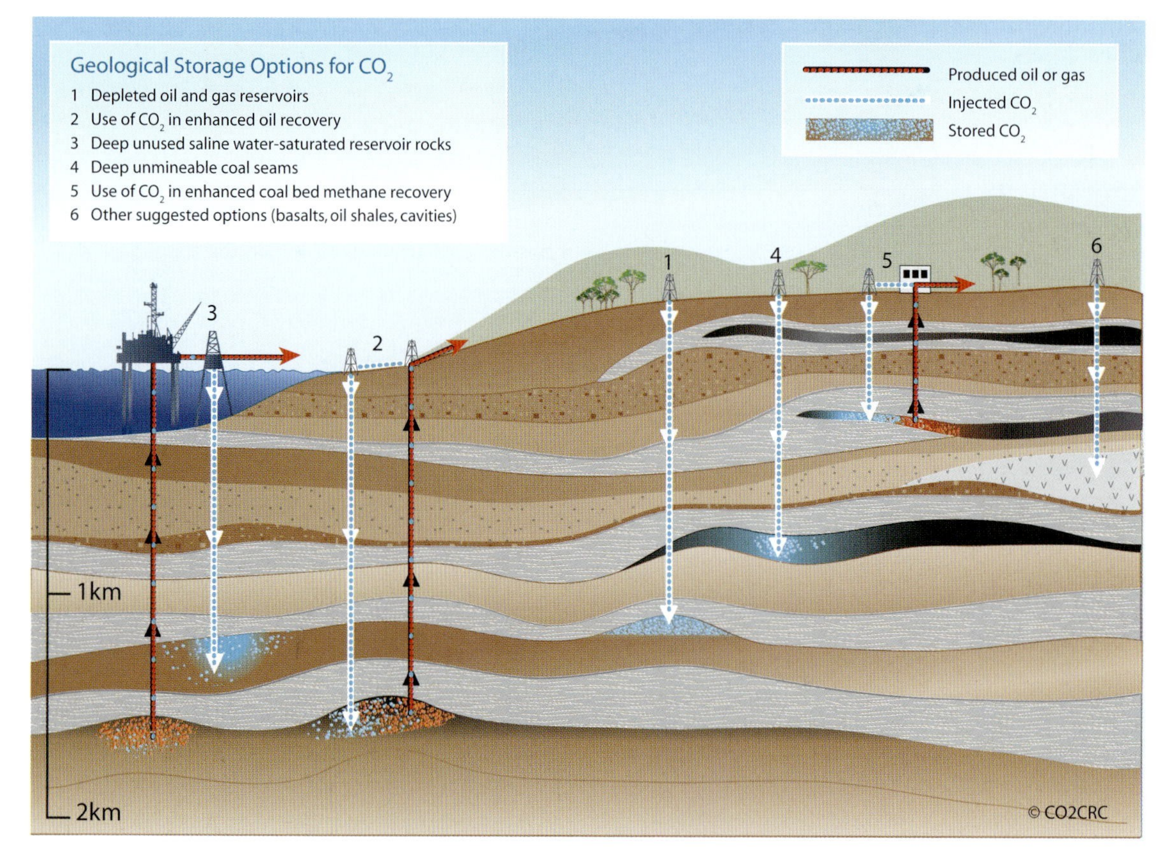

Figure 8.1 This now-classic diagram illustrates the range of options for geological storage of carbon dioxide. (Image: IPCC 2005)

Figure 8.2 Sediments such as the cross-bedded desert sandstones shown here from central Australia can extend over tens of thousands of square kilometres of a basin.

Figure 8.3 The environment under which sediments are deposited has a major influence on whether rocks are potentially suitable for storing CO_2. The ripple marks in this central Australian sandstone clearly shows that it was deposited in a shallow water environment possibly on the margins of a lake, to produce a well sorted porous and permeable sandstone.

- a source of sediments such as an eroding mountain range or coastline
- a sink or depression or basin, into which the sediments can be transported then trapped.

Sediments can accumulate in a range of environments and climatic zones within the sedimentary basin, including rivers, deltas, beaches, lakes, deserts, swamps, coral reefs and the deep ocean (Figure 8.2). Each environment produces particular sedimentary bodies (facies) that are a major influence on whether or not they are likely to be suitable for CO_2 storage (Figure 8.3). Sedimentary basins and their infilling sediments can extend over thousands of square kilometres and range in thickness from a few hundred metres to thousands of

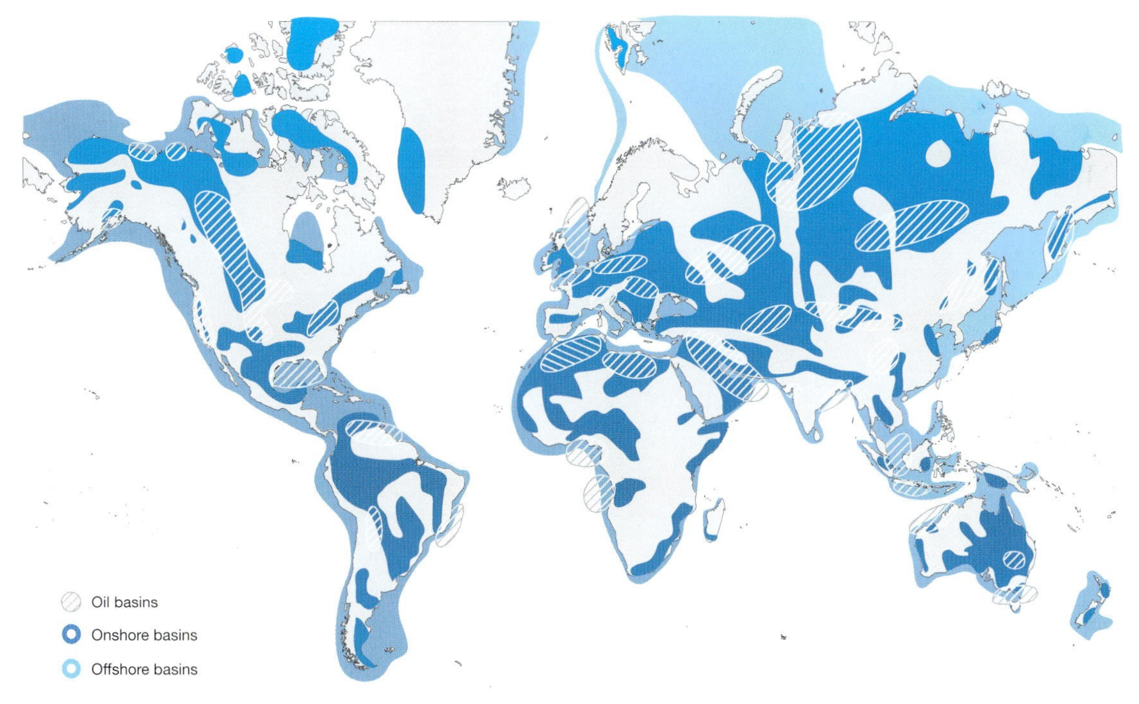

Figure 8.4 Sedimentary basins occur on all continents and continental margins. They host many resources and represent the starting point when searching for rocks that might be suitable for storing CO_2. (Adapted from IPCC 2005)

metres. The sediments within the basins can be as young as a few thousand years and as ancient as 3 billion years, though younger sediments are far more likely to be suitable for storage than ancient sediments.

Sedimentary basins have been the preferred habitat of man for thousands of years. Initially the basins provided the water, the forests and the game that constituted the primary food source. Subsequently, sedimentary basins were the preferred site for agriculture and for building materials. They also provide fossil fuels (oil, gas, coal) on which man has become increasingly reliant (Figure 8.4). So, civilisation has developed a high level of dependency on sedimentary basins as places to live and to grow food and as sources of water, building materials and energy supplies. It is therefore no accident that most of our major cities and towns and our major sources of anthropogenic CO_2 are located within sedimentary basins. From the perspective of CCS this is fortunate given that sedimentary basins offer by far the most promising sites for CO_2 storage.

This is not to say that all sedimentary basins are suitable. Some may be too old and the rocks too altered (too metamorphosed) to be suitable. Some are too faulted and folded and too geologically complex. Others are too shallow; too small; or inaccessible because they are in mountainous terrains or too deep in the ocean. But even excluding the sedimentary basins that are not suitable, there are still a great many that are potentially suitable for storing large volumes of CO_2 in many parts of the world. Sedimentary basins that host oil and gas deposits are particularly promising for CO_2 storage, although the need to avoid contaminating oil or gas resources with CO_2 is an important issue that has to be considered as part of any storage project.

Features of a sedimentary basin that may make it suitable for storage

Porosity and permeability

What are the features that make a sedimentary basin, or an area within a basin, suitable for storage? One of the key starting points is the presence of reservoir rocks – the sort of rocks in which oil or gas (Figure 8.5) or groundwater commonly occur: porous rocks that will hold water, analogous to the manner in which a paper towel holds water within tiny holes (pores), or a sponge holds water within much larger pores (Figure 8.6). A common misconception regarding storage of CO_2 (the same misconception often applies to oil and to groundwater) is that the fluid is stored in underground caverns. Whilst there are some instances of oil and gas being found in deep cave-like structures, this is very much the exception rather than the rule. Reservoir rocks, such as sandstones, which are often used as building stones, might look quite solid, but within the framework of sandstone grains (Figure 8.6) there are many tiny spaces or holes (pores). The smaller the pores and the denser the framework, the harder it usually is to force a fluid (water or CO_2) into the pores and it can take a great deal of time, perhaps thousands of years or more, for fluids to flow just a few kilometres in a rock that is 'tight'.

The ease with which a fluid can flow into and within a rock, is a measure of the permeability of the rock. A rock with pores that are connected, allowing a fluid to readily flow into (and through) the rock, has a high permeability (Figure 8.6). Rocks that have pore space but where the pores are not joined up has a low permeability. A reservoir rock that is suitable for storage will have lots of pore space (a high porosity) to hold the CO_2 and a high permeability to allow the CO_2 to flow into the pore space. Most reservoirs, whether for oil or gas or groundwater or for use in CO_2 storage, are

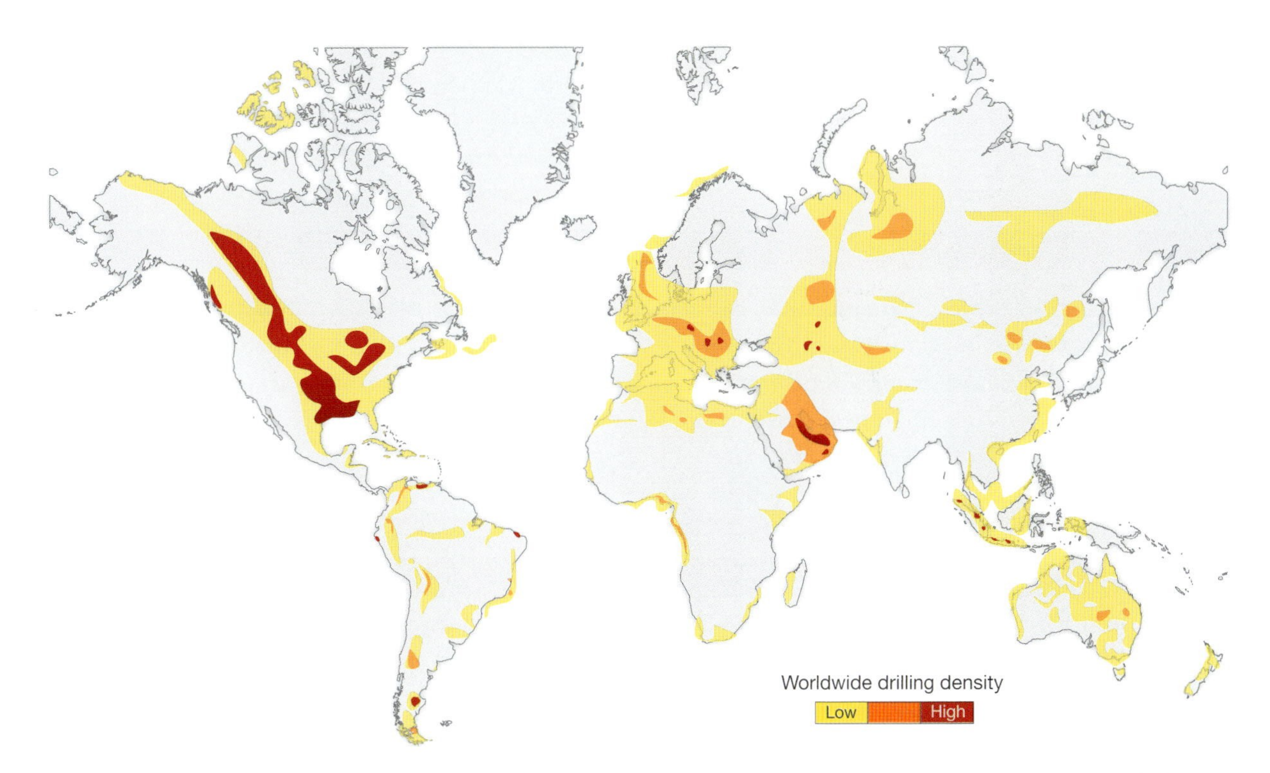

Figure 8.5 The abundance of drilling activity in sedimentary basins is usually an indication that there are oil and gas fields in the basins and porous and permeable rocks potentially suitable for storage of CO_2. (Image adapted from IPCC 2005)

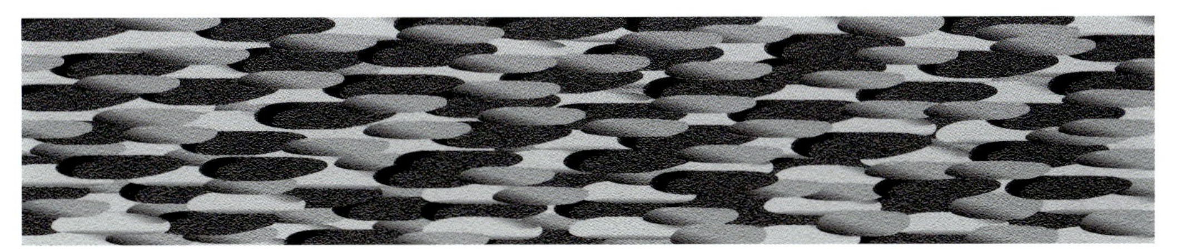

Figure 8.6 Rocks suitable for storage of CO_2 are likely to have a high porosity (abundant pore space between the grains) and high permeability (the pores are joined together) so that fluids such as water or CO_2 can readily move into the pore space. An overlying rock composed of finer grains and low porosity and permeability forms a seal rock which prevents upward movement of the CO_2.

fine grained porous and permeable sandstones, with the sand grains composed of stable minerals, notably:

- quartz grains in mature sandstones
- reactive, or less stable minerals or rock fragment in immature sandstones
- carbonate grains in limestones.

The composition of the grains is significant in that CO_2 may chemically react with grains to form carbonate minerals in an immature (reactive) sandstone but will not react to any significant extent with a mature (unreactive) quartz sandstone.

In addition to the physical or chemical nature of the reservoir, a further requirement is that it occurs sufficiently deep below the earth's surface that the CO_2 can be stored as a dense fluid rather than as a low density gas. At atmospheric pressure, 1 tonne of CO_2 occupies a volume of more than 500 cubic meters, whereas at high pressure (8 bars) 1 tonne of CO_2 may occupy a volume as little as 1.5 cubic meters (350 times less space). In other words, 350 times more CO_2 can be stored at a high pressure than can be stored in the same amount of space at normal atmospheric pressure.

High pressures occur in deep rocks mainly as a result of the weight of the overlying rocks and the water within those rocks. Together, the rock (lithostatic) pressure and the water (hydrostatic) pressure (Figure 8.7) can compress the CO_2 gas

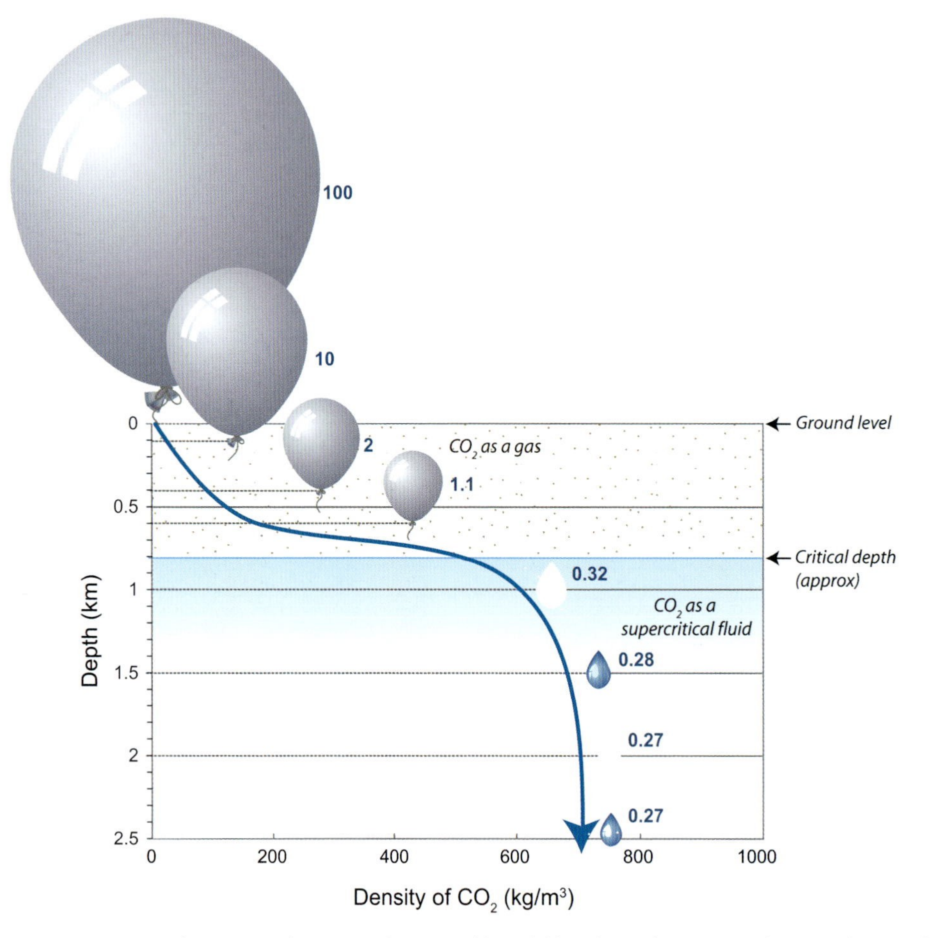

Figure 8.7 As pressure is increased on CO_2, it becomes dense and liquid-like. If it is then injected into rocks at a depth of 800 metres or more, it will remain in that liquid-like state, with a density several hundred times that of CO_2 at the surface.

to a dense liquid-like CO_2. Alternatively, if the CO_2 is already in that dense state when it is injected, then the pressure in the rocks can maintain it in that state.

The density of CO_2 is also strongly influenced by temperature – the higher the temperature, the less dense the CO_2. The temperature in the earth's crust increases with depth as a result of tectonic or igneous activity or the decay of radioactive minerals within the crust. Therefore we also need information on sub-surface temperatures to more accurately forecast the optimum depth for CO_2 storage. Using information from wells drilled for oil or water, we know how pressure and temperature changes with depth. While these changes are not uniform (they vary from place to place), in most parts of the world CO_2 needs to be stored at least 800 m below the surface in order for it to remain in a dense liquid-like state. Below 800 metres, while the CO_2 volume continues to decrease with depth, the volume change is quite small and any additional benefit derived from going deeper is limited, particularly as the deeper the well for injecting the CO_2 the greater the cost of drilling.

In addition to needing a porous and permeable reservoir rock at the right depth to contain the stored CO_2, it is also essential to have a suitable geological setting that will prevent the CO_2 from leaking out of the reservoir into adjacent rocks, and from there potentially into groundwater resources, and ultimately rising to the surface. We know that nature traps gases, from the existence of many massive natural gas deposits that are trapped below the surface. Similarly, fresh groundwater resources exist that are not affected by overlying or underlying salty aquifers – evidence that adjacent parts of the existing crust can be sealed from each other. When CO_2 is injected into a water-bearing geological formation (Figure 8.8), the CO_2 has a lower density than the water and it will therefore tend to rise. However, if there is a barrier above the

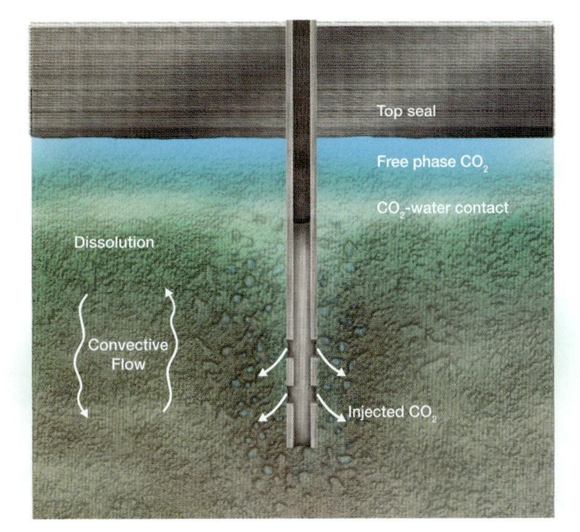

Figure 8.8 When CO_2 is injected into a rock such as a deep saline aquifer (DSA) from a deep drill hole, some of the CO_2 will go into solution, but initially much of it will rise through the rocks until it is trapped below an impermeable layer (the top seal), where it will continue to collect, pushing down the CO_2-water contact and spreading out below the seal.

reservoir, such as a very fine grained low permeability layer, a mudstone for example, then the CO_2 will be trapped within the reservoir. This low permeability unit is called a caprock or seal.

However, the trapping mechanisms can be much more complex than this. If, for example, the reservoir itself is tilted, then the CO_2 may remain within the reservoir interval, but continue to move upwards within the reservoir itself. If the rocks overlying a geological formation in which the CO_2 is trapped is fractured, then CO_2 can potentially leak out, which is why areas with very active faults would in general not be seen as good locations for CO_2 storage.

There are a number of ways that trapping mechanisms prevent the CO_2 escaping. The simplest form of trapping is structural trapping, where there is folding of the reservoir rocks and the seal, to form a dome or anticline (Figure 8.9a). Faulting of the rocks can produce a lateral impermeable barrier which prevents upward or lateral escape of CO_2 (Figure 8.9b). In some instances, the original geometry of the rocks,

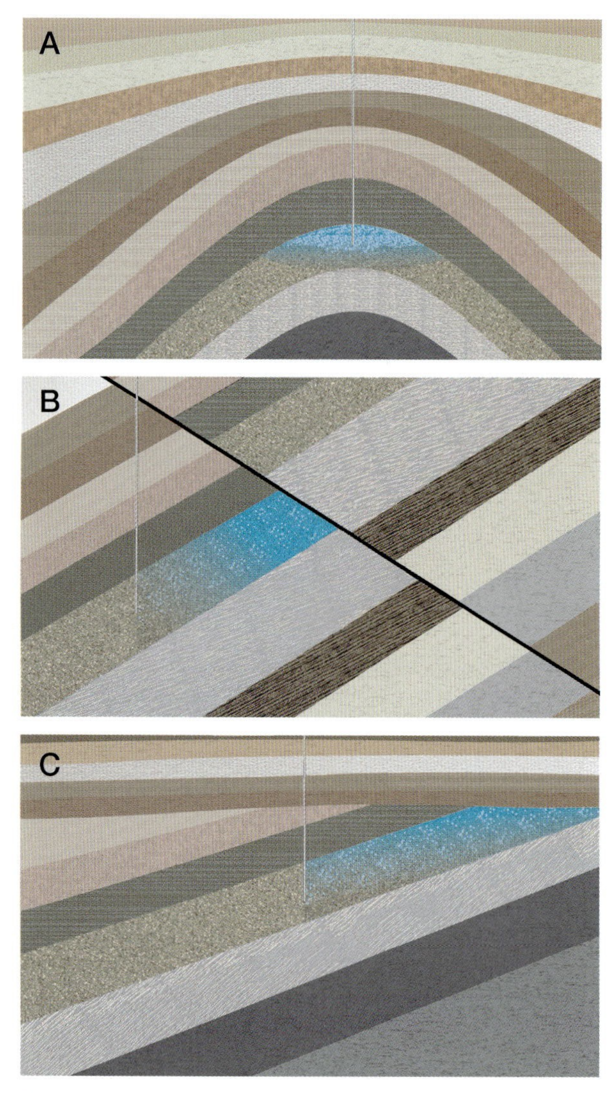

Figure 8.9 There are various geological settings that can result in trapping of CO_2 including: (A) in an anticline; (B) as a result of displacement along an ancient fault resulting in an impermeable seal being displaced against the porous rock with the CO_2; (C) as a consequence of the wedging out of the porous and permeable formation against an unconformity, in this case an old erosion surface overlain by impermeable rocks.

(resulting from the environment in which the rocks were initially deposited), provides a lateral impermeable barrier to CO_2 escape, resulting in what is called stratigraphic trapping.

In addition to structural and stratigraphic trapping, CO_2 can also be trapped as a result of hydrodynamic processes and/or chemical processes. Residual trapping can occur within a rock when there are two fluids such as water and CO_2. As the fluids move through the pores of the reservoir, the water will move across the surface of the grains bypassing small amounts of CO_2 in the middle of the pores. Even if you were to try to produce the CO_2 from a well, the flowing water in the formation would travel around the CO_2 trapped in the middle and you would only produce water; the CO_2 would be permanently trapped within the grains. The extent to which this occurs is dependent on the geometry of the grains and the 'connectivity' between the pores.

Another form of trapping, solubility trapping, occurs when CO_2 is dissolved in the formation water (the water in the pores of the reservoir rock). This is analogous to the manner in which CO_2 is dissolved in soda water or in some natural mineral waters. Over time, the proportion of stored CO_2 that is dissolved increases, unless for some reason there is a drop in pressure and then, just as when the top is taken off a bottle of soda water and the CO_2 fizzes out of the top of the bottle, so CO_2 can come out of solution forming in some instances a natural geyser of CO_2, such as those found in some volcanic areas. However, in the deep sub-surface, a sudden and unexpected decrease in pressure is unlikely to occur. Indeed the reverse tends to happen in the longer term: over time, as more CO_2 goes into solution, the water (plus CO_2) becomes progressively denser and begins to sink down into the basin. In other words, over time, and that time is hundreds of years or longer, the chance of CO_2 escaping to the surface does not increase, but diminishes.

This has been illustrated through some very elegant modelling which shows how first, fine fingers of CO_2-rich water move downwards in the porous and permeable reservoir and ultimately, most of the CO_2-rich water sinks down into the reservoir (Figure 8.10). As always, the full story is a little more complicated than

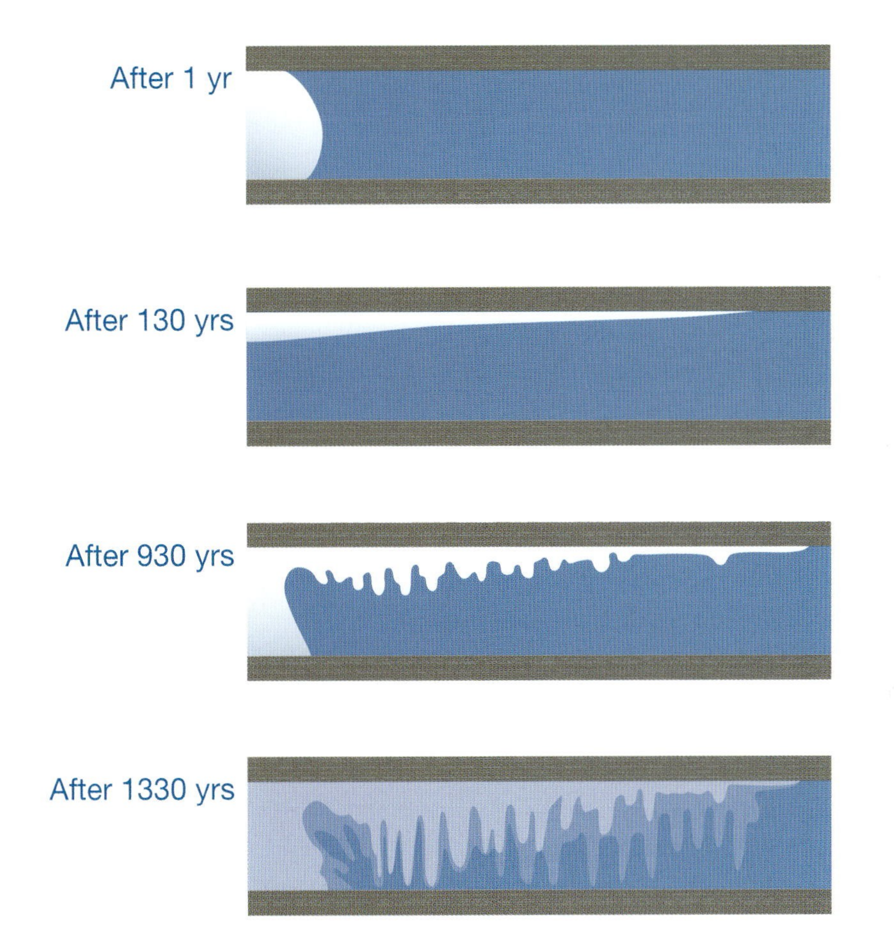

Figure 8.10 Some elegant computer modelling of the behaviour of carbon dioxide in a porous water saturated rock shows that initially the CO_2 moves up to the top seal and then spreads out laterally. Over time, progressively more CO_2 goes into solution, the brine becomes denser and descends into the deep porous rocks within the basin. (Image after Jonathan Ennis-King)

this, as the solubility of CO_2 in water also depends on temperature and pressure. Nonetheless, in general, the likelihood of stored CO_2 escaping to the surface appears to decrease rather than increase over time.

CO_2 can also be stored in the subsurface through mineral trapping. This occurs when CO_2 comes in contact with some types of comparatively unstable rock grains (usually calcium or magnesium silicates) and reacts with them to form stable calcium or magnesium carbonates. Once this happens, the CO_2 is essentially locked up for geological time as stable carbonate minerals. However, these reactions occur quite slowly, over many thousands of years or more. Consequently while mineral trapping does add to the security of geological storage of CO_2, it occurs slowly, on geological time scales. Work is underway to try to speed up the process, with a view to making it more relevant to human time scales.

Storage of CO_2 in depleted oil and gas fields

How often does the required combination of reservoirs and seals and a favourable geological setting, occur? An obvious site for geological storage is in depleted oil and gas fields: they generally have proven geological traps (such as

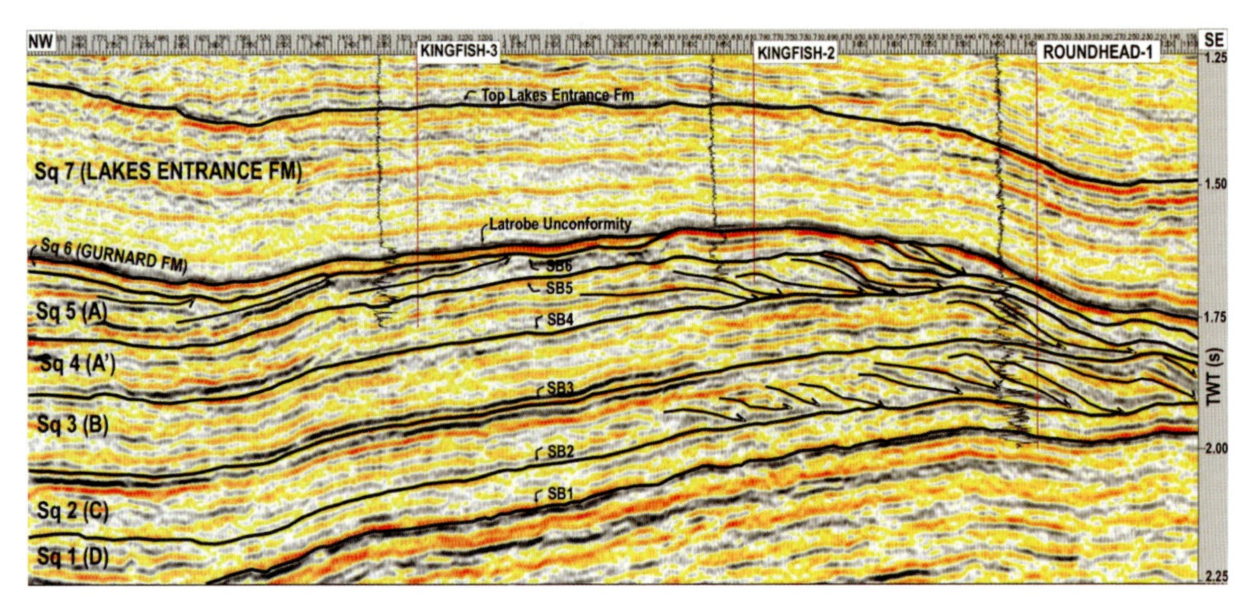

Figure 8.11 This seismic image from the Gippsland Basin in south eastern Australia, shows thick porous sediments dipping to the north-west under a regional unconformity surface (marked on the diagram as the Latrobe unconformity) formed as a result of a change in sea level and related erosion. Over this surface are draped relatively flat lying sediments. There are a number of reservoir-and-seal formations within the sequence which could provide excellent CO_2 storage opportunities, but care would need to be taken to ensure no adverse impact on some of the existing oil and gas fields in the basin. (Image C Gibson-Poole)

anticlines), good reservoir rocks and effective seals that have held oil or gas for millions of years. Their geology is also likely to be well known which means there is less likelihood of stored CO_2 behaving in unexpected ways because of unanticipated geological complexity (Figure 8.11).

There can be an added benefit from the injection and storage of CO_2 into depleted oil fields, in that it can result in an increase in the amount of oil that can be produced – referred to as enhanced oil recovery (EOR). After initial (primary) production of oil, further (secondary) production can be obtained by injection of water or steam or gas. Maximum output from some fields can be obtained through tertiary production, usually fields with fairly heavy and viscous oils, with late-stage injection of CO_2.

Enhanced oil recovery with CO_2 is now underway in more than 100 oil fields, mainly in the United States and Canada. In most oil fields, a maximum of only 40% of the oil in place can be recovered by conventional primary production. Secondary recovery can be used to increase this by 10–20% and this can be further increased by an average of 13% through tertiary recovery via the use of CO_2, which swells the oil and reduces its viscosity, making it easier for the oil to flow (and be pumped) to the surface (Figure 8.12). The CO_2 is then separated from the oil and reinjected, until finally it is no longer feasible to extract more oil, and most of the CO_2 then remains trapped in the reservoir. At the present time, of the order of 50 million tonnes of CO_2 are used each year in EOR operations in the United States and Canada. The majority of the CO_2 used is from geological sources, but there is increasing interest in using anthropogenic CO_2 from major stationary sources including gas separation plants, fertiliser plants and chemical plants. In the United States CO_2-EOR is seen as an important driver for getting large scale CCS operations underway. The benefit is of course that the extra cost of CCS can potentially be more than offset by the value of the extra oil produced, with up to 4–5 barrels of additional oil produced for every tonne of CO_2 injected.

LABARGE (RANGELY, SALT CREEK, MONELL)

The ExxonMobil Labarge natural gas field (Wyoming, USA), composed of approximately 66% CO_2 (with some hydrogen sulphide and helium), commenced gas production in 1986. The CO_2 is separated from the natural gas stream at the Shute Creek processing plant, the world's largest CO_2 separation plant, using a combination of solvent based technology and a new controlled freeze technology. Initially CO_2 production was 4 Mtpa CO_2 but since 2010 has been between 6 and 7 Mtpa CO_2. The CO_2 is sold to a number of enhanced oil recovery (EOR) operations. The Chevron Rangely field has been injecting LaBarge-derived CO_2 at rates of more than 2 Mtpa since 1986 with more than 25 Mt now permanently stored in the field. A limited monitoring and verification program is underway, including gas seepage studies, and to date there is no evidence of leakage of injected CO_2. The Salt Creek oil field operated by Anadarko (Wyoming, USA) began CO_2 enhanced oil recovery operations in 1993. To date more than 8 Mt of Labarge-derived CO_2 have been injected with an additional 22 Mt planned over the life of the project.

An outstanding example of CO_2–EOR is the Weyburn Project. Carbon dioxide from a coal gasifier in Beulah, North Dakota is transported 325 kilometres by pipeline to the Williston Basin in Saskatchewan Canada and used there for CO_2–EOR. Approximately 3 million tonnes of CO_2 is injected each year for EOR at Weyburn. The Rangely oil field in Colorado is also

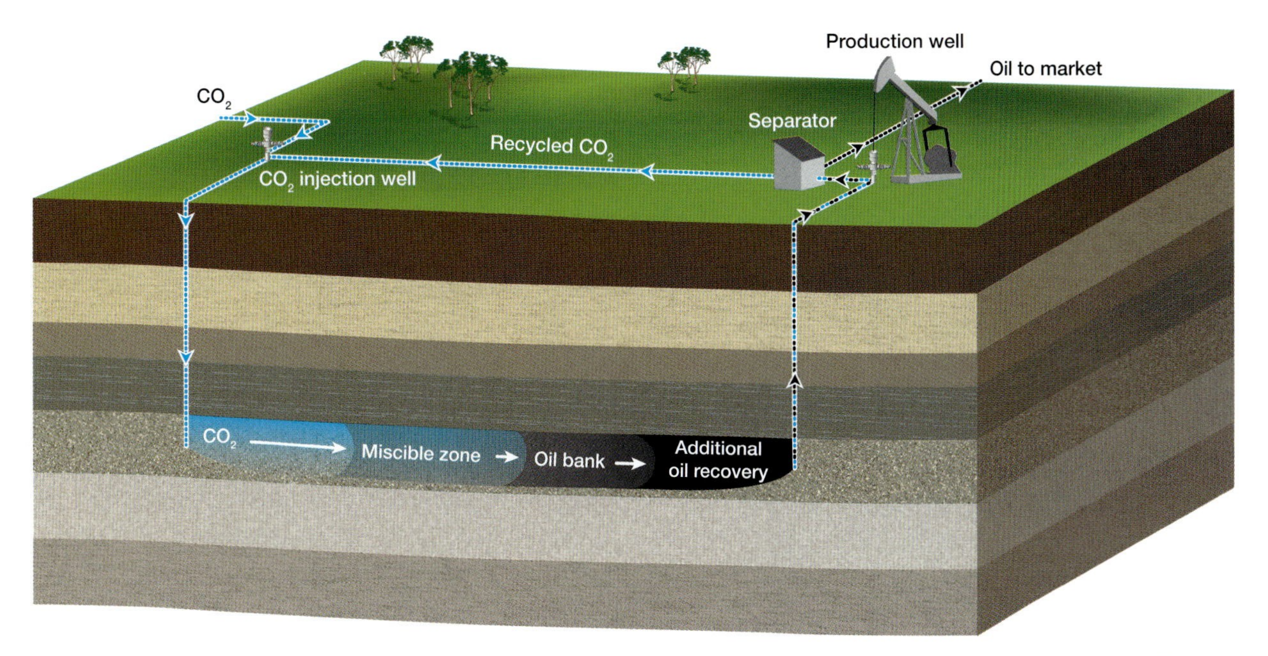

Figure 8.12 Enhanced oil recovery using CO_2 (CO_2-EOR) has been underway in North America for almost 50 years. While the primary purpose of CO_2-EOR is not CO_2 storage, over time CO_2 storage does occur as progressively more recycled CO_2 is trapped in the oil-bearing rocks. EOR has also provided some valuable lessons in terms of handling CO_2 at the surface and the sub-surface. (Image adapted from IPCC 2005)

successfully carrying out a large scale CO_2-EOR operation, with the CO_2 derived in part from anthropogenic sources.

We are likely to see more CO_2-EOR in the future but some argue against such operations, seeing them as leading to yet more use of fossil fuels and therefore more CO_2 being emitted to the atmosphere. Enhanced oil recovery in itself is unlikely to lead people to use more oil. Instead it is a substitute for conventionally produced oil. Its importance to CCS is that it provides experience of CO_2 injection and storage and a pathway to accelerated deployment of CCS.

In North America, CO_2 –EOR could be an important opportunity for accelerating CCS. Further opportunities may also exist in Brazil, the Middle East, China, South-East Asia and the North Sea. The light oils found in Australia and some other parts of the world do not readily lend themselves to CO_2–EOR, and scope for deployment in those countries currently appears to be limited.

Vello Kruuskaa of the Advanced Resources Institute (ARI) has recently suggested that improved oil recovery can be obtained if the CO_2 is injected early in the production phase rather than at the late stage. Globally CO_2–EOR may not be a major storage opportunity, but it is nonetheless one that is likely to be used wherever possible, particularly in the initial phases of deployment of CCS, because it provides an offset to storage costs through the production of more oil.

Enhanced gas recovery (EGR) using CO_2 (CO_2–EGR) also presents some opportunities that may ultimately prove to be more generally applicable than EOR, but it remains an option that is not commercially deployed to any extent at the present time. The principle is simple enough – as a gas field is depleted and pressure in the field drops, CO_2 can be injected into the field at the base of the gas column to maintain the pressure and 'push' more of the natural gas out. Challenges include ensuring that there is no significant mixing of the CO_2 with the remaining natural gas and knowing the optimum time to inject the CO_2.

The largest injection of CO_2 into a gas field to date has been into the In Salah Gas Field in Algeria where 1–2 million tonnes of CO_2 are injected into the deep part of the field each year, not necessarily with a view to EGR, but primarily for storage of CO_2 separated from the natural gas as part of the gas production process. In the Netherlands sector of the North Sea, CO_2 has been successfully injected for several years into a gas-bearing structure known as K12B, with a view to testing CO_2 –EGR.

In Australia, a small scale injection of CO_2 (66 000 tonnes CO_2) into a depleted gas field as part of the CO2CRC Otway Project (see Otway project, Chapter 9), has provided encouraging insights into the prospects for CO_2 storage in a depleted gas field. For example it was found that while not all of the 'space' resulting from gas production was refilled by CO_2, of the order of 60% was. If this value applies to other gas fields then ultimately hundreds of gigatonnes of CO_2 could be stored in depleted gas fields around the world, equivalent to decades of total global stationary emissions. This suggests that there will be greenhouse gas benefits arising not only from the increased use of natural gas as a lower carbon fuel, but also from the increased CO_2 storage opportunities arising from access to depleted gas fields.

As gas demand increases and more gas fields are developed, there will be greater opportunity for more CO_2 storage in depleted fields. But as gas demand increases it is also likely that more high-CO_2 fields will be developed and at least some of the storage space will need to be used for storage of the natural gas-associated CO_2. Nonetheless, CO_2 storage in depleted oil and gas fields will be

SLEIPNER PROJECT

In 1991 Norway became the first country to institute a tax on offshore CO_2 emissions, including on CO_2 emissions from offshore oil and gas production. At that time, the Norwegian oil company Statoil and its partners were finalizing the Sleipner West gas field in the Norwegian sector of the central North Sea, which contained a high percentage of CO_2 (9.5%) in the natural gas stream and had emissions of more than 1 Mtpa. In order to reduce those emissions (and the liability for a carbon tax), beginning in 1996 Statoil began separating CO_2 from the natural gas produced from the Sleipner field. The CO_2 is removed from the stream using solvent technology. It is then transported to a nearby, but separate, offshore rig where it is compressed and injected into the Utsira Formation. The Utsira Formation is a deep saline aquifer that occurs at a depth of about 1000 metres below the sea floor and above the natural gas-bearing formation.

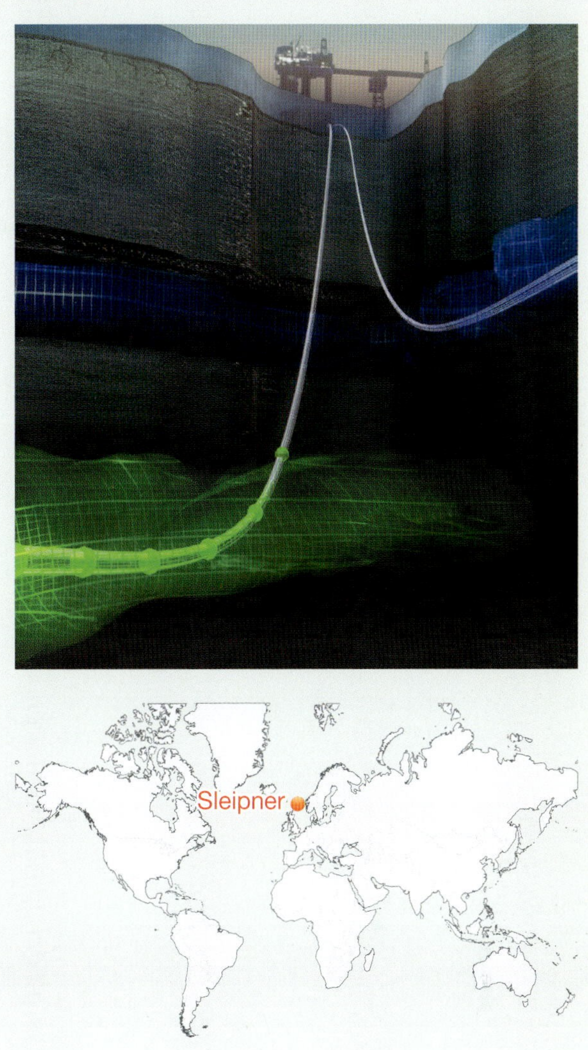

Since 1996, Statoil has injected CO_2 at a rate of approximately 1 Mtpa of CO_2, with a total of approximately 16 Mt CO_2 now stored. A major program of seismic monitoring has been undertaken by the project which has served to demonstrate the effectiveness of seismic monitoring for CO_2 storage and also to ensure that there has been no leakage of CO_2 from the Utsira Formation. (Image: courtesy of Statoil)

a valuable future opportunity, with or without EOR or EGR.

Storage in deep saline aquifers

Important though depleted fields may be, there appears to be a much larger and more widespread storage opportunity in deep saline aquifers (DSAs). Deep saline aquifers are deep, porous and permeable rocks (usually, but not always sandstones) that are saturated with water that is too salty or in some other way unsuitable to use for drinking water or agriculture (Figure 8.13). DSAs are found in many parts of the world, extending from near the surface to kilometres below the surface and underlying millions of square kilometres of the earth's surface. They constitute massive rock volumes and contain vast quantities of pore water which can range in salinity from slightly salty to more salty than seawater.

A characteristic of most DSAs is that the movement of the groundwater through the rocks is extremely slow – perhaps just metres of lateral

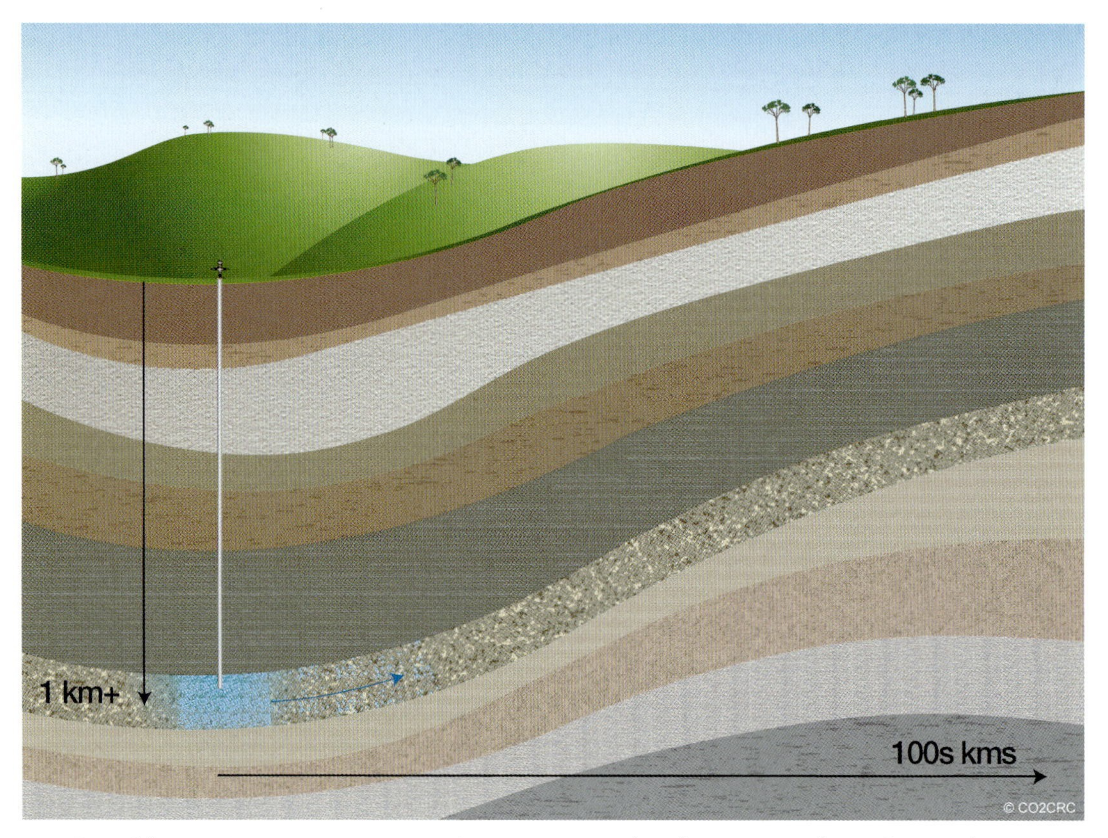

Figure 8.13 One of the most important opportunities for CO_2 storage is thought to occur in deep saline aquifers (DSA) where the CO_2 is trapped below the top seal but very slowly migrates laterally within the DSA, with progressively more CO_2 becoming trapped.

movement a year. Therefore, in some large sedimentary basins it can take a million years or more for the formation water to move across the basin through the DSA from the catchment area to a distant discharge point.

The principle of CO_2 storage in a DSA is that in part CO_2 will dissolve in the formation water and be entrained in that water as it slowly moves through the basin. As pointed out earlier, over time the brine charged with CO_2 becomes denser and will tend to move deeper into the basin. However, much of the CO_2 is also trapped – residually trapped – in the pores, as the free CO_2 is moved along with the formation water (Figure 8.14). It is thought that residual trapping is likely to be a very important mechanism in DSAs, but estimating the proportion of the CO_2 that will be dissolved and the proportion that will be residually trapped, is difficult. Experiments are underway by the CO2CRC and others to better determine the relative importance of those various trapping mechanisms, using a series of experiments involving injection then production of water from the same well, followed by injection of CO_2 and then mixtures of both water and CO_2.

Theoretically, movement of CO_2 into a shallower freshwater aquifer could occur in the 'near field' (i.e. in the vicinity of the injection site) if the injection pressure was to exceed the physical strength of the overlying impermeable seal rocks, or the CO_2 was to migrate up a fault. Pressure will increase in a closed container (such as an anticline or a depleted gas field) but this can be controlled by limiting the amount of CO_2 injected so that the pressure field stays below the known strength of the confining geological units. It is also possible to relieve the pressure by extracting water from the reservoir.

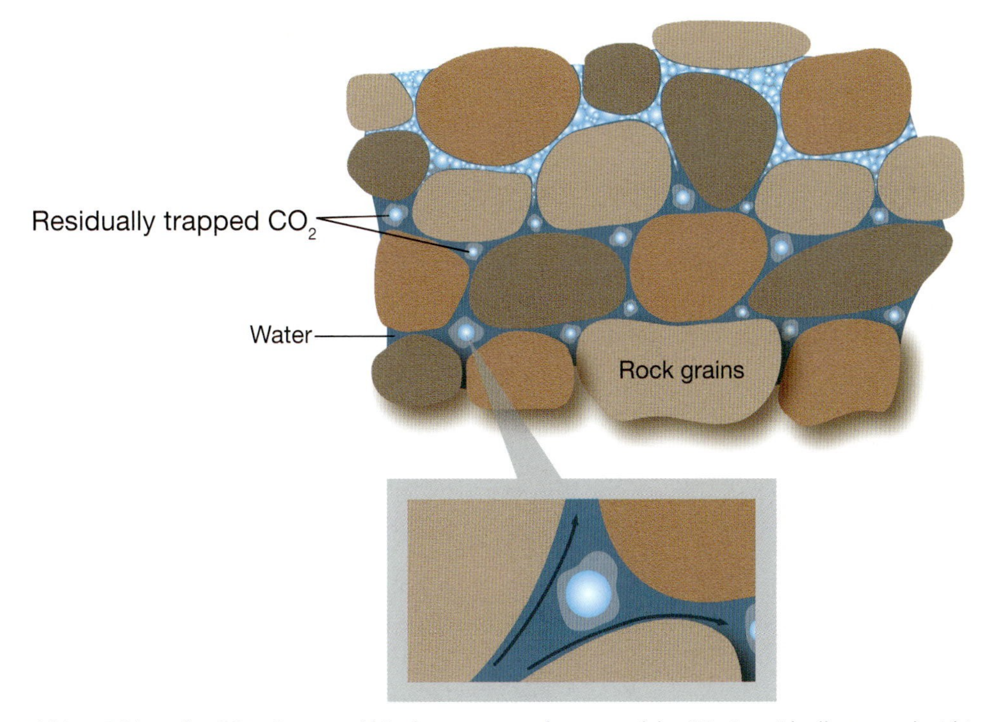

Figure 8.14 Within a DSA as the CO_2 migrates within the porous rock, some of the CO_2 is residually trapped within the pore space between the grains. Once trapped in this way it is permanently stored.

In the case of open or partly open systems, researchers at the Lawrence Berkeley National Laboratory (LBNL) have found that pressure build up is relieved naturally by movement of 'native' formation waters into regions some distance from where the CO_2 is injected. But it is also necessary to be careful in dealing with the 'far field' in that the pressure can build up many kilometres from the injection site and potentially displace brine into freshwater. Steps can be taken to ensure this does not happen, by carefully monitoring the pressure field and also, if necessary, by controlling the pressure through the drilling of pressure relief wells. DSAs hold great promise for CO_2 storage on a very large scale, but there must be monitoring of the DSA to ensure that formation pressures remain within acceptable limits and that there is no contamination of fresh water aquifers.

Confidence regarding storage in DSAs, is provided by the experience of the Sleipner Project in the Norwegian sector of the North Sea. There, one million tonnes of CO_2 per annum, separated from natural gas, is injected into a DSA made up of a thick sandstone with very saline formation water, at a depth of about 800 metres below the sea floor and overlain by an impervious layer of mudstone (Figure 8.15). Injection at Sleipner has been underway since 1995 and since that time, approximately 16 million tonnes of CO_2 have been stored without any leakage of CO_2 from the DSA and without any excess pressure build up. Further, the work at Sleipner has demonstrated that it is possible to successfully monitor the migration of CO_2 within a DSA, using seismic surveying techniques. Storage of CO_2 in DSAs is likely to be very important in the future.

Storage in coals

The discussion so far has dealt primarily with coarse grained reservoir rocks (mainly sandstones) overlain by fine grained cap rocks

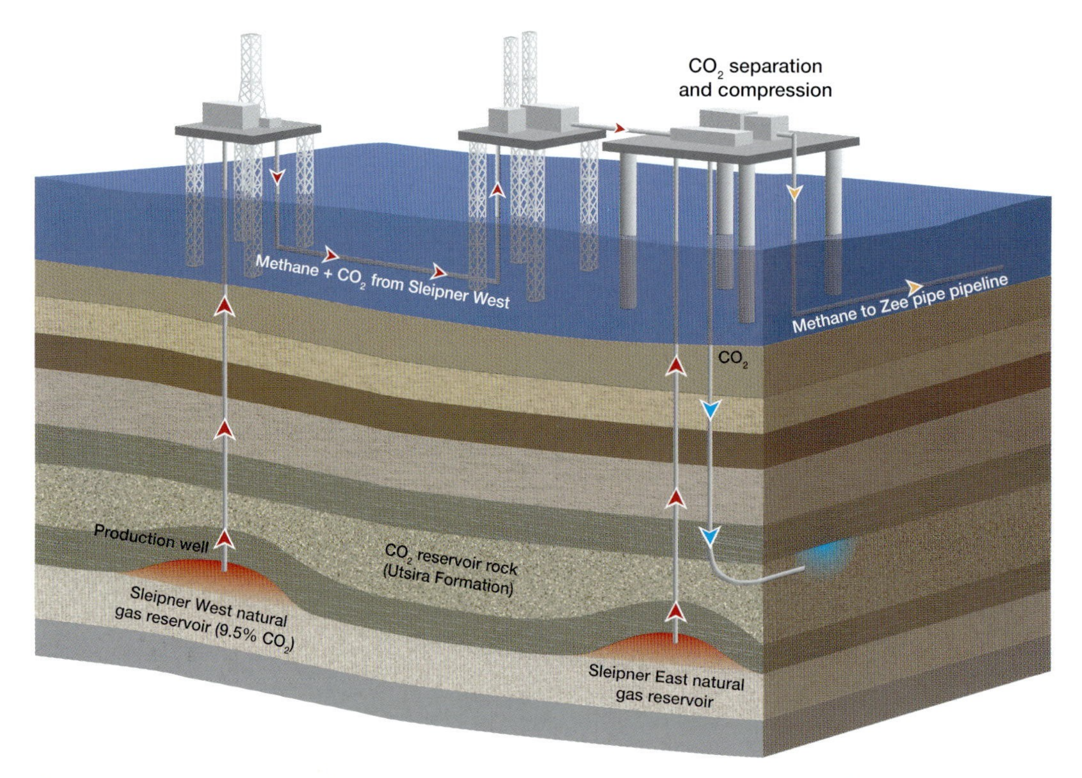

Figure 8.15 The Sleipner Project of Statoil and its partners in the North Sea was the world's first large scale CO_2 storage project. Natural gas (methane plus 9.5% CO_2) from the Sleipner West Field is treated, the CO_2 separated out, then compressed and injected in a DSA (the Utsira Formation) at a depth of about 800 metres below the sea floor. To date approximately 16 million tonnes have been geologically stored at Sleipner.

(seals), but coals can also naturally store CO_2 in fine cracks (cleats) and micropores and also by adsorption onto the fine carbon-rich particles that make up the coal. Some coals have naturally high concentrations of relatively pure CO_2 (with little or no methane) that has been trapped in the coal for geological time.

Not all coals are suitable for storing CO_2. One example is shallow coals that are economically significant and that may be mined in the future; if large amounts of anthropogenic CO_2 were to be stored in the coal, its future use as a fuel would be pre-empted. Another example is coals that are too deep to be mined: although CO_2 might be stored in them, often these coals are likely to be too strongly compacted and have too little permeability to be able to effectively inject CO_2 into them. Therefore it is a matter of looking for coals that are not so shallow that they could be mined in the future and not so deep that they are too compacted; 'Goldilocks coals' are needed that are 'just right' for storage!

Increasingly, the gas industry in many parts of the world is looking to coal bed methane (CBM) as a major source of natural gas. Production of CBM is not as straight-forward as conventional natural gas, as there is little permeability in the original coal and it is usually necessary to create permeability by fracturing the coal and sometimes the adjacent rocks (Figure 8.16). Gas flow is then stimulated by pumping out the groundwater, which usually means lowering the water table. Vast quantities of methane are being produced in the United States and Australia using this approach and it is likely that there will be even greater production in the future. But there is opposition from groups concerned at the

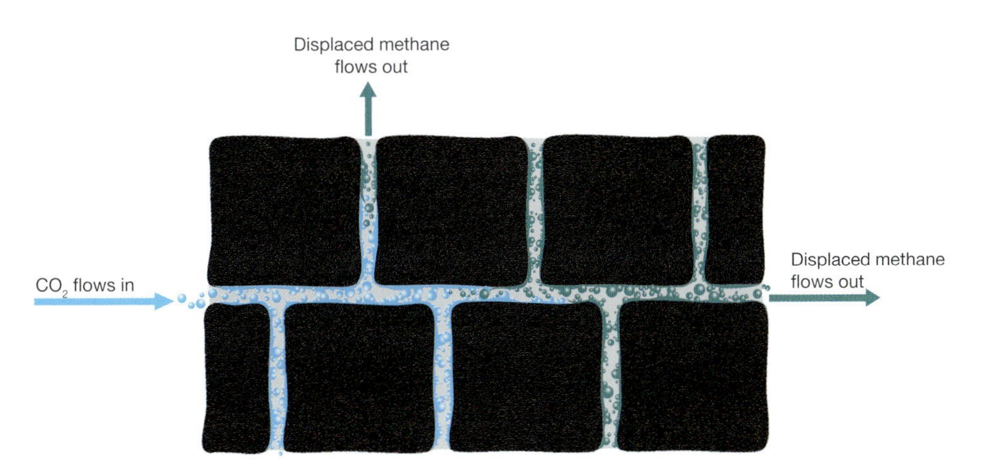

Figure 8.16 Injection of CO_2 into permeable coals displaces methane and the CO_2 is adsorbed on the surface of the coal in the cleats (the fine cracks within the coal).

environmental impact of CBM production, particularly the lowering of the water table.

What then is the connection between CBM and storage of CO_2? Enhanced coal bed methane production (ECBM) can result from injection of nitrogen into the coal. However, CO_2 can also be used for ECBM, with the added benefit of storing the CO_2. Research suggests that for every molecule of methane displaced from a coal seam, two molecules of CO_2 can be adsorbed by the coal.

A number of projects have endeavoured to put CO_2–ECBM into practice, with the most significant test to date being that undertaken by Burlington Resources as part of the Allison Project in New Mexico. There, CO_2 was injected into coal seams over several years, resulting in storage of 277 000 tonnes of CO_2 and enhancement of the methane produced by the equivalent of one additional molecule of methane for every three molecules of CO_2.The test was suspended because of loss of permeability (making it difficult to inject the CO_2) and because of the overall cost of the project and the rate of return. The falling price of natural gas had a major impact on the economics of the project, but, purchasing CO_2 was also an additional cost on the Allison Project. At the present time, CO_2–ECBM is not economic, in that the financial benefit derived from enhancing CBM production is less than the added cost of obtaining and injecting CO_2, but this may change in the future, depending on the carbon price and the gas price.

There are technical challenges in terms of injectivity – many coals expand when CO_2 is injected into them, thereby destroying whatever permeability was originally there. Indeed bearing all this in mind, it could perhaps be argued that coal may be more satisfactory as a seal than as storage reservoir, and in some geological settings this may well be so.

It is also difficult to make a judgement on what is and what is not an unmineable coal seam. While conventional CBM production potentially allows the coal to be mined in the future, this is not the case with CO_2–ECBM and, because of this, there is reluctance on the part of some regulators to consider CO_2-ECBM at this time. Overall, it is unlikely that storage in coal will constitute a major opportunity globally to decrease emissions, but there may be special circumstances where it may make economic and environmental sense.

It has also been suggested that in situ coal gasification (ISCG) can be linked to CCS. ISCG is a technically feasible technique by which coals are 'combusted' 1–200 metres (or deeper) below the surface, to produce gas (methane) and potentially other by-products. The gasification process is largely controlled through the engineered flow of groundwater into or out of the area while gasification is underway. It may be possible to link CCS with this process, but only if suitable deep reservoirs and seals exist in the area to allow for 'conventional' CCS. It is unlikely that the 'chamber' created by the ISCG process can be used for long term storage of CO_2, given that the chamber is likely to be shallow and the rocks adjacent to it fractured.

Figure 8.17 The prospect of being able to store CO_2 in basalts is receiving some attention. In parts of the north-west United States, large areas are overlain by hundreds of metres of the Columbia River Basalts. (Adapted from U.S. Department of Energy and National Energy Technology Laboratory 2010)

Storage in basalts

Volcanic rocks – basalts – are also possible storage targets. They occur in volcanic regions, where they are erupted to form lava flow on the flanks of volcanoes and in adjacent valleys. Others form as flood basalts, quietly erupted from multiple fissures at the earth's surface and covering vast areas such as in the north-west United States (the Columbia River Basalts; Figure 8.17) and in India (the Deccan Traps; Figure 8.18).

When molten lava cools, lots of tiny bubbles (vesicles) form within the lava flow, and the CO_2 can potentially be stored in these holes – provided there is sufficient permeability (from fractures and cracks) to allow injection of the CO_2 into them. We know that some flood basalts, such as those in the north-west United States, are important regional aquifers, suggesting that they can be quite permeable, although in some instances the aquifers may in fact be interbedded sandstones rather than the actual basalts. Within the lava flows there can be very low permeability layers composed of glassy lava that can act as a caprock or seal. The storage potential of basalts is enhanced by the fact that they are likely to be reactive to CO_2, forming stable carbonate minerals in a relatively short period of time.

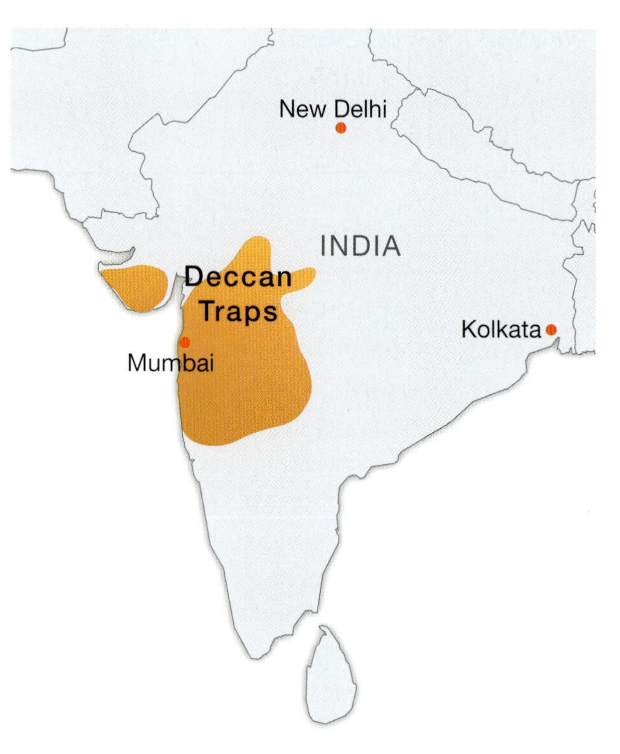

Figure 8.18 At this stage, India appears to have few opportunities onshore for storage of CO_2 in sedimentary basins. Because of this, attention has turned to the Deccan Traps, a thick and very extensive sequence of basalts in central west India, which may have some storage prospects. (Adapted from Woods Hole Oceanographic Institute)

There has only been limited work undertaken on storage in basalts, although some work is now underway at Wallula, Washington State and also at the Hellisheidi geothermal power station in Iceland, with a view to determining the storage potential of basalts. In most parts of the world, it is likely that sedimentary basins will offer far more storage potential than basalts, nonetheless they do warrant consideration, particularly in areas with no suitable sedimentary basins.

In addition to the more obvious option for potentially applying CCS to very extensive well known basalt occurrences such as those in the Pacific north-west of the United States and in India, basalt formations are also quite widespread in other parts of the world. For example, researchers from Columbia University have suggested that the proximity of basalts in the north-eastern United States (both onshore and offshore, including those off the coasts of New York, New Jersey and Massachusetts) to major stationary sources of CO_2, provides a potential opportunity to apply CCS. However, a great deal of work remains to be done on the nature, thickness and distribution of these basalts before the concept can be progressed.

Storage in serpentinites

A further option for storage may exist in serpentinites. These are rocks composed of magnesium silicates and other related minerals, that are quite reactive to CO_2 and in which the CO_2 can be stored as carbonate minerals as a result of a chemical reaction between the CO_2 and the rock. However, the geology of serpentinites is usually far more complex than that of basalts, in that they are frequently steeply dipping and commonly folded and faulted. It is likely that both their porosity and permeability, about which little is known, are low, which means that injecting CO_2 directly into serpentines may be difficult. Matter and Kelemen have suggested that serpentinites could be hydraulically fractured and then heated to get the chemical reaction with CO_2 underway, before the reaction itself starts to generate heat and further speeds, finally permanently fixing the CO_2 in the rocks as carbonate minerals.

Serpentinites are very common deep in the earth's mantle, at a depth far greater than could ever be contemplated for CO_2 storage. However, there are parts of the earth where they occur at quite shallow depths, outcropping in well defined geological zones. It is these areas, including some parts of eastern Australia and the western United States, which have received attention.

There have been no significant field trials of the concept this far and, for the present, in situ reaction of CO_2 with serpentinites is probably a distant option. An alternative approach may be to mine and crush the serpentinites and then react them with the CO_2 at elevated temperatures to form carbonate minerals, and as a potential by product, extract serpentinite-hosted metals of commercial significance (Figure 8.19). This technology has been studied by a number of people, but the costs appear to be very high. In addition, the energy intensity of mining, transporting and crushing the serpentinite is so high that the overall carbon reducing benefits of this form of mineral sequestration are questionable. Nonetheless storage in serpentinites and related rocks cannot be totally dismissed as a possibility for the future.

Assessing storage capacity

We know that geological storage of CO_2 is feasible at scale through studying natural analogues, through the experience of the oil and gas industry (including EOR), through large scale projects such as Sleipner, In Salah and Weyburn (see Sleipner project, this chapter; In Salah project, Chapter 6, Weyburn project, Chapter 6), and through research projects such as the CO2CRC Otway Project, the Ketzin Project in Germany,

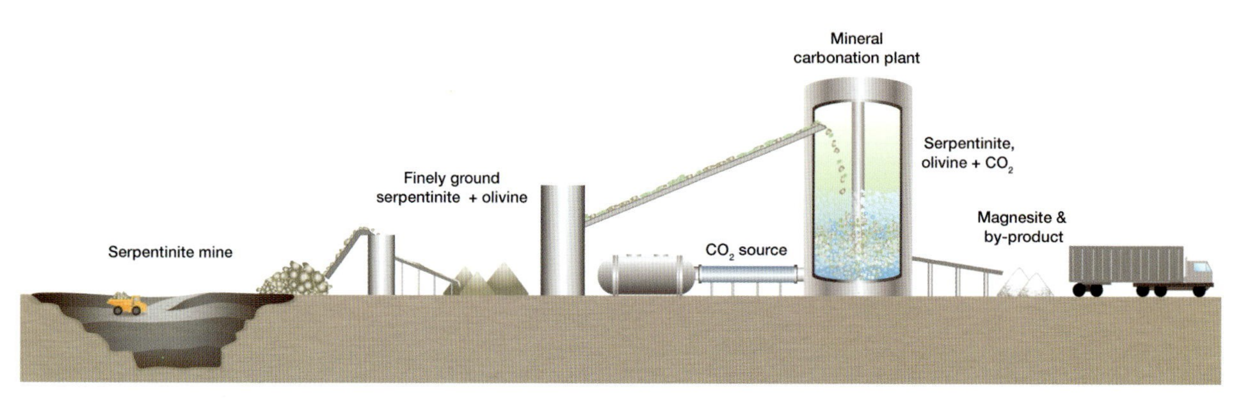

Figure 8.19 Serpentinites and related rocks occur in various places at or near the earth's surface. Because they will react with CO_2 to form stable carbonates, they offer an opportunity for mineral sequestration. Under normal circumstances, the reaction is likely to be slow and various schemes have been devised for mining and crushing the serpentinite to increase the rate of reaction and possibly obtain some by-products. For the present, none of these schemes appear to be economically feasible because of the high cost of mining, crushing and transportation, but research continues.

and a number of Regional Partnership Projects in the USA. But if we were to add up the total amount of CO_2 stored through all of these projects, it would constitute only a tiny fraction of the amount of anthropogenic CO_2 emitted each year. So what then are the opportunities for large scale geological storage of CO_2 and what is the magnitude of that opportunity?

In order to answer these questions, it is necessary to have knowledge of the geology of the area (region, country, continent) that is to be assessed for storage potential. Given that scientists have been undertaking geological mapping for the past 200 years, it may come as a surprise to learn that in many areas, knowledge of the geology is nowhere near good enough to make an accurate assessment of storage capacity (see Box 8.1). The reason for this is that while we have a fairly good idea of the geology of the top few hundred metres of the earth, our knowledge of the geology from around 800 metres to 3000 metres (the interval of particular interest for storing CO_2) is quite incomplete, other than in areas where there has been extensive oil and gas exploration.

Our knowledge of the deep geology is most comprehensive in areas where there are deep mines (usually areas that are not prospective for CO_2 storage) or areas where there has been a lot of oil and gas exploration, which also happen to be areas where the characteristics of the rocks are likely to be most favourable for storage. The identification of areas suitable for CO_2 storage is done in the first instance by compiling all the available geological information, so that the rocks of the basin can be 'characterised'. This includes obtaining information on, among other characteristics:

- the size and location of the basin
- the thickness and type of sediments in the basin
- whether the basin is geologically simple or complex
- whether it is a basin in which there are frequent earthquakes.

This then allows the basin to be graded in terms of the likely suitability (or unsuitability) for storage. A number of countries and states have already undertaken this preliminary assessment. Subsequently, having identified promising rocks, we need to know the volume of rock and the pore space in which the CO_2 might be stored. We also need to know the depth of the geologically favourable rocks and have some idea of their temperature, so that we can estimate the density of stored CO_2. Together, this information

BOX 8.1: DETERMINATION OF STORAGE CAPACITY

The determination of CO_2 storage capacity provides an estimate of the amount of CO_2 that can be stored in subsurface geologic formations. Because of inherent geological uncertainties in storing CO_2 kilometres beneath the surface it is not possible to know the exact details of the rocks and therefore it is not possible to calculate the precise amount of CO_2 that can be stored. Storage capacity estimates rely on the integrity, skill and judgment of the geologist and the level of confidence in the capacity depends on the complexity of the subsurface and the availability of data.

Saline formation capacity

Capacity is calculated using a volumetric equation developed by the United States Regional Carbon Sequestration Partnerships, the Carbon Sequestration Leadership Forum, and other authors.

$$G_{CO2} = A * h * \phi * \rho * E * g$$

Where

G_{CO2} (Mt) = Storage capacity
A (km^2) = Area of the region being assessed
h (km) = Gross thickness of the saline formation
ϕ (%) = Average porosity of the saline formation over thickness h
ρ (Mt/km^2) = Density of CO_2 evaluated at reservoir pressure and temperature
E (%) = CO_2 storage efficiency factor
g (%) = Shape factor for the geometry of the oil and gas trap (not generally used for DSAs)

The storage efficiency factor (E) adjusts total gross thickness to net gross thickness, total area to net area and total porosity to effective (interconnected) porosity actually containing CO_2. Without E, the equation above presents the total pore volume or maximum upper limit to capacity. Inclusion of E provides a means of estimating storage volume for a basin or region with the level of knowledge (uncertainty) in specific parameters determining the type of CO_2 storage capacity estimated.

In many parts of the world DSAs are poorly known and it is often necessary to estimate some of the parameters such as thickness and porosity. In addition, the poorly known parameters are often given as a range (for example 1–10% porosity) which obviously has a major impact on the estimate of the storage capacity.

CO_2 storage capacity estimations in depleted (or near depleted) oil and gas fields are generally easier than estimates for DSAs because there is typically a greater amount of data associated with oil and gas fields and hence they are better characterised. However, another parameter must be added to the equation above for oil and gas fields to account for the geology of the oil and gas traps in that the storage efficiency factor for depleted oil and gas reservoirs also reflects the fraction of the total pore volume from which oil and/or gas has been produced and that can be filled by CO_2. It also needs to account for irreducible water saturation, as well as an estimate of the remaining irreducible oil and gas saturation and recovery factor. More complex mechanisms, such as dissolution of CO_2 into the oil and/or water and saturation levels related to reservoir drive processes, may also be included in the storage efficiency factor.

CO_2 storage in coal seams differs from storage in oil and gas reservoirs or DSAs, in that the trapping mechanism is the adsorption of CO_2 onto the coal and not storage of CO_2 in rock pore spaces. The assessment of coal seam storage capacity requires additional knowledge of a coal's adsorption capacity at a given depth and temperature, and will vary depending on the quality (rank, grade and type) of the coal. Competition for access and utilisation of coal resources for mining, coal seam gas extraction or *in situ* gasification must also be considered to ensure that coals are not rendered unusable in the future by their use for CO_2 storage.

provides an indication of what the theoretical storage potential might be.

However, as much of the pore space might not actually be available or accessible for storing CO_2, most studies now assume that only a very small percentage of the total pore space will actually be available for storage. The rock may be comprised of say 40% pore space but perhaps as little as 4% or less of the pore space will actually be available. The problem is that this storage 'efficiency factor' is not accurately known and only now is work underway to establish it with confidence. For the moment, the conservative approach of assuming a low efficiency factor is appropriate.

There is a distinction between the maximum amount of CO_2 that can be theoretically stored (technical storage potential represented by the total pore volume) and the amount of storage space that is actually likely to be available and accessible for CO_2 storage (Figure 8.20). The theoretical storage potential is usually a large number and the 'actual' potential significantly smaller. This same trend is shown by all mineral and energy resources, where initially a resource (whether minerals or hydrocarbons) will be very large, but as the deposit is studied in greater detail, it becomes evident that a much smaller proportion of the mineral or oil can be extracted, because the remainder is too deep or too difficult or too expensive to extract, or because it is not accessible for cultural or environmental reasons. So it is with storage potential, in that as we get to know progressively more about a geological storage formation and decrease the level of uncertainty, we become more confident about the storage capacity, but the estimate of the storage capacity actually decreases as a consequence of this assessment process. This is explained as a resource pyramid (Figure 8.20): the total pore volume (aka the total available pore space) is represented by the large volume of the pyramid, starting out at the base of the pyramid (where we may have limited information); the volume decreases as we approach the apex of the pyramid where we have much more information and therefore a higher degree of confidence in a much smaller volume.

It is also important to point out that the storage potential can also increase, or decrease in the future as the cost of drilling or of other operations vary, as we learn how to use the rocks more effectively and also, potentially, as the price of carbon (i.e. the financial benefit derived from storing CO_2) goes up or down – the higher the price of carbon, the greater the storage potential becomes.

National assessments of storage potential

Some studies have been undertaken to assess the storage potential of specific regions or countries. An example is provided by the recently completed study of Australia's storage potential by the National Storage Mapping Task Force, which considered oil and gas fields and DSA (aquifer) storage (Figure 8.21). In the case of depleted oil and gas fields, the Australian storage capacity was estimated by the Taskforce at 16.5 Gt CO_2, most of it located offshore (where most of Australia's large gas fields are located). In the case of DSA (aquifer) storage, there is far less certainty about the numbers because of the lack of information. This uncertainty is recognised and it is proposed to address it through a program of deep drilling and seismic surveys. However, at a high level of confidence, the Australian storage potential ranges from 33 Gt to 226 Gt CO_2. In addition, CO2CRC had earlier estimated the storage capacity of unmineable coal seams as 9 Gt CO_2.

If all of this storage capacity is compared with rates of stationary emission; for the eastern half of the Australian continent (where 90% of the Australian population lives), the storage capacity

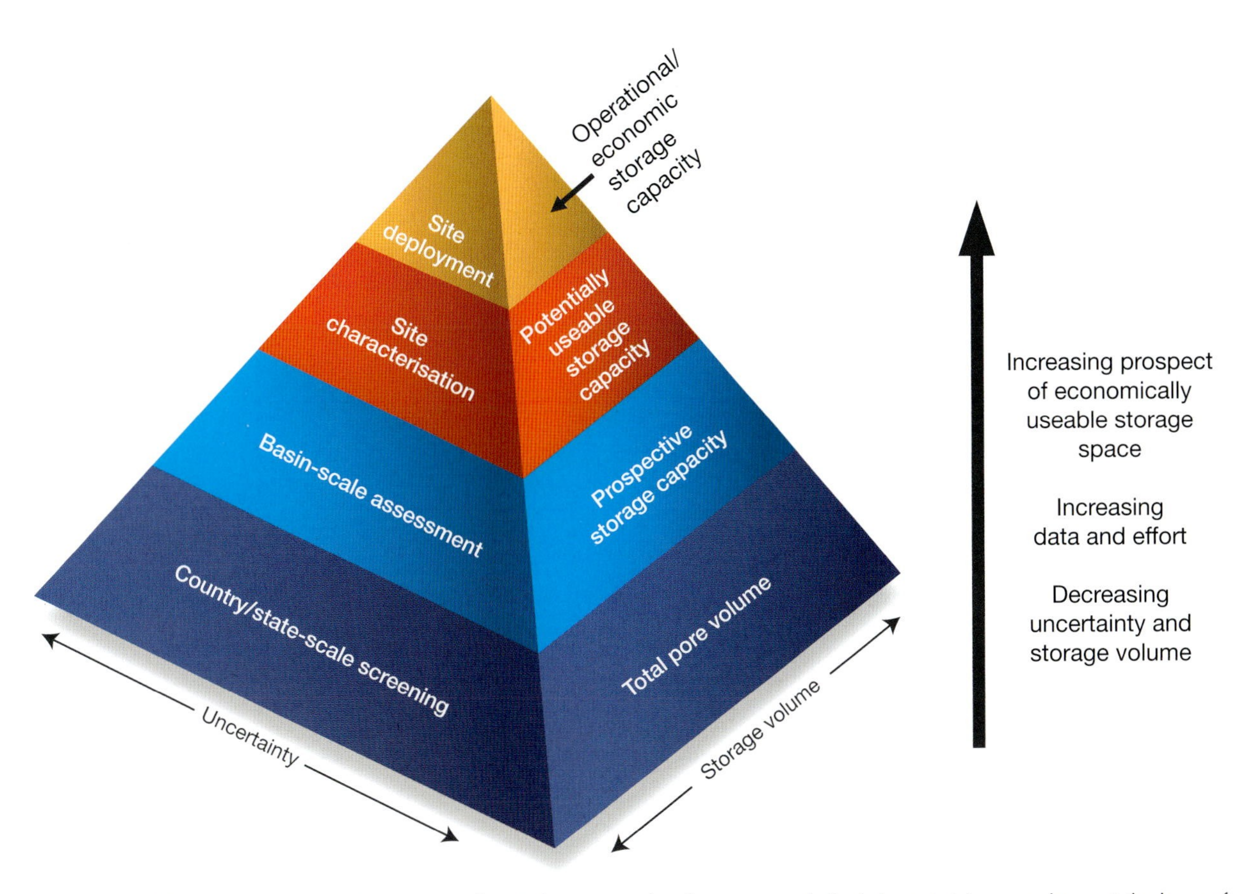

Figure 8.20 The resource pyramid is a way to depict the progression from a speculatively large total pore volume at the base of the pyramid which may or may not be suitable for storage of CO_2, through to an economically exploitable (bankable) but smaller pore volume that can be used for storage. (Adapted from Carbon Sequestration Leadership Forum 2005)

there is regarded by the taskforce as sufficient to store 70–450 years of emissions at a storage rate of 200 million tonnes of CO_2 per annum (approximately the current rate of emissions). For the western half of the continent, the capacity there is sufficient for 260–1120 years of emissions at 100 million tonnes of CO_2 per annum. In other words, Australia has more than enough storage potential to meet its storage needs for the next 100 years and beyond.

The storage potential of North America has been assessed in a series of continent-wide studies, backed up by regional studies. The excellent Carbon Sequestration Atlas of the United States and Canada, published by the National Energy Technology Laboratory (NETL), summarises the spatial distribution of oil and gas reservoirs, unmineable coal seams and deep saline aquifers, as well as more speculative storage opportunities such as basalt formations and organic rich shales. More detailed assessments are then provided for regions and basins. Oil and gas fields (depleted and currently producing) are estimated to have a storage capacity of 82.4 Gt CO_2. Unmineable coal seam capacity is estimated at 156–173 Gt CO_2 and aquifer (DSA) capacity is estimated at 919–3378 Gt CO_2. Based on a compilation of 4365 stationary sources, the Atlas estimates current total stationary emissions as 3.8 Gt CO_2 per annum for the United States and Canada (Figure 8.22). Comparing this with the capacities, suggests a storage potential equivalent to approximately 20 years of emissions using depleted oil and gas fields, approximately 40

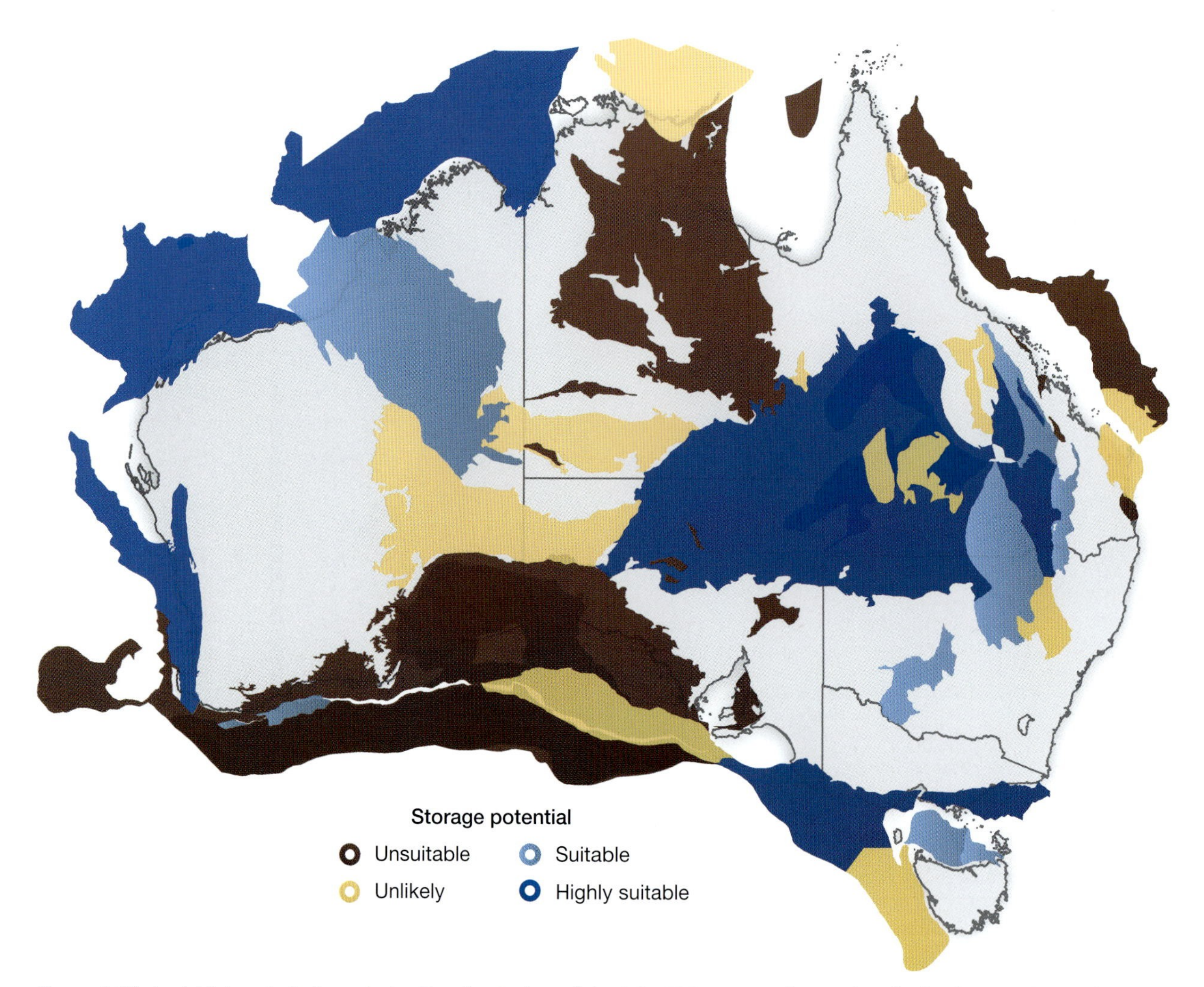

Figure 8.21 An initial analysis through the Geodisc Project of the CO2CRC suggested many hundreds of years of Australian storage potential capacity. More recent analysis by the National Storage Mapping Taskforce has supported this by mapping a number of areas underlain by DSAs with high storage potential. These are likely to be sufficient for storage of hundreds of years of CO_2 at current rates of emissions. (Adapted from Carbon Storage Mapping Taskforce 2009)

years using unmineable coal seams and of the order of 240–890 years for DSA (aquifer) storage. These preliminary figures suggest that for the USA and Canada, the storage potential is likely to be sufficient for the next 100 years and well beyond, at current rates of emissions.

The Geocapacity Project of the European Commission, has completed a pan-European assessment of CO_2 storage potential (Figure 8.23) that has estimated the following storage capacities: oil and gas fields 20–32 Gt CO_2, coals: 1–2 Gt CO_2, and aquifers (DSAs) 96–326 Gt CO_2.

This gives a total estimated capacity of 117 Gt CO_2. Western Europe's total emissions are 3.2 Gt per annum, of which 1.9 Gt per annum is from major stationary sources. Therefore Western Europe's per total storage capacity appears to be sufficient to provide, very conservatively, 60 years of storage based on current annual stationary emission of 1.9 Gt CO_2 and at the upper end of the estimate enough storage for 180 years. Therefore it is likely that the Western European storage capacity is equal to 100 years or more at current rates of emission.

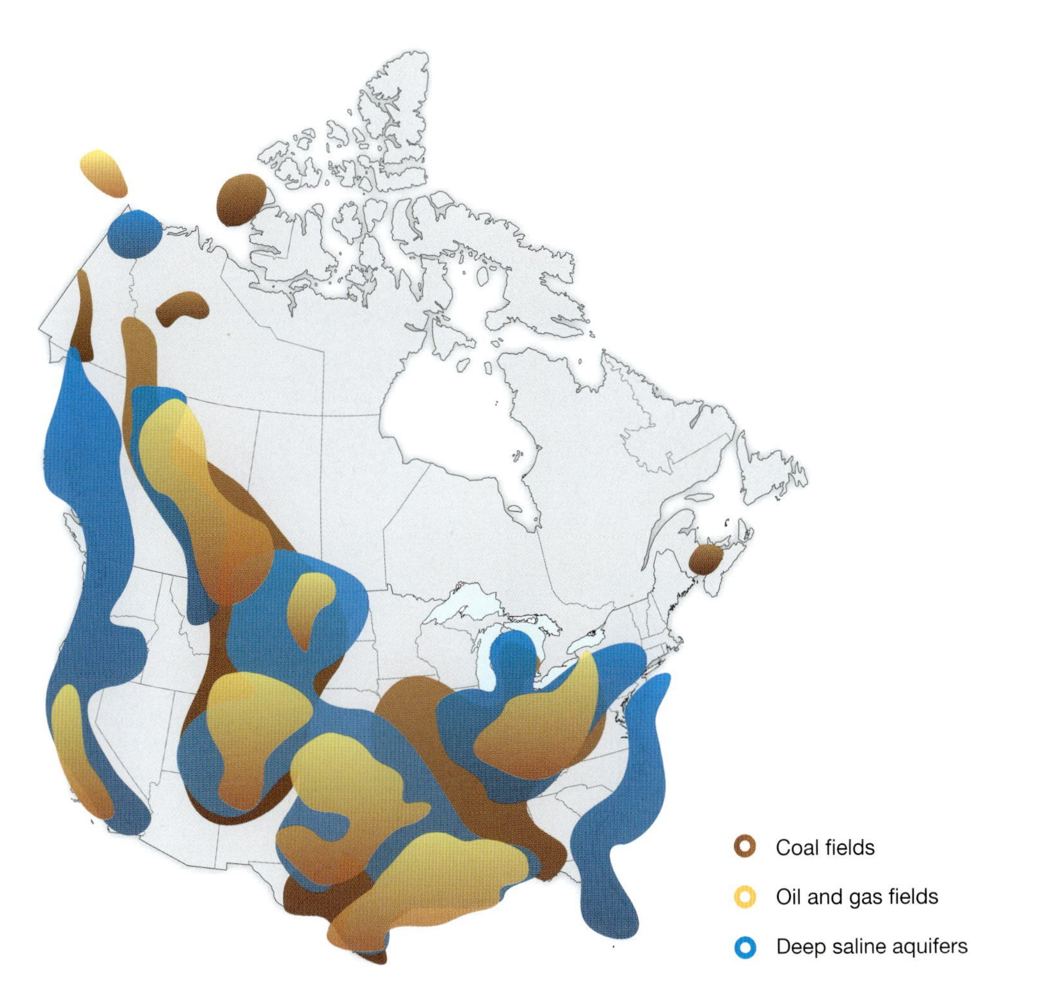

Figure 8.22 Through the Regional Partnerships Program, a series of excellent regional maps have been developed for the United States and Canada showing the distribution of unmineable coal seams, oil and gas reservoirs and deep saline aquifers and their CO_2 storage potential. Some areas have two or more of these storage options available. Much of the capacity is found in mid-west and central western states, extending into western Canada. (Adapted from U.S. Department of Energy and National Energy Technology Laboratory 2010)

The Geological Survey of China has identified 50 basins as being prospective for CO_2 storage with a total capacity of the order of 1455 Gt CO_2, almost all of this (1435 Gt) in DSA systems. Oil and gas fields have an estimated storage capacity of 7.8 Gt and 12 Gt is the estimated capacity for unmineable coals. According to the International Energy Agency, China's major stationary sources emit approximately 3.7 Gt CO_2 per annum, suggesting that China has storage capacity for 390 years. Even if this is over optimistic, it certainly suggests that China's storage capacity is likely to be sufficient for the next 100 years at current rates of emission.

Therefore, in four major regions of the world, there appears to be sufficient CO_2 storage capacity to meet all likely needs for large scale mitigation of stationary emissions for the next century, which is consistent with the conclusions of the 2005 Special Report of the IPCC.

Beyond these regions, assessments of storage capacity have not been undertaken to any great extent. We can surmise that countries of eastern South America, some of which have significant oil production, are likely to have large CO_2 storage capacities, but the western part of the continent is likely to have only limited capacity.

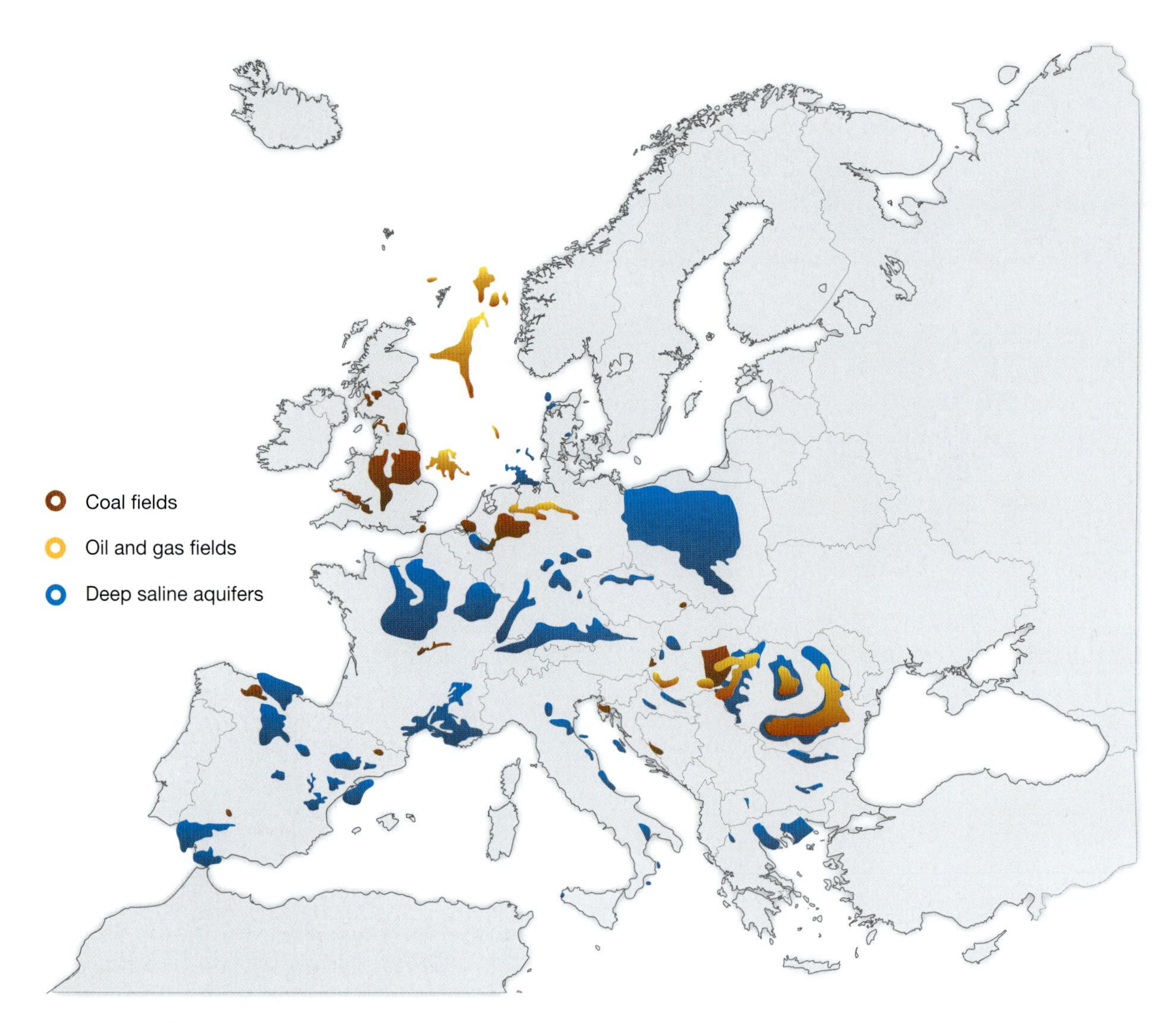

Figure 8.23 A European compilation of onshore storage capacity has been developed by the EC-supported Geocapacity Project to show the distribution of oil and gas fields, coals and DSAs. Much of the capacity occurs in north-western Europe, particularly underlying the central North Sea. (Adapted from EU Geocapacity Project 2009)

Much of central and southern Africa at this stage appears to have only limited onshore storage capacity. However, regions such as Nigeria, North Africa, large parts of Russia, the Middle East and South-East Asia, all regions with major oil and gas production, are likely to have a high potential for CO_2 storage. Conversely, areas of the Western Pacific such as Japan and Korea and the Philippines as well as Southern Asia (India, Sri Lanka) may have only limited capacity, much of which will be mainly offshore or, more speculatively, in basalts. So a global picture emerges where fossil energy-rich countries are likely to have a significant storage potential, whereas fossil energy-poor countries are likely to have at best, modest storage potential.

Conclusions

Bearing all this in mind, what can be said about the global geological storage potential? This is a

Table 8.1: Estimated storage potential (billions of tonnes CO_2)

	Oil/gas fields	Unmineable coals	Aquifer (DSA)	Estimated annual emissions from large stationary sources	Indicative total storage potential in years based on current emission rates
USA/Canada (onshore)	82	156–173	919–3378	3.8	300–1000
Australia (onshore & offshore)	17	9	33–226	0.3	200–800
W Europe (mainly onshore)	20	1	96	1.9	60
China (onshore)	8	12	1435	3.7	400
Global (IPCC, 2005)	675–900	3–200	1000–10 000	13	100–1000

difficult and at times controversial question, with estimates varying enormously, depending on some of the underlying assumptions. The 2005 IPCC Special Report on CCS provides global estimates of the 'technical potential' for oil and gas fields (675–900 Gt CO_2), unmineable coal seams (3–200 Gt CO_2), DSAs (1000–10 000 Gt CO_2).

Work done since the 2005 IPCC Report has tended to support the upper range of the IPCC global estimates for depleted oil and gas fields and DSAs, but storage in coals remain problematical and there is currently no global value for storage in basalts or serpentinites. There is obviously a significant level of uncertainty about the numbers. However, even accepting this, it is evident that the global potential of DSAs is far greater than that of any other storage option, followed by depleted oil and gas fields and then a relatively modest global potential for coals. Even though these numbers require far more scrutiny, it is worth bearing in mind that global CO_2 emissions are of the order of 30 Gt CO_2 per annum of which 13 Gt are from stationary sources (i.e. sources potentially amenable to CCS). Thus it appears that conservatively the global geological storage potential is likely to be adequate to mitigate 100 years of global stationary CO_2 emissions at existing emission rates, and perhaps far more (Table 8.1). CCS will not be universally applicable to all areas and all countries because of the variability of the Earth's geology, but many major population centres and a large proportion of the major stationary sources are generally found in sedimentary basins, where there are likely to be storage opportunities. In the longer term, perhaps long distance transport of CO_2 by pipeline and even by ship, may help to provide truly global opportunities for storage of CCS, even for regions which do not have suitable geology.

9 HOW DO WE KNOW CCS WILL BE EFFECTIVE?

Undertaking good science and obtaining credible and reproducible research results are critical first steps in the development and deployment of any new technology. The previous chapters have outlined the opportunities for deployment of CCS in the light of the fact that there are no technology 'show stoppers'. But beyond this, it is also critical to ascertain what uncertainties there are and risks that might arise from the application of the technology, so that people to be assured appropriate steps are taken to manage risk. Equally important is a third step of effectively communicating to the public the science and technology of CCS.

It is evident from the discussion so far, that we know how to separate CO_2 from emissions, and how to transport that CO_2. We also know that CO_2 can be geologically stored and that the global storage potential is very large. But how do we know that CCS is safe? What are the risks involved in CCS? We assess risk in our everyday lives usually without being conscious of doing so. Should we cross the road now or wait until the traffic slows down? Should we ski down this exciting but steep slope, or take the gentler safer but less exciting way down? Is it ok to overtake this car now or should we wait?

The nature of risk assessment

When major projects or activities are planned, it is necessary to formally evaluate a wide range of risks, including possible economic loss, environmental effects, health and safety issues, political risk and risk to reputation, to name but a few. This risk assessment process is important for designing a project and for determining whether a project should go ahead or not. In the initial stages of a project, risk assessment is used to:

- decide if a project is likely to be economically feasible
- make sure the project will be able to meet government regulations
- reassure the banks so that finance can be obtained on reasonable terms
- judge whether the public and local community are likely to be in favour of or against the project
- and decide what action may need to be taken to ensure that the project is acceptable.

Members of the public will have widely differing perceptions of project risk, based on their past experience, their level of knowledge and understanding, proximity to the project and the extent to which they or their family will personally benefit from or be disadvantaged in some way by the project.

Risk assessment (Figure 9.1) is a widely used formal process which considers two fundamental issues: first, what is the chance of a particular event occurring and second, what will be the likely consequences or the impact of that event. When there is a long history of carrying out a process or

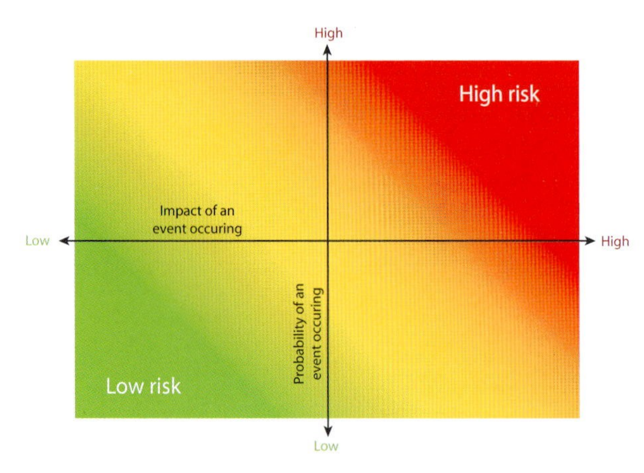

Figure 9.1 Risk assessment is based on the likelihood of an event occurring and the impact of that event, were it to occur. Events which fall in the green zone are unlikely to hold up a proposed CCS project while those within the red zone would be likely to hold up a project. For example, issues such as insufficient funding, strong community opposition and uncertainty regarding liability could all be located in the red (high risk) zone. Conversely issues such as availability of a drilling rig, adequately trained staff or approval of the regulator may be in the yellow or green (low risk) zone.

activity such as operating a piece of equipment, or transporting a liquid, then there is sufficient experience to draw on to say that the chances of a particular event occurring, are high or low. There is also likely to be sufficient experience to be able to say that the consequences of the event are likely to be major or minor.

Whilst much of the attention regarding risk has focused on storage, the evaluation of risk has to be applied to the entire CCS process. Perhaps because CO_2 capture is a process similar to many other chemical processes with which industry and regulators are familiar, it is seldom considered to involve significant risk beyond 'normal' industrial risks which are already covered by long standing health safety and environmental regulation and procedures. There are of course some unique issues such as health or environmental risks which might arise from the use of large quantities of amine solvents, but for the most part there are well established precedents for handling risk associated with capture.

In the case of pipelines, there are some differences between natural gas pipelines and CO_2 pipelines (CO_2 pipelines are, for example, inherently safer because CO_2 is neither explosive nor flammable). In general the public is aware of the existence of thousands of kilometres of gas pipelines, for which there are well established pipeline regulations as well as health and safety procedures. It will be necessary to adapt some of the existing regulations to the needs of a large CO_2 pipeline. For example CO_2 is more corrosive in a pipeline than methane and it has particular physio-chemical properties. One of the hazards of CO_2 previously discussed is that it is heavier than air and consequently there is a risk of exposure to high levels of CO_2 and potential asphyxiation.

Dispersion models (Figure 9.2) predict how the CO_2 will disperse . in the event of a pipeline rupture. Such models can be used to predict the area that will be dangerous because of a high concentration of CO_2 and for how long the danger might last. Extensive dispersion modelling has determined that should a pipeline rupture occur, the level of CO_2 sharply decreases away from the pipeline and that the risk in populated areas can be mitigated significantly by having valves at frequent intervals along the pipeline, in order to limit the amount of CO_2 that can leak out. In addition 'crack arrestors' ensure that if a fracture were to develop its impact would be very limited. Through proper design and by following safety regulations, CO_2 transport can be done safely and the safety record for transporting up to 50 million tonnes of CO_2 per annum in the United States is proof of this.

As pointed out previously, the formation of very cold dry ice when CO_2 is released from a pressurised condition, can cause frostbite or cold burns before the dry ice turns into a gas and disperses. While it is essential to undertake a risk assessment of all aspects of the transport of CO_2, it is worth reiterating that while some

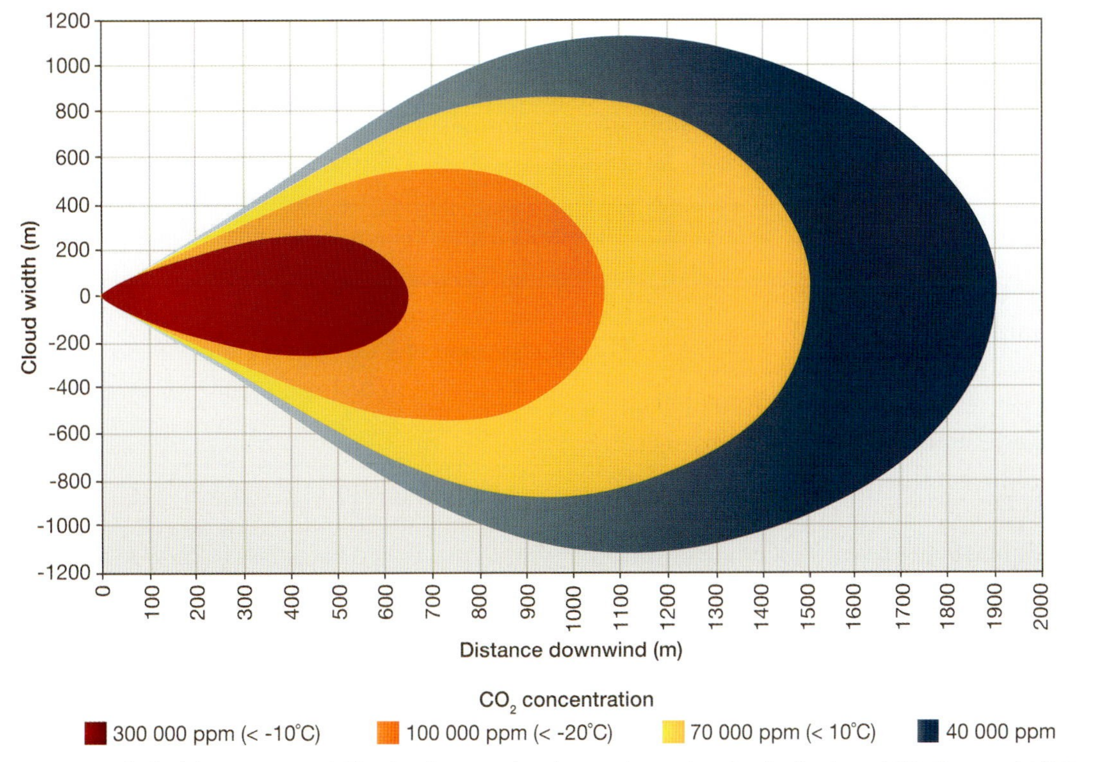

Figure 9.2 A great deal of dispersion modelling has been undertaken to determine the distribution of CO_2 that would if there were to be a pipeline leak. In this example a large leak is modelled with the wind direction having a significant impact on CO_2 distribution.

modifications will be required on a project-by-project basis, the mechanisms for ascertaining and managing risk associated with CO_2 transport are well known.

Geological risk

For most people, geological storage is the least understood component of CCS and therefore it is likely to be perceived by the public as carrying the greatest risk. It is critical to address the issue of just how much risk is involved in geological storage and the rest of this chapter focuses on this. The concept of geological storage of CO_2 was only developed about 20 years ago and therefore by scientific standards it is still quite a young concept. But it is based on an enormous amount of knowledge of the subsurface, acquired over the past 200 years through the work of geologists, particularly those working in the oil and gas industry over the past 50 years.

The concept of geological storage is also based on examples of comparable processes (such as natural gas storage) and natural analogues (natural accumulations of CO_2 in the Earth's crust). In addition it draws on experience of operational storage projects, enhanced oil recovery using CO_2 and a number of pilot and demonstration storage projects. In other words there is already a wide body of knowledge on which we can draw, in order to assess the risks arising from geological storage of CO_2.

Existing natural gas storage facilities

Natural gas (methane) has been stored in underground reservoirs since 1915 and worldwide there are now more than 600 underground gas storage facilities, at depths ranging from a few hundred metres to 2000 metres or more. Their purpose is to provide 'peak loading' so that the demand for natural gas can always be met including at the times of high demand. At times when the demand is low, the storage reservoir is then refilled with natural gas.

Many of these gas storage facilities are close to and even under major cities. For example, there is a major natural gas storage facility immediately under the site of the 1936 Olympics in Berlin, which provides extra energy security for Berlin should there be a disruption in the supply of gas from its regular providers in Western Europe or Russia. The many hundreds of natural gas storage sites in the United States are located in a diverse range of environments and have a total storage capacity of 4000 billion cubic feet (bcf) of gas equivalent to the volume occupied by 200–300 million tonnes of stored CO_2. The lona natural gas storage facility in Victoria, Australia has its storage reservoir at a depth of about 1,600m (approximately the same depth at which CO_2 is being stored in the nearby Otway Project) and is a depleted gas field. There are many gas storage sites in the Paris Basin region of France. Given the number of natural gas storage facilities around the world, their safety record is very good, particularly given that natural gas is both flammable and explosive. There is every reason to believe that CO_2 storage will be at least as safe, and probably even safer than natural gas storage.

Natural accumulations of CO_2

Some geological structures contain large quantities of CO_2. Some natural gas fields have a CO_2 content which can range up to 70% CO_2 or more; the higher the CO_2 content, the less likely the field will be commercially viable. For example the biggest natural gas field in South-East Asia and one of the biggest in the world, is the Natuna gas field in Indonesia which has 46 trillion cubic feet (TCF) of natural gas reserves. But the field also contains an average of 70% CO_2, so approximately 153 TCF (8650 million tonnes) of the gas is CO_2. Consequently it is likely that the field, vast though it is, will never be exploited for natural gas, unless the co-produced CO_2 can be re-injected and geologically stored. But the point is also that Natuna is an example of natural long term storage of CO_2 on a scale similar to that which will need to be put in place for large commercial CCS projects.

There are many other examples of natural accumulations of CO_2 such as those found in the western United States, Italy and southern Australia. In many cases the CO_2 appears to be of volcanic origin. In some instances, there are towns and villages on or near the CO_2 –rich structures, despite the fact that CO_2 in the natural environment can constitute a hazard at times. In some volcanic areas of Italy, naturally occurring CO_2 leaks into the basement of houses and has caused a number of deaths.

Some volcanoes emit large volumes of CO_2, on occasions explosively, causing property damage and loss of life. Examples include Vesuvius in Italy, Popocatapetl in Mexico and Merapi in Indonesia. The CO_2 is thought to be derived from sedimentary limestones 'cooked up' by the high temperature lavas. The result can be massive outbursts of CO_2. For example, in a recent event Popocatapetl emitted more than 38 000 tonnes of CO_2 a day, the equivalent of two 1 GW power stations.

Perhaps the most dramatic example of a large natural CO_2 leak was that which occurred at Lake Nyos in Cameroon, central Africa, in 1986. Lake Nyos is a deep lake in the crater of a volcano with cold stagnant water at the bottom of the lake and layers of warmer water above. Because of the stratified nature of the lake, CO_2 leaking from the deep volcano accumulated in the cold bottom waters of the lake and for

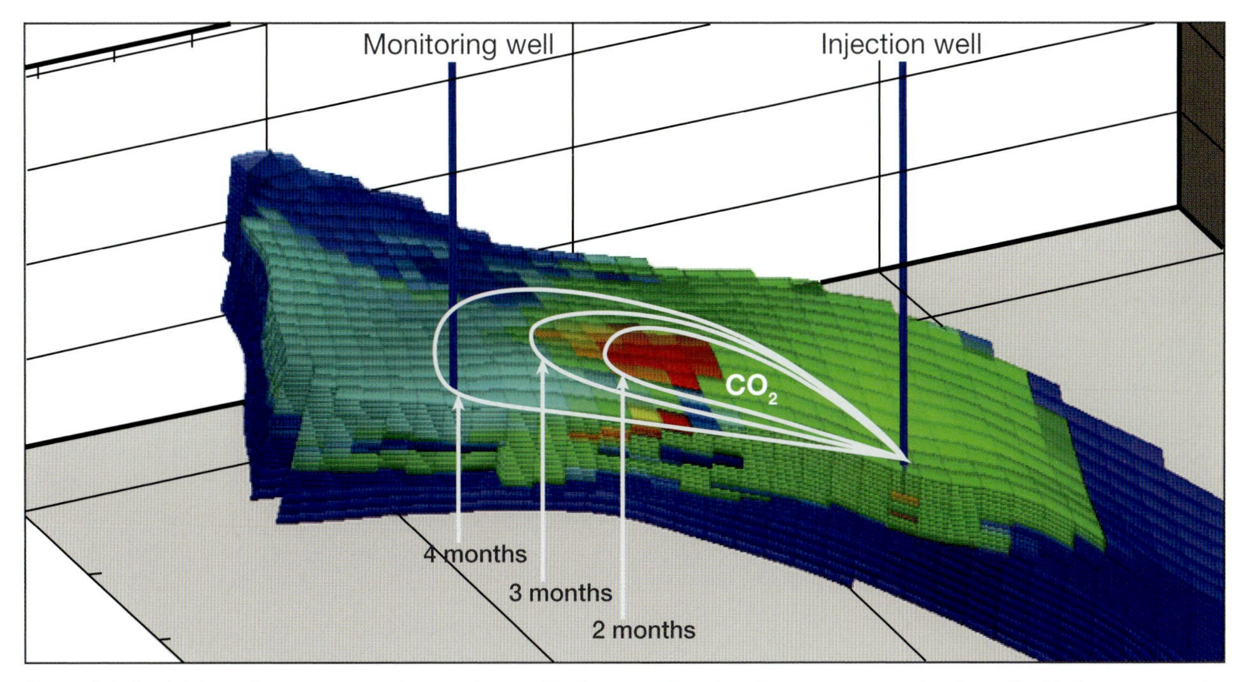

Figure 9.3 Sophisticated computer models can be used to forecast the subsurface movement of carbon dioxide in porous and permeable rock. In this model obtained for the CO2CRC Otway project, the movement of CO_2 from the injection well was modelled and then confirmed by obtaining CO_2-charged water samples from the monitoring well. These proved that it took four months for the CO_2 to laterally migrate approximately 300 metres within the storage formation.

reasons that are not clear (but possibly caused by a large landslide on the flanks of the crater lake), there was a massive disturbance of the water column, allowing the dissolved CO_2 in the deep cold water to escape, which decreased the pressure and allowed even more CO_2 to come out of the solution. The result was an enormous release of millions of tonnes of CO_2 in a very short time. This massive cloud of CO_2 flowed down the flanks of the volcano, into the valley below and over pastures and villages, killing thousands of people as well as livestock. There is no likelihood of a large scale CCS project contemplating CO_2 storage in an active volcano or volcanic lake. Lake Nyos is in fact a very poor analogue for a CCS project.

Knowledge derived from large scale commercial CO_2 storage projects

A number of large scale commercial CO_2 storage projects and enhanced oil recovery (EOR) projects are underway at the present time and these greatly contribute to our level of confidence in the effectiveness and safety of geological storage. Acid gas injection (which includes a number of gases in addition to CO_2) has been successfully underway in western Canada for the past 20 years. In addition a number of smaller scale research and demonstration projects have been undertaken in the past 10 years in Japan, the United States, Germany, France and Australia. These projects have provided great insights into the behaviour of CO_2 in the deep subsurface and also have offered the opportunity to model and measure that behaviour (Figure 9.3)

In the CO2CRC Otway Project, 66 000 tonnes of CO_2-rich gas has been injected at a depth of 2 km, and monitored using a highly innovative sampling system developed by Barry Freifeld at the Lawrence Berkeley National Laboratory. This project has assisted with the understanding of the chemistry of deep groundwater changes over

SOME PILOT PROJECTS

VATTENFALL (SCHWARZE PUMPE) PROJECT + KETZIN

In mid 2008 Vattenfall inaugurated the Schwarze Pumpe pilot project, the world's first oxy-fuel capture project. The 30 MW plant, located in Spremberg, eastern Germany, next to a 1600 MW lignite-fired power plant and costing €50 million, can capture up to 75 000 tonnes per year with a purity of more than 90% CO_2. Initial plans for a related large scale storage project have been put on hold due to access issues, but a smaller scale trial is planned using the Ketzin site. The Ketzin site is a CO_2 storage project operated by the CO2SINK consortium and sponsored by the European Commission. CO2SINK is using three wells (one for injection and two for observation) and since 2008 has injected nearly 50 000 tonnes of food grade CO_2 at a depth of 650 metres. Because the site was previously used as a natural gas storage facility (in a shallower formation), it is already well characterised, reducing the need to collect significant amounts of new geological data. A very extensive program of geophysical and geochemical monitoring has been undertaken by the Project. Although Ketzin and Schwarze Pumpe were originally separate projects, in 2011 a portion of the CO_2 from the capture project was trucked and injected into the storage project, successfully demonstrating integrated CCS.

AEP MOUNTAINEER PROJECT

Alstom and American Electric Power (AEP) jointly installed a retrofit post-combustion capture facility to AEP's 1300 MW coal-fired Mountaineer power plant in New Haven, West Virginia (United States). The capture plant, using Alstom's chilled ammonia solvent technology, was applied to a 20 MW equivalent slip stream, capturing at a rate of approximately 100 000 tonnes of CO_2 per year. Storage is at a depth of about 2500 metres in the Rose Run and Copper Ridge sandstone formations. The pilot project is the World's first fully integrated post combustion capture and storage project, but the trials are currently suspended for financial reasons.

CO2CRC OTWAY PROJECT

The CO2CRC Otway Project in south-western Victoria is Australia's first demonstration of the geological storage of CO_2. The CO_2, which is sourced from a natural gas well producing 79% CO_2 and 21% methane (by weight), is transported approximately 2 km by pipeline to the injection site. The first stage of the project, commencing in 2008, successfully injected 66 000 tonnes of CO_2 (with methane) into the main sandstone reservoir interval within a depleted natural gas field, at a depth of 2 km. This was accompanied by a very extensive program of monitoring including atmospheric, soil, hydrological and sub-surface monitoring; numerous seismic studies and tracer monitoring; and extensive fluid sampling from a dedicated monitoring well. As part of Stage 2, in 2010–2011 a second well was drilled and a series of experiments were undertaken, involving injection and back-production of small amounts of food grade CO_2 and also various tracers, at a depth of 1500 metres. By measuring the difference between how much CO_2 was injected and how much was produced, the experiment determined how much CO_2 can be residually trapped in a heterogeneous sandstone reservoir. Additional experiments are planned for the Otway facility.

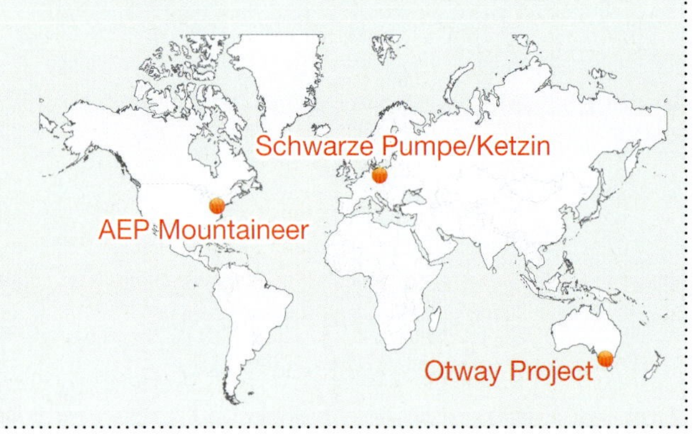

time as CO_2 is injected, and the vertical and lateral migration of the injected CO_2 over time. This sort of information is critical to give regulators the confidence to approve geological storage projects.

Location-specific risk assessment: characterising the site

We therefore have a lot of experience of large scale gas storage, including CO_2 storage, plus knowledge of how CO_2 behaves when it is geologically stored. CCS is not a journey into the unknown! At the same time CO_2 storage represents an anthropogenic perturbation of the natural geological system (the geosphere) and it has to be recognised that the geology of every location is different. So how can we be sure that nothing will go wrong? For instance, how can we be sure that some of the stored CO_2 will not contaminate aquifers used for drinking water, or leak to the surface?

The starting point for being confident that a particular geological location will provide a safe storage site is to fully understand the rocks at the site and in surrounding areas (see Chapter 8 for detailed discussion of this). Some of the questions that we should ask in undertaking risk assessments for suitable storage sites are:

- Are there suitable rocks present? This generally (though not exclusively) means rocks in a sedimentary basin and preferably a geologically stable and structurally simple sedimentary basin.
- Are there suitable reservoir rocks, overlain by suitable caprocks or seals that will prevent CO_2 leaking out of the reservoir and into shallower rocks and ultimately to the surface?
- Is there a suitable geological structure?
- Are the groundwaters fresh or salty? And are they connected to an aquifer that is used for drinking water?
- Are there recorded earthquakes in the region and if so, were they major or minor?
- Are there any inactive faults and what is their orientation and could we in some way produce some movement on the fault through the injection of CO_2.
- Are the rocks likely to be strong enough that CO_2 can be injected without fracturing them?

These and many more questions need to be asked, and answered satisfactorily, in order to characterise the site. Once it is decided that a site is potentially suitable for storage, detailed seismic surveying is undertaken and unless there is existing drill core, the site is likely to need a well to be drilled and samples of core obtained from key depths. This also provides rock samples that can be analysed to ascertain the conditions under which the rocks were formed and to determine their porosity and permeability. All this information can then be used to develop a 'geological model' of the site and adjacent areas. This in turn can be used to develop a 'reservoir model' to predict:

- how CO_2 will move through the rocks
- how it will react with both the minerals in the rock and the groundwaters
- what the shape and extent of the CO_2 'footprint' (plume) in the subsurface will be
- how far the plume might spread in the future.

This sort of information also provides the basis for forecasting the rate at which we can inject CO_2 into the storage formation and how much CO_2 in total can be safely stored? Characterising the site is therefore the first crucial step in gaining confidence that a site will be suitable for storage.

But all this is only the start, for in addition to building up an accurate picture of the geology, it is also necessary to develop a risk register of what might happen. Carbon dioxide can pose a risk to the environment by forming carbonic acid, and if it were to leak it would change the pH of soils and groundwater, which can in turn impact on plants and animals. CO_2 can leach trace metals out of some geological formations also impacting on the environment. Therefore we must ensure containment of the CO_2 in the storage formation, in order to minimise the possibility of CO_2 escaping not just in decades, but in hundreds of years.

Therefore as part of the risk assessment process it is necessary to identify the ways in which leakage may occur (Figure 9.4). This can include leakage through a seal that is more permeable than anticipated, through faults in seals, or through existing wells. Again, good characterisation of a site is vital but it is also essential that the wells are constructed to a high standard of engineering to reduce the risk of leakage from them. It is also very important to know whether any pre-existing oil or water wells exist and if they do, what condition they are in, for leakage from old wells or from incorrectly engineered new wells is one of the more likely sources of leakage. Fortunately it is one of the easier mitigation pathways to identify and remediate. It is also necessary to assess the chances of a structural trap at a storage site having less capacity than originally thought and as a result the CO_2 may migrate into an

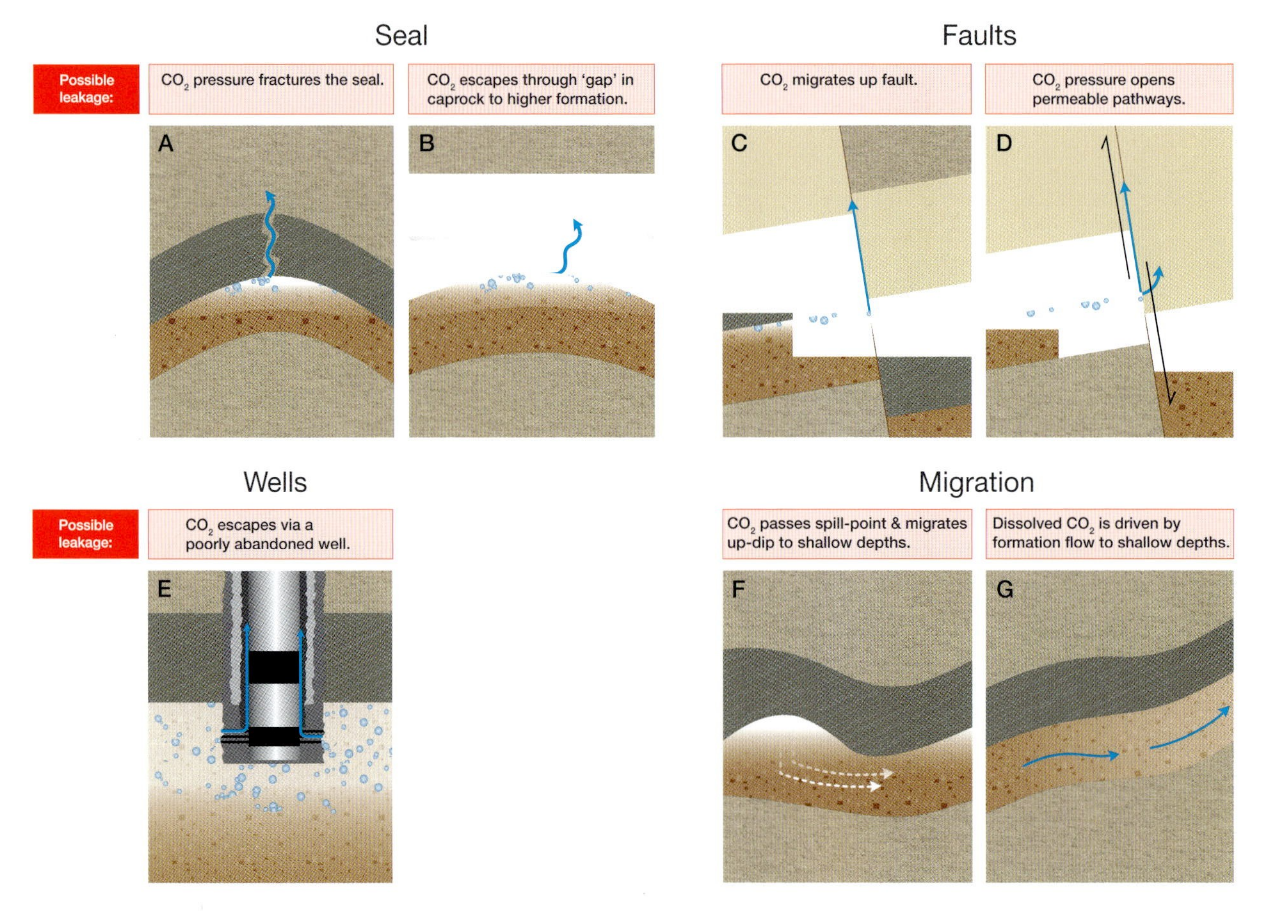

Figure 9.4 A number of hypothetical CO_2 leakage or mitigation pathways are shown, including: (A & B) unexpected failure in the seal; (C&D) unanticipated migration along faults; (E) escape via a poorly engineered abandoned oil or gas well; (F) migration as a result of filling a structure beyond its storage capacity; (G) unanticipated groundwater movement carrying the CO_2 to shallow depths. Various mitigation strategies can be applied, ranging from extraction and reinjection of the CO_2 into a new structure, filling the leaking zone or fault with a sealant such as a gel or foam and, in the case of an abandoned well, filling and replugging the well with CO_2–resistant cement. If it is just a small leak that poses no danger, then ongoing monitoring may be the most appropriate course of action.

adjacent structure, or move in a direction other than predicted and migrate into another geological unit.

The risks of earthquakes

A frequently asked question is what would happen if an earthquake occurred in the vicinity of a storage site? What we do know is that there would not be a cataclysmic escape of CO_2, because contrary to the images shown in disaster movies, there are seldom open faults running deep down into the earth into which people fall, or from which CO_2 might escape! There could conceivably be some migration of CO_2 through new faults or fractures, but it is likely that the volumes of CO_2 would be small. We also have evidence from the Nagaoka storage project in Japan which was hit by a strong earthquake (6.2 on the Richter Scale) in 2004. What impact did the earthquake have on the 10 000 tonnes of CO_2 geologically stored at the site and how much CO_2 escaped? The answer to both of these questions is none!

What is the chance of CO_2 injection triggering an earthquake? It is possible to trigger microseismic activity by injecting fluids into the subsurface. In most cases, the magnitude of the activity is so small that it cannot be felt and only registers on very sensitive instruments. Nonetheless there are some well substantiated cases of induced earthquakes. For example the injection of chemically-contaminated fluids into the Rocky Mountain Arsenal Well to the east of Denver in the 1960s, triggered a series of small earthquakes. More recently, injection of water as part of a geothermal project triggered a series of small earthquakes in the Basel region of northern Switzerland in 2007.

Steps can be taken to minimise the chances of this happening at storage sites, by identifying the location and orientation of existing faults and by ensuring that the pressure used to inject the CO_2 is less than the pressure that would trigger movement on the fault. Many old faults are in fact 'sealing faults' and act as a barrier to fluid movement; in thousands of oil and gas fields it is the sealing faults that are responsible for trapping the oil or gas. It is very important to map all faults as part of the site characterisation process, but their presence does not necessarily mean that the site is unsuitable for storage.

The risk to groundwater

The other concern relates to the possibility of CO_2 leaking into groundwater. Again the answer largely lies in carrying out comprehensive site characterisation, including demonstration that at least one, and preferably more than one impermeable seal, occurs between the deep formation in which the CO_2 is stored and aquifers that are used to provide water for humans or animals or agriculture (Figure 9.5).

But in spite of the most meticulous site characterisation, the best engineering and the most careful operations, because we are dealing with a subsurface system, can something unexpected still happen? The risk assessment process should identify the possibility of the 'unexpected' occurring and the likely consequences if it were to occur.

Despite best practice being followed, if the CO_2 migrates out of the storage formation, what can then be done? First, it is necessary to assess whether or not the CO_2 will be trapped by another seal (in which case there may be no need to take any action) or be likely to leak to the surface? If it is the latter, then it is necessary to evaluate the magnitude of the leak and the potential hazard posed by the leak. If the amount of CO_2 involved is very small, then the action may be to merely monitor the leak but take no immediate action. If there is a large scale

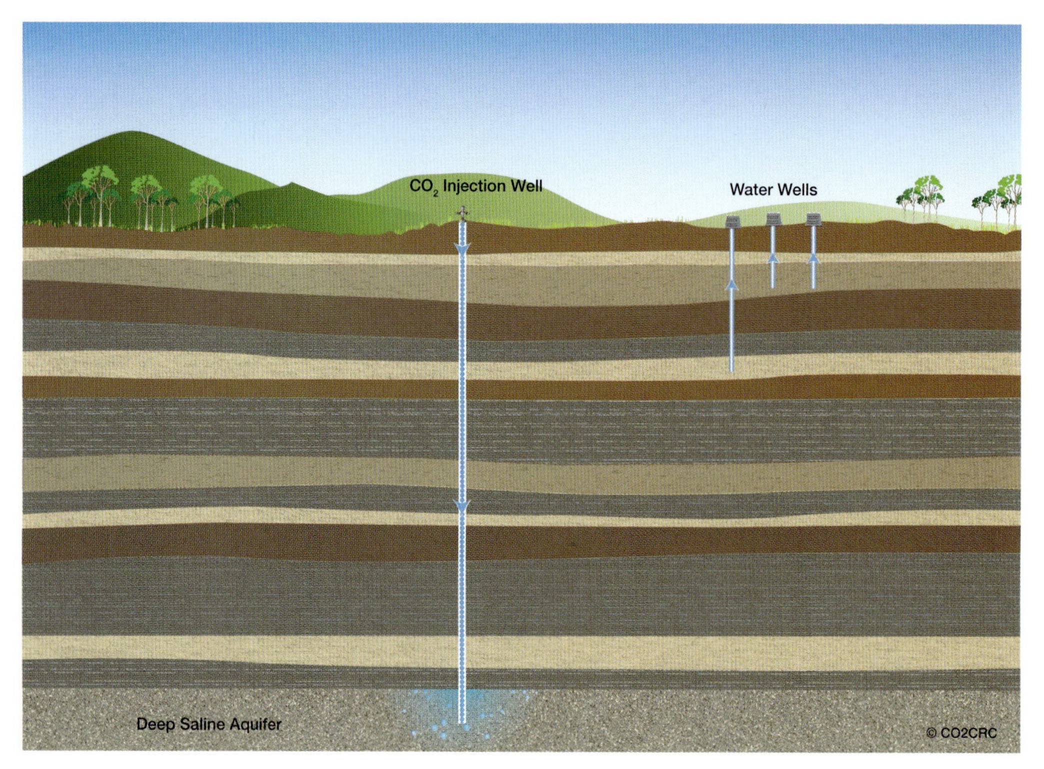

Figure 9.5 Illustrated here is an excellent storage site with a DSA overlain by several seals. If there were to be any leakage through the lower (primary) seal, the secondary (shallower) seals would prevent leakage to the surface.

leak, then immediate remedial action may be necessary, such as decreasing the pressure in the storage formation or, if the leak is from a well, recementing or in some other way remediating the well.

Monitoring

Forecasting what might happen when CO_2 is stored at a site is a very important part of the planning process for a CCS project. But an equally important component is to have a program of monitoring of the site. Monitoring is an integral part of providing confidence to the regulators and the community that the project is operating safely. In order to do this, two types of monitoring are necessary (Figure 9.6):

- **integrity monitoring** is carried out to confirm that, as predicted, the CO_2 is remaining within the geological storage unit

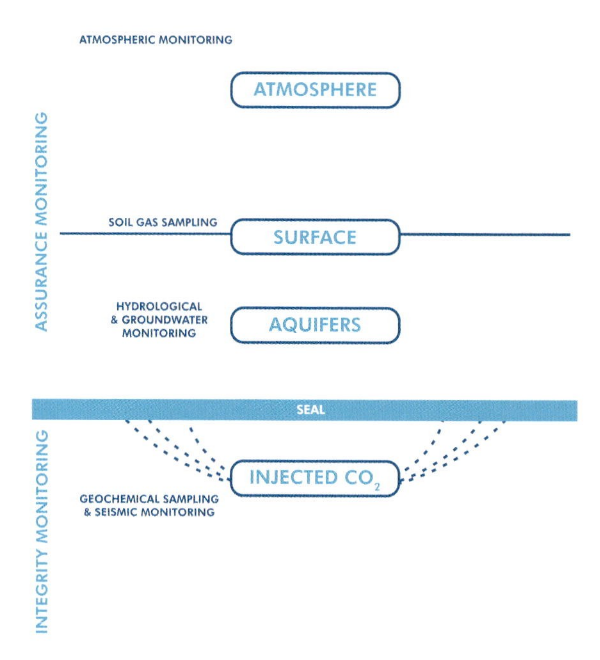

Figure 9.6 It is important to carry out both deeper integrity monitoring to confirm containment of CO_2 in the storage formation and shallower assurance monitoring to confirm that there is no CO_2 leakage into shallow aquifers, soils or the atmosphere.

CRANFIELD PROJECT

The Cranfield oil field in southern Mississippi (United States) was discovered in the 1940s and was depleted around 20 years later after recovering approximately 22% of the oil in place. In 2008 Denbury Resources (the field operator), along with SECARB Regional Carbon Sequestration Partnership (RCSP), began injecting CO_2 into the Tuscaloosa Formation for enhanced oil recovery at an initial rate of approximately 400 000 tonnes CO_2 per year, increased in 2009 to 1 Mt CO_2 per year, with the intention of increasing this to 1.5 Mt per year in phase 3. The Tuscaloosa Sandstone is a regionally extensive unit with a deep saline aquifer (the storage interval) at a depth of about 3000 m.

Using monitoring technologies such as tracers, downhole sensors, u tube sampling devices, well logging and multiple seismic technologies (including 4D and vertical seismic profiling), the Texas Bureau of Economic Geology and collaborating organisations has carefully monitored the injection of 2.5 million tonnes of geologically sourced CO_2 with the intention of also injecting up to 250 000 tonnes of anthropogenic CO_2 from a power plant in the region. (Image courtesy Susan Hovorka, BEG)

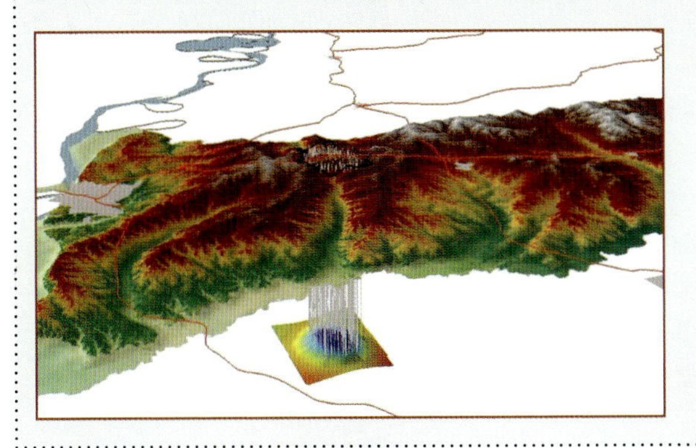

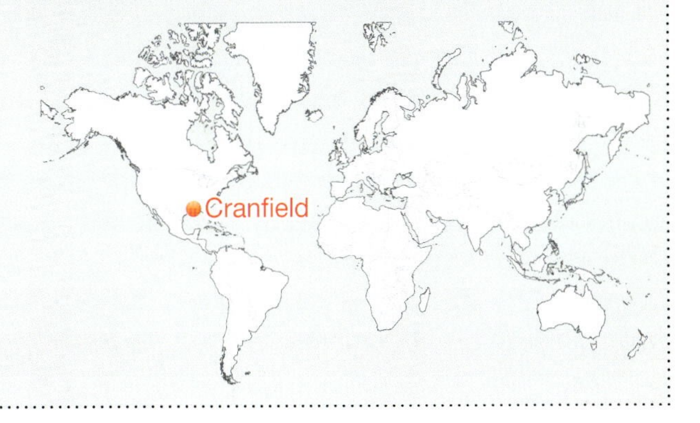

- **assurance monitoring** is carried out to assure the regulator and the community that CO_2 is not leaking into potable aquifers, the soil or the air, or posing some environmental or health hazard at or near the surface.

Integrity monitoring

Integrity monitoring is important for confirming the behaviour of the CO_2 within the storage reservoir. In much the same way that ultrasound can be used to image internal organs in the human body, we can use sound waves to image rocks deep below the surface. This is done using explosives, or heavy 'drop plates' or large scale vibrator trucks to generate a sound wave, which is then reflected off the various rock layers (Figure 9.7).

By using sensitive processing techniques it is then possible to detect the difference between rocks containing water and those containing CO_2 (because the velocity of sound through

Figure 9.7 Seismic trucks put shock waves into the underlying bedrock in order to image the deep rocks (and also in some circumstances the CO_2) as part of integrity monitoring.

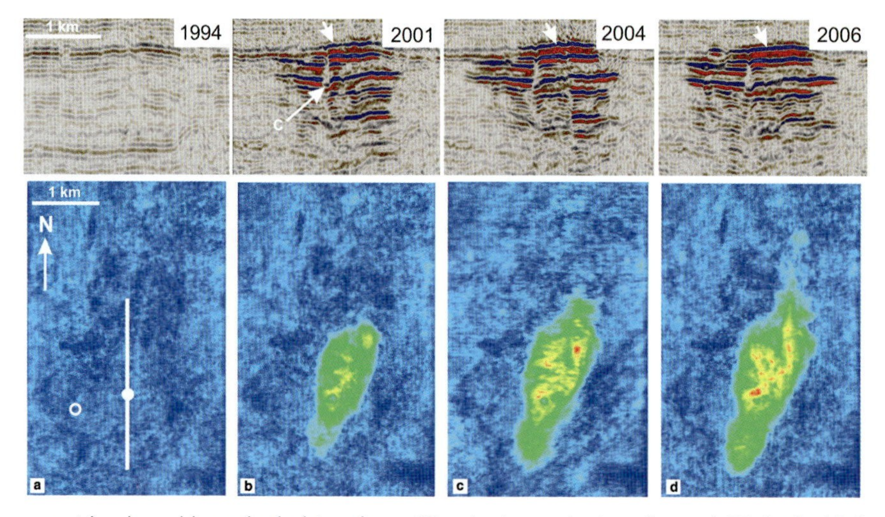

Figure 9.8 Sleipner provides the gold standard of time-lapse 3D seismic monitoring of stored CO_2 in the Utsira Formation, North Sea. Top panels show a vertical cross-section through the plume; bottom panels show a plan view of the Utsira reservoir and the plume footprint migrating over time. (Image courtesy of Statoil)

water is different to that of sound through CO_2). By carrying out very detailed seismic surveys, a 3D picture of the subsurface and the stored CO_2 can be developed. By carrying out the same survey at various time intervals, a 4D picture can be developed, showing how the CO_2 behaves over time. In other words, it is possible to see how the CO_2 is migrating within the storage formation and whether or not it is migrating out of the storage formation and into shallower units. The Sleipner Project has been able to build up an exceptional picture of CO_2 storage through seismic surveys and has become the 'gold standard' for seismic monitoring (Figure 9.8).

However, the method is very expensive and it is not always suitable (or practical) for onshore storage. Because of this, work is underway to develop new methods of detection to show that the CO_2 is contained within the storage formation. Methods under development or consideration include detecting changes in the gravitational field due to CO_2 injection, the use of fixed seismic systems and the application of satellite-based methods.

For example, using a satellite system (InSAR) it has been possible to detect the small scale surface uplift and tilt associated with CO_2 injection at the In Salah project in Algeria. To monitor changes in the chemistry of the fluids within a storage formation, it is necessary to have a monitoring well that can also be used for sampling the fluids for later analysis. This has been used very effectively in a number of projects for providing insights into the rate of migration of the CO_2 and also to provide a picture of how the chemistry of the groundwater changes over time when CO_2 is injected (Figure 9.9). For example, at the CO2CRC Otway Project , a monitoring well made it possible to obtain water and gas samples from 2km depth which showed that it took four months for CO_2 to migrate 300 metres within the storage formation and that the water became more acidic as the CO_2 front approached the monitoring well. In the case of this depleted gas field, it also provided an accurate picture of how the CO_2 behaved when it met residual methane. Therefore using a combination of techniques, it is possible to develop an accurate picture of how stored CO_2

Figure 9.9 Water in a storage formation (DSA) can be sampled using a sophisticated u-tube sampling system developed by Lawrence Berkeley National Laboratory. The system shown here is at the CO2CRC Otway site.

behaves at depth and to confirm that it is behaving as expected.

Assurance monitoring

Assurance monitoring is concerned with detecting any CO_2 that might escape from the storage formation and again it is possible to use a range of technologies for doing this. This can include 3D and 4D seismic surveys to detect migration of CO_2 through a seal. But in addition shallow wells can be used to monitor the chemistry of the groundwater for any changes, such as changes in the acidity of the water or increases in some trace metals. The soils can be monitored for excess CO_2 and atmospheric monitoring can also be undertaken to detect any suspicious anomalies in atmospheric concentrations of CO_2 in the vicinity of the storage site.

One of the difficulties that arises with atmospheric or other forms of assurance monitoring, is to know whether a change is the result of leakage of CO_2 or the consequence of the natural variability of CO_2 in the groundwater, the soil or the atmosphere (Figure 9.10). For example the atmospheric concentration of CO_2 just above the surface shows enormous variability on a daily, seasonal and annual basis and it is difficult (but not impossible) to detect the CO_2 signature from a small leak above all this natural variability. It is therefore very important that, before a CCS project gets underway, baseline surveys are undertaken to determine the background concentration of CO_2 and the natural variability in that concentration. In addition, the stored CO_2 may have subtle differences in composition compared to naturally occurring CO_2 at the site, which can help to identify the increase in concentration from stored CO_2 that may have leaked rather than being just a natural variation in the concentration (Figure 9.11). It is also possible to 'label' the

Figure 9.10 Soil sampling for CO_2 anomalies is a fairly simple field operation; but interpretation of the data can be complex due to the natural variability of the CO_2 content of soils.

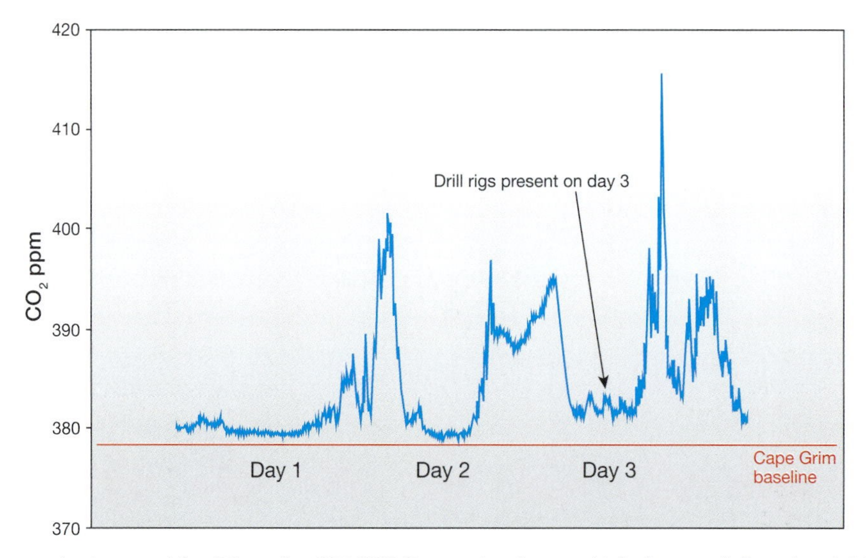

Figure 9.11 The atmospheric record for CO_2 at the CO2CRC Otway site shows a high degree of diurnal variability due to the photosynthetic cycle. However, by careful analysis it is possible to see anomalous anthropogenic events producing locally higher CO_2 values. (Data source: David Etheridge, CSIRO/CO2CRC)

stored CO_2 using trace chemical compounds that can be detected in trace amounts in order to confirm whether any detected CO_2 is natural or stored. Tracers have been successfully used in a number of small projects but may not be practical for large scale storage.

The regulatory regime

It is possible not only to predict how a potential storage site will work but also, through monitoring, to confirm that it is working as expected.Another key step needed to support public confidence in CCS is an effective regulatory regime. A number of countries and states now have put in place regulations to ensure that storage sites are effectively and safely monitored and managed (Figure 9.12).

Australia has for some time had Federal regulations in place for dealing with offshore CO_2 storage and Victoria has them in place for onshore and nearshore storage. Other Australian states have, or are now close to having their own onshore regulations. In the United States, the Environmental Protection Agency has released discussion documents on regulations under which CO_2 storage would be dealt with in much the same way as deep brine disposal is already effectively regulated. In Canada, the Province of Alberta is advancing its draft regulations, with a view to having them submitted to the government by late 2012. In Europe, the European Commission has issued a number of directives dealing with regulation of CCS projections.

Most of the regulations and draft regulations follow the same pattern: they aim to ensure the orderly and effective use of the subsurface, and in a manner that ensures that there is no danger to the public, no long term adverse impacts on the environment or on other earth resources (such as groundwater or petroleum) and no financial liability falling on the public in the future. An underlying principle in all the regulations is to minimise and manage any risk.

Most governments are approaching the regulations through a series of steps, the first of which is to define the condition under which an area would be granted to a company to explore for suitable storage sites. This is done in much the same way as oil and gas exploration licences are granted, to ensure that the

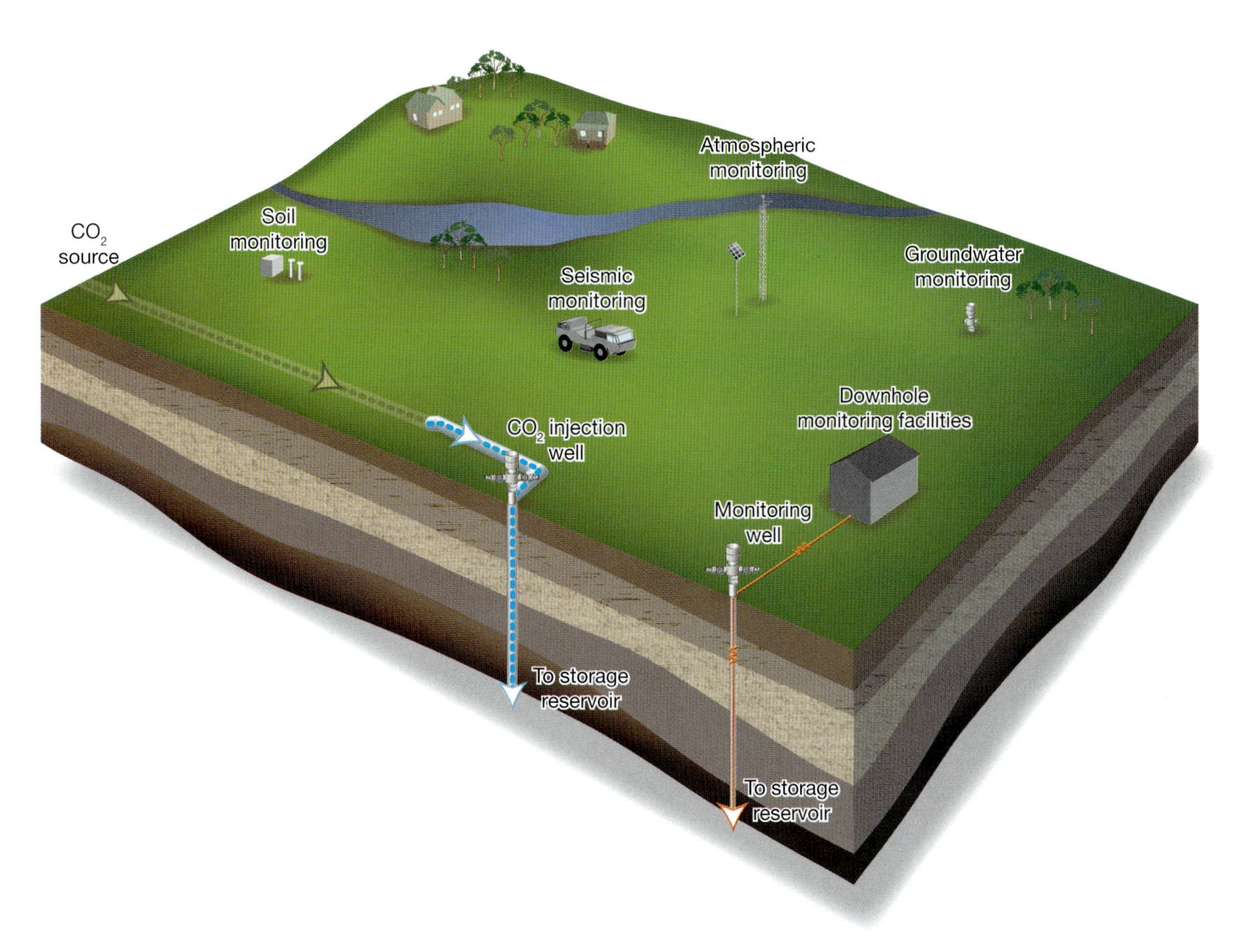

Figure 9.12 A comprehensive site monitoring system is needed for onshore CCS projects, such as that established for the CO2CRC Otway Project, involving atmospheric, soil, groundwater and subsurface monitoring.

operator has the necessary qualifications and resources to successfully take the project forward. Once a suitable site has been identified, the project proponent has to satisfy the government, via the regulator, that the site is indeed suitable for storing an amount of CO_2. This will usually involve very detailed site characterisation, extensive modelling and a comprehensive risk assessment, before a storage permit is granted. However, before injection can actually commence, it will be necessary to undertake baseline surveys, develop detailed engineering plans (including plans for remediation should it become necessary) and also detailed plans for monitoring and verification, as well as for closure and future abandonment of the site.

Once underway, the regulator has oversight of the storage operations, ensuring that the approved operational conditions (such as the injection pressure) and the agreed monitoring program are adhered to. If, during the program, the behaviour of the stored CO_2 deviates from the anticipated behaviour (such as the storage plume migrating in a manner not forecast by the reservoir models, or an unanticipated increase in formation pressure) or there are indications of a leak, then the regulators may decide to change the operational conditions or even suspend the operation until the deviations from anticipated behaviour have been thoroughly investigated. If a significant leak were to happen, the operator would be expected to remediate it and the project could even be closed down. However, under normal circumstances, the project would proceed until all the approved CO_2 had been injected. In some circumstances, the regulator could agree to injection of additional CO_2 if the operational history indicated that it would be safe to do so and there was adequate additional storage capacity.

Once injection has ceased, most regulations will require ongoing monitoring by the project for some years after closure, to confirm that the CO_2 (and the site) is behaving as expected, and that any migration of CO_2 within the reservoir, has remained within the limits originally approved by the regulator. If the CO_2 behaviour does not conform to the approved conditions, the regulators may require that monitoring of the site be continued beyond the normal period and, in extreme circumstances, that remedial action be taken.

Ultimately it is expected that all well designed and effectively regulated sites will get to the stage where the CO_2 (and the site) is behaving as expected with no indication that there will be any problems with sites in the future. Most regulatory systems anticipate that at this point there will be a transfer of liability from the project (the company) to the Government. Why should the project or company not continue to hold the liability indefinitely? Why should it be relieved of responsibility if there was a leak in say 200 years or even 1000 years time? The answer to this is clear if we look at the longevity of companies and organisations. A few Japanese trading companies have been active for almost 1000 years; some European companies have been in existence since the Renaissance some 600 years ago. But the vast majority of companies, including the largest banks, have generally been in operation for less than 100 years. In other words, there can be no expectation that the CCS company will still exist in say 200 years time. Therefore no company can offer assurance to the public that it would be around to fix a problem, were one to arise. But we can be confident that some form of government will be around in 100, 200 or even 1000 years to take on responsibility.

Does this not then mean that the company has been relieved of its responsibilities, in that it would not have to deal with a future problem? First it is important to recognise that the transfer of liability would only occur when it could be confidently predicted that risk of any future adverse event was very low indeed. Second, it is up to the government to take over the long term (post-closure) monitoring, but the government can require that a financial bond be lodged by the project or a fee paid per tonne of CO_2 stored, to cover the cost of any ongoing future monitoring or remediation. It should also be borne in mind, as pointed out earlier, that the perception of risks has to be tempered by the fact that risk is the likely consequence of an event occurring and also the potential impact of such an event. There is no such thing as a risk-free activity and even deciding to take no action carries a risk. Deciding to take no action to limit CO_2 emissions undoubtedly carries a risk and any possible (and manageable risk) arising from CCS has to be balanced against the (probably unmanageable) risk of the concentration of CO_2 in the atmosphere rising to dangerous levels as a consequence of the inadequacy of steps taken to decrease CO_2 emissions.

A 'social licence' for CCS?

So there is a basis for being confident about geological storage of CO_2 as an effective and safe mitigation option because:

- we have effective subsurface technology
- we understand the way in which CO_2 behaves
- we can monitor and verify that behaviour
- we can be confident that an appropriate governmental regulatory regime has been, or will be established.

But there is one additional element that needs to be considered to ensure that CCS is an effective mitigation option and that is to ensure that it has the 'social licence' to operate. Crucial to achieving this is to listen to and effectively communicate with the community at the local level (in the immediate vicinity of the project), at the regional and national level (where policy makers and politicians make the decision to support a project, or not), and at the international level where the impact of international bodies (whether UN-type bodies or NGOs) can profoundly influence the media to be in favour of or against a CCS project. At the moment few people know about CCS and even fewer understand it. There is therefore inevitably some uncertainty, even fear, about this unfamiliar technology.

The experiences of CCS project proponents to date have been mixed when dealing with local communities. There are examples of communities being opposed to CCS projects. For example a number of communities in the United States were opposed to the siting of the FutureGen Project (see FutureGen project, Chapter 11) in their district. Conversely other communities were eager to host FutureGen because it brought new jobs and perhaps other benefits. Projects in Europe have had a mixed response. Some have been accepted and others such as Barendrecht in the Netherlands strongly opposed. Offshore projects have encountered the least problems, perhaps because they are 'out of sight and out of mind' and because there is no local community.

In Australia, the experience with the CO2CRC Otway Project has been very positive. That is not to say everyone is supportive, but the community as a whole is, and indeed even has a sense of pride in a project in their area that has acquired international status. The key to achieving acceptance is to start talking (and listening) to the community before there is a real project. It is also important to ensure that a completely open approach is taken in explaining the project to the community, including welcoming them to the site when operational conditions allow and ensuring that they know who to go to (preferably somebody who lives in the immediate vicinity) if problems do arise.

Beyond the community in the immediate vicinity, a CCS project can be extraordinarily valuable for educating and informing people more widely about CCS. Being able to see a real project, talk to the plant operator and see the facilities is usually more valuable than reading about CCS. But obviously it is also important to take CCS to a broader audience through the media, not with a view to lobbying or advertising the value of CCS, but to present the facts and let people make up their own minds. Hopefully this book will contribute to that process of communication.

10 THE COST OF CLEAN ENERGY

What will it cost to deploy CCS at a large scale? This is extraordinarily difficult to answer for CCS (or for large scale geothermal or solar thermal for that matter). In fact before it is possible to attempt to answer it, it is usually necessary to pose a number of additional questions such as: Using what technology? At what scale? When? Where? New build or retrofit? What is the composition of the emissions? What discount rate is applied? The problem is that people (and perhaps especially politicians) like a simple answer and preferably a single number that can be readily communicated to the public. They want to be able to easily work out how much more expensive 'clean' (via CCS or some other technology) electricity will be than some other form of 'clean' electricity, or compared to 'dirty' electricity, so that the most cost effective approach to mitigation can be developed. This is a reasonable objective of course, but not easy to achieve. For the fact is that all clean energy technologies show a high degree of variability in cost depending on, among other things:

- location
- the impact of variability of supply, or the value put on the certainty of electricity supply (or the cost of intermittency)
- the cost of externalities including of course the cost of emitting carbon to the atmosphere.

Some argue that these cost issues can be largely resolved by putting a price on carbon and that is discussed in the next chapter, but this chapter is concerned with looking at the direct cost of CCS and how those costs might compare with other clean energy technology options.

The interplay of costs

The cost of a new build 500 MW power plant using coal-based integrated gasification combined cycle (IGCC) might be \$3–\$4 billion or more, whereas the cost of a 500 MW conventional pulverised coal-fired plant might be of the order of \$1–2 billion. But the relative cost of CO_2 capture for the two systems will be in the reverse order: lower cost for IGCC using pre-combustion capture and higher cost for post combustion capture on a pulverised coal unit. So, there is a trade off between the capital cost of the power plant and the cost of the capture system plus the ongoing operational expenditure.

Similarly there is a trade off between new build and retrofit. A new power station with CCS offers opportunities to greatly improve efficiency by, for example, using an ultra supercritical system, or by more effectively using waste heat, or both. On the other hand the cost of borrowing money may be high and, if the old power station is fully paid for, with the only cost being for the retrofit capture plant, then the total cost of retrofit (taking into account the cost of money) might be cheaper than new build. The efficiency of a power station is reduced by the deployment of a

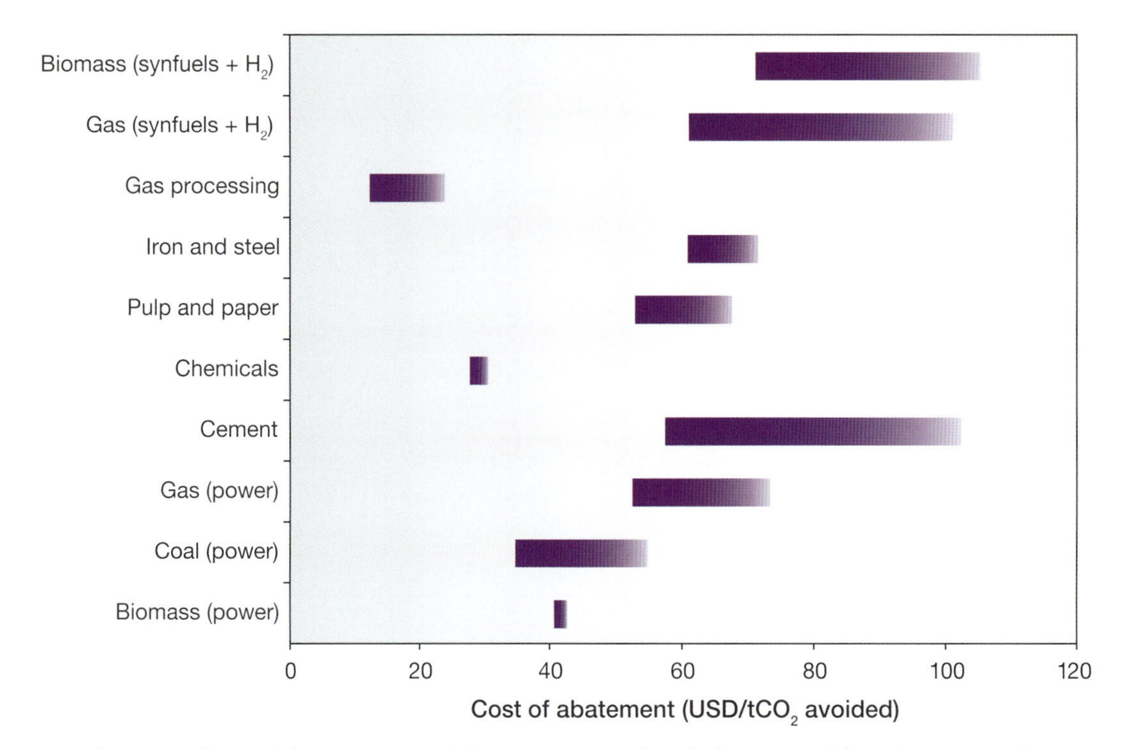

Figure 10.1 The potential cost of abatement using CCS varies enormously with the source of the CO_2. Lowest CCS costs are for abatement of CO_2 emissions from gas separation and production of chemicals such as urea. Mitigation for power is seen (perhaps optimistically in some cases) to be in the mid range and biomass and synfuels at the higher end. However, these values should be treated with caution because the cost range is often very large and in addition it appears to not fully take into account the value (cost benefit) of the synfuels. (Data source: IEA 2009)

capture plant because energy is required for the capture process. Thus, a power plant can either produce less marketable energy (and therefore less profit) because some energy is required for the capture process, or more fuel can be used to power the capture process and make up for the losses. Not only would this require an expansion of the power plant's generating capacity, but the extra fuel itself would be a significant cost to the power plant, particularly if more expensive, but cleaner, fuels such as gas are used instead of coal.

CCS costs are also affected not only by the 'starting point' (how much CO_2 is in the original emissions) but also by the desired 'finishing point' (how pure is the stream of CO_2 to be stored). The purer the CO_2 to be stored, the higher the cost of capture and separation, but the trade off of a purer stream is that costs are saved by not having to compress, transport and store 'non-CO_2' gases such as nitrogen. It can be even more complicated than this if there are toxic compounds in the gas stream that must be taken out before transport and storage, which in turn can drive costs higher.

Can we bring down capture costs by making compromises so that we do not capture all of the CO_2? Well of course the less CO_2 caught the lower the cost, but this is defeating the purpose of CCS. Nevertheless in the case of post-combustion capture, it may make sense to turn off the capture plant for say two hours a day when electricity demand is at a peak and electricity costs are an order of magnitude more than they are at off peak times. This might mean 10% less CO_2 is captured but could result in the overall cost of capture decreasing by perhaps 20 or 30%.

The costs of capturing CO_2 emissions from non-power sources

What about the cost of capture for non-power sources? Separation of CO_2 from natural gas, prior to putting it into an liquefied natural gas (LNG) plant, is a necessary part of gas processing, and produces a pure stream of CO_2 at a relatively low cost (Figure 10.1). This is why CCS has first been applied to gas projects such as Sleipner and in the future will be applied to the Gorgon LNG Project in Western Australia. CO_2 from gas processing represents 'low hanging fruit' for the deployment of CCS, but even so it comes at a cost. For example the CCS component of the Gorgon LNG Project adds an additional $2 billion to the cost of the project, which obviously impacts on the profitability (or the viability) of a project when competing with other producers who are not mitigating their CO_2 emission.

The emissions streams of iron and steel plants, and fertiliser plants, also have high concentrations of CO_2, which means that capture can potentially be applied at a relatively low cost. So why is it not being done? Again as in the case of LNG, the answer is simply that iron, steel and fertilisers are sold into a global price competitive market and therefore CCS, which adds to the cost, is unlikely to be applied until there are policy or other drivers to make it happen (this is discussed in the next chapter). In addition, as technologies progress from research through demonstration to large scale deployment, costs come down (Figure 10.2), though by how much is somewhat speculative.

So, given all these uncertainties and variabilities, is it possible to arrive at any sort of price for deploying CO_2 capture and separation? There have in fact been a number of attempts to determine the levelised cost of capture by a range of organisations, including the Intergovernmental

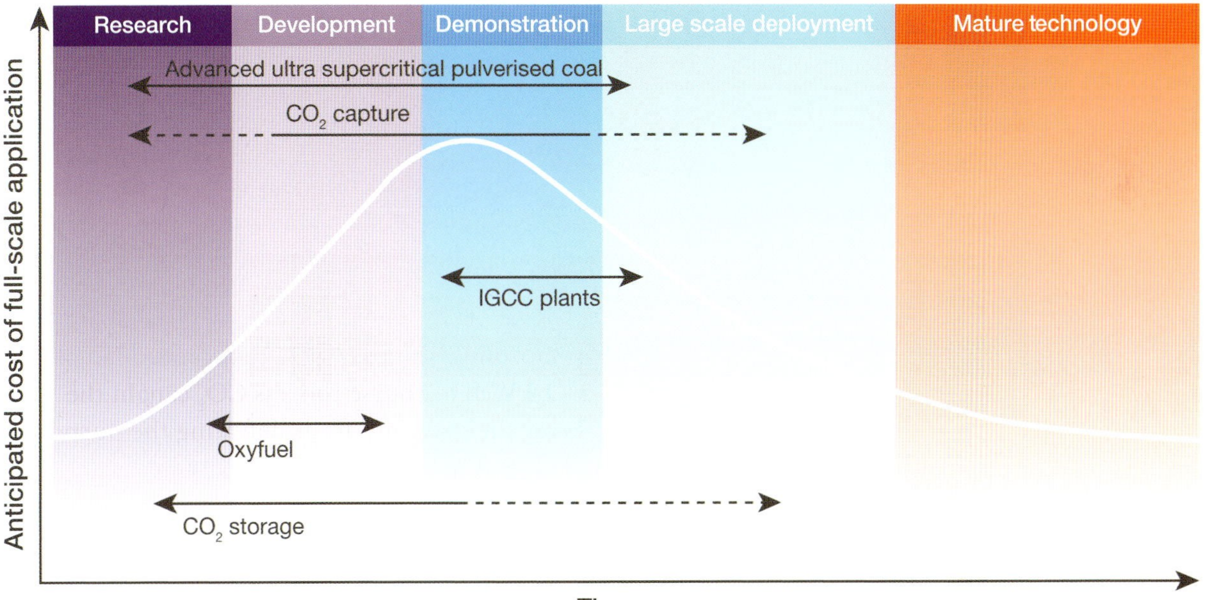

Figure 10.2 The cost curve for new technologies typically shows maximum costs at the early demonstration stage before declining significantly as the technology matures. The length of time it will take to progress along the cost curve varies within the CCS chain. Transport is already a mature technology; components of storage are relatively mature; large scale capture technology is quite immature and therefore there may be some way to go before costs come down significantly. (Adapted from CAETS 2010)

Panel on Climate Change (IPCC), the International Energy Agency, the Global Carbon Capture and Storage Institute, CO2CRC, the Massachusetts Institute of Technology, and the Electric Power Research Institute

In 2005, the IPCC produced a series of costs (based on 2003–2004 figures) from which it appears that capture costs are lowest for IGCC, intermediate for post combustion capture and highest for natural gas combined cycle (NGCC). A 2011 study by the Organisation for Economic Cooperation and Development (OECD) and the International Energy Agency shows the same relativity of costs for capture in that IGCC is lowest, NGCC highest, and post combustion for a pulverised coal-fired power station intermediate. Capture costs for an oxyfuel system are somewhere between IGCC and post combustion.

For simplicity, representative costs are sometimes expressed as a single number. However, single numbers should be treated with extreme caution, because they have a high level of uncertainty attached to them, though the relativity appears to be consistent. But it is very important to bear the full cost in mind. IGCC may offer the cheapest capture option but if (as appears to be the case) IGCC is a far more expensive power plant option than pulverised coal or NGCC, then any benefit in the lower cost of capture, may disappear in higher capital and financing costs.

Similarly while NGCC appears to be expensive in terms of capture costs per tonne of CO_2, natural gas only produces half the CO_2 emissions per unit of electricity. In addition NGCC power plants are relatively cheap to build, so the total cost of the NGCC plant with capture may be cheaper than some of the other options. But again it has to be remembered that fuel costs are double for gas than for coal. Swings and roundabouts!

The indicative capture costs for 2005 and 2011 provided by various studies suggest that costs have increased significantly over the past few years, because materials such as steel and concrete have increased in price over the period (an issue shared by many other clean energy technologies), as have labour costs. However, it is important to remember that construction costs for conventional power stations (and other power systems) have also risen significantly in the last few years.

Costings for large scale capture technologies also have to reflect the level of uncertainty (or risk) regarding the final cost. When something has been built many times before, the manufacturer or project developer has a high degree of confidence the plant will work efficiently and they will know the likely cost. In this case, a contingency of only 10–20% is applied to cost estimates. If, on the other hand, it is the 'first of a kind' (and a large scale capture plant on a large power station would be 'first of a kind') then the company or project will factor in a high level of uncertainty which will be reflected in a higher cost. In this case, a contingency of 40% or more may be applied.

The only way to determine the real cost (and bring down the cost) is to build and operate some large scale capture plants. 'Learning by doing' is really the only way of going forward and decreasing the financial risk that is currently a major inhibition to taking CCS and other large scale clean energy options forward.

What about the cost of fitting a capture system to an existing power plant? Retrofit versus new build: which is cheaper? In a recent study commissioned by the IEA Greenhouse Gas (IEAGHG)Program and also in a study by CO2CRC, it is evident that a number of factors have the potential to make retrofit more expensive, including the fact that old power plants are not designed to take a capture unit, may have a higher energy penalty and lower efficiency, and may have an operating life that is shorter than that of a new plant. Despite these factors, the IEAGHG study concludes that '*for a range of conditions that might be encountered in practice it appears that the costs of electricity from power plants retrofitted with CCS may be lower than from new build power stations with CCS. Lower costs of CO_2 capture in new build power plants compared to retrofit may be offset by the higher capital cost of the base power plant itself*'. This suggests that some older and dirtier power stations could cost effectively be made into cleaner power stations.

A decision on whether or not to 'retrofit' is likely to be highly dependent on the policy settings including the taxation system, age and type of power station, its location, the availability of land on the site for major new capture facilities and so on. Despite these complications, it is important that at the least, given the billions of dollars involved, the option of retrofit is thoroughly investigated for existing plants. Thousands of them are in existence around the world, all producing large amounts of CO_2 and likely to be doing so for many years to come, unless action is taken to limit their emissions through CCS.

Transport and associated costs

What about the cost of CO_2 transport and the associated compression costs? Here there is a far better basis for developing costings, such as the North American experience of transporting up to 50 million tonnes of CO_2 per annum over the past 40 years or so, using several thousand kilometres of dedicated CO_2 pipelines. Added to this is the experience gained from the construction and operation of hundreds of thousands of kilometres of natural gas pipelines under a range of conditions. As pointed out previously, the amount of CO_2 being transported and the scale of the transport operation has a massive impact on cost per tonne of CO_2. For example a 20 cm diameter gas pipeline costs around $400 000 per kilometre, whereas a 100 cm diameter gas pipeline will cost approximately $3 million per kilometre but is capable of transporting 25 times as much CO_2 as a 20 cm pipeline. The cost of laying a pipeline depends not only on the diameter of the pipeline, but also on whether it is onshore or offshore, whether the topography is flat or mountainous and whether the area is densely or sparsely populated. In addition to the cost of the pipeline, and depending on the length and route of the pipeline, it will be necessary to have compressors or pumps every 2–300 km, to maintain the pressure of the CO_2 within the pipe. Each compressor, depending on size, costs perhaps $40 million capital cost and several million a year in operational costs. A study of transport costs by CO2CRC based on a notional 200 km pipeline (half onshore and half offshore), indicated a capital cost of around $800 million, with operating costs of up to $30 million a year for transporting five million tonnes of CO_2 per annum.

In 2005, the IPCC provided some indicators of the comparative cost of onshore and offshore pipelines. Using a pipeline with a capacity of 5 million tonnes of CO_2 per annum (based on 250 km of 'normal terrain conditions'), costs are estimated at US$2.20–3.40 per tonne of CO_2 transported onshore compared to US$3.60–4.30 per tonne of CO_2 transported offshore The absolute values are not significant to present-day projects as they relate to 2004 costs, but they do suggest that CO_2 pipeline transport offshore costs around 50% more than transport onshore, which will obviously flow on to the overall cost of a CCS project.

The IPCC also compared the cost of pipeline transport to ship transport, concluding that for distances greater than 1000 km, shipping transport is cheaper, though this does depend on water depth, harbour facilities and a range of other variables. According to Teekay, ship transport is feasible on a small scale and provides a flexible method of transport in the

700–1500 km range. In the future, cost may not necessarily be an insurmountable barrier to the long distance transport of CO_2: for example, from a country with limited storage potential to a country with has large scale offshore storage potential. However, for some time to come, most CO_2 will be transported by pipeline, at a cost estimated by the IPCC to be in the range of $1–8 a tonne CO_2, for a nominal 250 km distance. The large price range once again serves to highlight how location-dependent costs are.

Storage costs

Turning to the cost of storage, again the oil and gas industry provides a good basis for estimating costs. However, as before, it is important to stress the variability in costs depending on location. For example, a 2000-metre exploration or injection well drilled offshore in say 100 metres of water, may cost 10 times more than a 2000-metre well drilled onshore. Costs can also vary greatly from region to region. A 2000-metre onshore well drilled in an area of the USA where there is a lot of oil and gas activity would cost as little as $1 million, whereas a 2000-metre well drilled in a remote area of Australia may cost $5 million.

Once again this high degree of variability in costs can only be addressed by asking the question 'where?' when considering the cost of CCS. In 2005 the IPCC estimated that the cost of storage was in the range of $0.5 to $8 a tonne of CO_2. This cost range would now be regarded as being low. Since that time, we have developed a better understanding of the influence of the geology, particularly the permeability, of the rocks on storage. If the permeability is low, then it is difficult to inject the CO_2 and this has to be overcome either by drilling more injection wells, or by using horizontal wells to provide a greater injection interval, or a combination of both. Alternatively the rocks can be fractured to create permeability. But these measures to overcome the barrier of low permeability add greatly to the cost of storage compared to the cost in an area where the rocks have naturally high permeability and where it is easy to inject large quantities of CO_2 using perhaps just one or two wells.

Researchers at the University of New South Wales have looked at this issue in some detail for CO2CRC, assessing the tradeoffs involved in injecting CO_2 into very permeable rocks which are some distance away (which therefore then involves higher transport costs), compared to using rocks with a low permeability near to the CO_2 source (which therefore have lower transport costs). They found that more often than not, it was better in terms of costs, to store in a more distant geological region with high permeability, rather than in a nearby region with low permeability. However, there are other influences, such as the depth to the storage interval and the temperature gradient within the earth's crust. The deeper the well, the more costly it is to drill and once again there may be tradeoffs between shallow rocks with low permeability versus deep rocks with high permeability, although in general, shallower rocks do tend to be more permeable than deeper more compressed rocks.

Subsurface temperature impacts on the density of the stored CO_2 and therefore on the amount of CO_2 that can actually be stored. In a 2010 analysis of the tradeoffs between permeability, depth and distance for a series of sedimentary basins in Australia, the National Storage Mapping Task Force concluded that the cost of transport plus storage could range from as little as $7 a tonne CO_2 in a favourable source-sink setting (such as the Gippsland Basin in south-eastern Australia), to $70 a tonne or more, with less favourable settings (and where CCS is therefore unlikely to be a viable option in most circumstances).

Another storage cost which needs to be considered, is the ongoing cost of monitoring of a storage site, not just during the injection phase but for perhaps decades after closure. Once again this cost will vary greatly with the site and whether it is onshore or offshore, in mountainous terrain or on a delta, in an arctic environment or a tropical environment. The IPCC suggests that monitoring costs are of the order of 10 to 30 cents per tonne of CO_2, although there is a great deal of uncertainty attached to those numbers. The best that can be said is that, overall, it is not expected that the cost of monitoring will add significantly to the cost of a CCS project. The caveat here is that of course this cost will depend not only on the location but also on the regulatory regime. Futhermore, the imposition of unrealistic requirements for monitoring could impose a significant additional cost on a project.

Indicative total costs for CCS

So, to return to the question of 'what will CCS cost?', it should be apparent by now, that it is not simply a matter of adding up the indicative numbers of capture plus transport plus storage, to arrive at the full cost for CCS. To do this could, on the one hand, produce quite unrealistically low numbers by taking an extraordinarily optimistic scenario. On the other hand, by using a worst case scenario, it could arrive at unrealistically high numbers. Despite this, many studies have courageously produced indicative total costs for CCS projects. In 2008, McKinsey and Company investigated costs using a 200 MW-plant demonstration project. They suggested that there would be a cost of €70 per tonne of CO_2 avoided for a 200 MW plant, and €45 per tonne avoided for a 600 MW plant. These costs were arrived at with the expectation that there would be significant cost improvements once industrial-scale roll-out took place. The figures of the IPCC from 2005 are consistent with these numbers, but the range for each component of the CCS system is so large that their aggregation suggests the total cost would range from an unrealistically low value of \$17 a tonne CO_2 avoided, to a high value of \$91 a tonne CO_2 avoided. In 2011 Paul Feron and Lincoln Paterson suggested an overall cost for a coal-fired power plant plus CCS in the range \$80–140 per tonne CO_2 avoided.

Cost estimates derived from operational CCS activities

All of these 'order of magnitude' numbers are useful in that the question of 'what does CCS cost?' is an entirely reasonable one and it is not satisfactory to respond by saying 'we don't know!' or 'It's too complicated to give an answer'. Nonetheless it is essential to seek to answer the question on the basis of specific projects, rather than on the basis of global costs. But as there is no operational large scale power-with-CCS project, we have few definitive numbers available to us, which is why it is important to continue with research and development, but it is especially important to get some large CCS projects underway, so that the real costs can be established and equally importantly, those costs can be brought down.

The best we can do at present is to try to use some numbers from operational CCS activities to provide a 'sanity check'. For example Frank Mourits reported in 2008 that the capital cost of the Weyburn enhanced oil recovery (EOR) project in the United States was \$1.3 billion. This obviously includes EOR-specific costs and it is unclear what they are, but given that 30 million tonnes of CO_2 will be transported 320 km and stored over the life of the project, it suggests a maximum capital cost of around \$40 a tonne of CO_2 transported and stored. A 2001 report of the National Energy Technology Laboratory (NETL) indicates a \$100 million capital cost for the pipeline and compressors for

Weyburn. Operational costs specific to transport and storage are not known, but could be in the order of $5 a tonne of CO_2. CO_2 is sold into the pipeline at an estimated $15 a tonne, which presumably covers at least the cost of capturing CO_2 from the Beulah coal gasification plant. So, using the example of Weyburn and some admittedly speculative numbers, the total cost of CCS attached to a coal gasification plant appears to be of the order of $60 a tonne CO_2 captured, transported and stored (it is not possible to translate this into a 'CO_2 avoided value).

A second example is provided by the Sleipner Project in the North Sea, which was established in the knowledge that if the CO_2 separated from natural gas was not sequestered, a tax of around $50 a tonne CO_2 would be levied, which obviously imposes an upper limit on the cost of storage, otherwise the company could have chosen to pay the tax, as a cheaper option. In fact, the IEA reported in 2010 that the cost of compression and storage was $16 per tonne CO_2. In some ways, Sleipner is the 'perfect' project in that the cost of separation is essentially zero as far as CCS is concerned (because the CO_2 has to be removed anyway before the natural gas can be put into the gas network). In addition the storage interval, the Utsira Formation, is a highly permeable reservoir which directly underlies the production platform, so there are no transport costs. The cost of compression at Sleipner is likely to be high because it is an expensive offshore operation, but nonetheless it does provide an indication of storage costs at $16 a tonne CO_2 avoided.

A third indicator, again from the gas sector, is provided by the Gorgon Project in Western Australia; where the storage project under Barrow Island is reported to have a capital cost of around $2 billion for injection of 3–4 million tonnes CO_2 per annum for 30 years. This suggests a capital cost equivalent to approximately $18 a tonne CO_2 injected, and whilst operational and other costs are not known, and transport and separation costs are essentially zero, this $18 figure is perhaps a useful indicative number for the capital cost of storage alone.

It is important to stress that these are very crude figures to the extent that they do not take into account cost increases that might have occurred since the projects were initiated, nor the cost of borrowings or depreciation, or operational costs. But putting these caveats aside, these real world examples suggest that a cost of $10–20 a tonne CO_2 for storage alone is a reasonable indicative value for a reservoir rock with good storage characteristics and a figure of $20–30 a tonne CO_2 for 320 km of transport (to Weyburn) or $6–10 tonne per 100 km. The gap in our knowledge of the real cost of projects is of course in the large scale capture area and as mentioned previously, this gap will not be filled until we undertake some large scale projects. In the absence of projects, it is reasonable to take a mid range cost estimate for post combustion capture (whether or not oxygen-rich combustion) of around $55 per tonne CO_2 avoided.

Therefore to sum this up, and bearing in mind all the caveats regarding the numbers quoted or derived from projects, a cost of $70–$85 per tonne of CO2 avoided is possibly of the right order for a hypothetical project involving:

- a coal fired power plant of 250–500 MW
- with post combustion capture
- transporting 2–3 million tonnes CO_2 per annum of the order of 100 km to a well characterised geological location, such as a depleted oil field, with excellent storage characteristics.

This does not take the cost of financing into account. There will be some circumstances where the costs will be lower than this and other circumstances where the costs will be higher!

Costing uncertainty

How can we do better than 'possibly of the right order' for costs?' Putting a price on uncertainty might be a step in the right direction, for not only does CCS involve technical uncertainties, but there are also many others such as variability in the cost of construction materials or labour; the speed (or slowness) of the approvals process; and the cost of debt. All of these and many other uncertainties, are of course factored into every major project, but what makes CCS costing more difficult is that a large scale, complete, CCS system has not yet been built. Added to these uncertainties are the enormous uncertainties regarding the financial returns from implementing CCS in the absence of clear long term policy (or price) drivers in most countries. We therefore have uncertainties compounding on uncertainties which results in any proposed CCS project having to place a higher premium on risk and uncertainty than most other large scale projects.

The normal process of developing costings for any major project, is to progress through a series of stage gates, with the information available becoming progressively more definitive and the engineering design becoming more detailed at each stage, until the project can confidently reach the final investment decision, at which time the project is either abandoned because the risk is too high and/or the likely rewards too low, or it is decided to proceed to construction because the rewards are likely to outweigh the risks. In other words, probabilities are attached to the estimates and a balanced decision is made. But in the case of CCS, because of the lack of major projects on which to draw on for experience and the external risks created by policy vacuums, there is a high level of uncertainty. This leads to a low level of confidence of financial success, which in turn has to be factored into the costings, thereby creating a premium of 20%, 30%, 40% or more on what the project costings might otherwise be.

It is possible that the CCS project costs cited by, for example, the IPCC or the IEA may not adequately cover the cost of uncertainty and that these costings may therefore be unrealistically low. In contrast, a project proponent will want to minimise the risk of failure and may therefore err on the side of caution, putting a high price on uncertainty. The project proponent's costing at this stage in the development of CCS may therefore be unduly high. First movers in CCS have no alternative other than to put a high cost on risk and uncertainty and this has to be reflected in the price of a CCS project, unless governments step in and decrease the risk by directly covering the cost of uncertainty, or funding additional research or pre-competitive studies.

Comparison costing

While this discussion serves to explain why it is difficult to say how much CCS will cost and why costs have such a large range, the fact remains that in mapping a way forward in greenhouse policy, decision makers will continue to ask how much it will cost. As indicated earlier, the best response is to seek to answer the question at the project-specific level. At the same time, at the higher level there is a need to somehow answer the underlying question of not only how much CCS costs but how much it will add to electricity costs and how the cost of CCS compares with other clean energy options.

Once again, these are complex questions to answer. In the case of the cost of electricity, the question has to be refined to whether it is cost at the power station, cost to the wholesaler or cost to the consumer? Whether all the CO_2 has to be captured or say only 90%, is also an important question that impacts on cost. And of course a

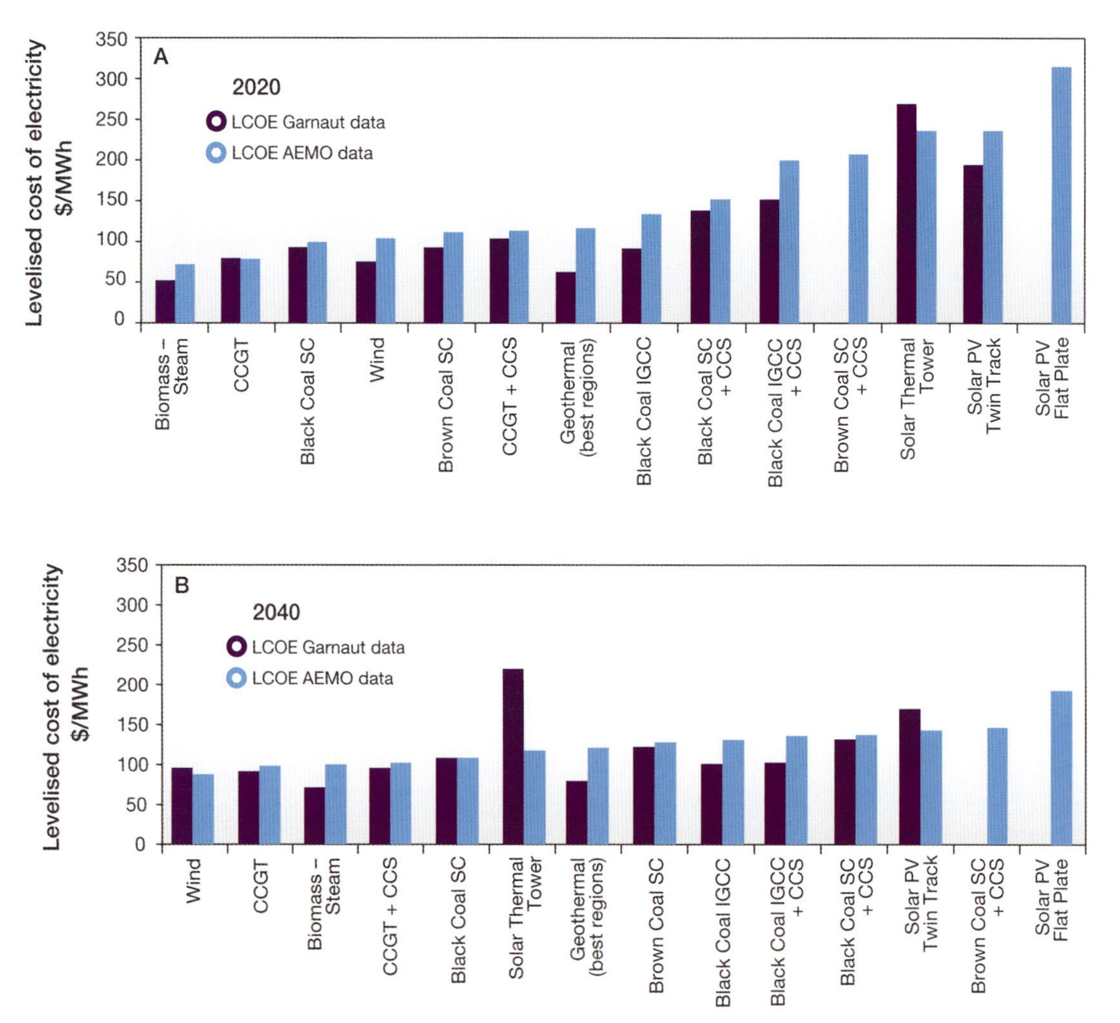

Figure 10.3 Using the levelised cost of electricity (LCOE) as a way of taking a uniform approach to comparing the cost of the various energy technologies, John Burgess for the Australian Academy of Technological Sciences and Engineering (ATSE) has modelled anticipated LCOE costs for (A) 2020 and (B) 2040. This modelling is based on data from Professor Ross Garnaut (an expert adviser to the Multi-Party Climate Change Committee and author of the Garnaut Climate Change Review, commissioned by the federal government) and the Australian Energy Market Operator. While the Garnaut data are generally lower than those of the AEMO, the trends are much the same for both, but with the striking difference of opinion regarding solar thermal tower technology in 2040 which Garnaut sees as the most expensive clean energy option. Wind is consistently the cheapest technology. The indicative cost of geothermal for 2020 may be rather optimistic. The various forms of CCS cover a range of costs but are below all solar costs in 2020, less so in 2040. Gas (CCGT) with CCS is considered the cheapest of the CCS options throughout. (Data source: Burgess 2011)

critical question is which fuel and the cost of that fuel? Comparing CCS costs with other clean energy technology costs is also difficult, because not every sector has the same approach to costings. Should the cost of necessary upgrading of the grid be included or excluded from the costs of a particular technology? How should water used by the technology be priced? The result is that most studies end up trying to compare apples and pears!

Studies in 2008 and 2011 by John Burgess, for the Australian Academy of Technological Sciences and Engineering (ATSE), provides power cost comparisons for a range of technologies. Costs quoted are for Australia and

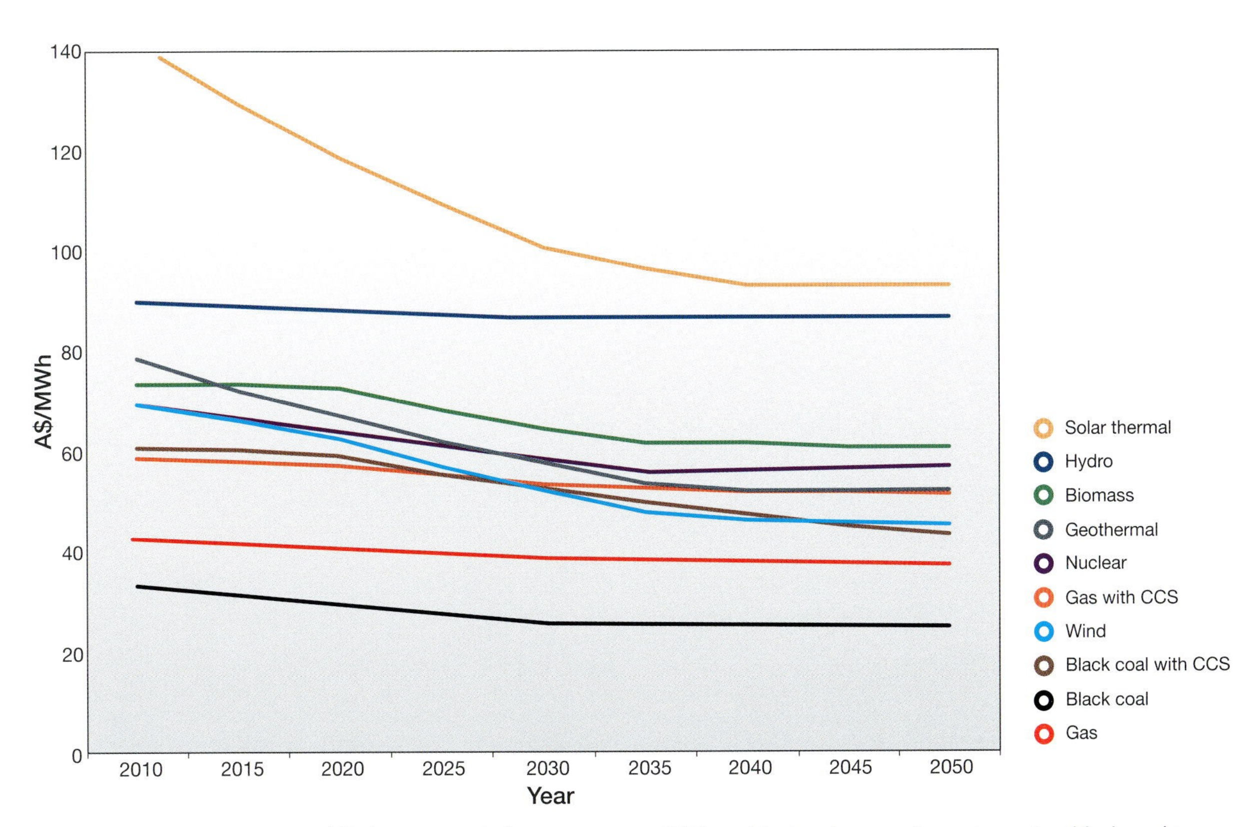

Figure 10.4 In an assessment of likely cost trends for energy up to 2050, and in the absence of a carbon price, black coal is seen as provided the lowest cost electricity, with gas the next cheapest. Solar thermal remains the most costly throughout, even though prices are expected to drop sharply to 2040. Hydro remains uniformly high with biomass next. Then follows a group of technologies all occupying much the same cost throughout the period. These include wind, geothermal, nuclear, and coal or gas with CCS. Therefore there is no reason to exclude any of these 'mid level cost' technologies from the mitigation portfolio at this stage. (Data source: The Cooperative Research Centre for Coal in Sustainable Development/CSIRO)

will vary significantly from country to country depending on fuel and other costs. Nonetheless the cost relativity of the various technologies is likely to remain much the same in most OECD countries.

The projected 2015 capital cost of the various technologies expressed by Burgess as cost per unit of generating capacity of a plant, range from a base case of $1200–$1300 per kilowatt for a combined cycle gas turbine with no CCS to $1900–$2300 per kilowatt with CCS. The capital cost for black coal (supercritical) power station with CCS is estimated by Burgess at $2900–$4500 per kilowatt compared to $2100–$2900 per kilowatt for wind. Solar PV ranges from $4600–$5700 per kilowatt depending on the type of PV system and geothermal is in the range $4000–$6300 per kilowatt. Over time all these capital costs are projected to come down as the technologies mature. Initially CCS with gas or coal is two to three times cheaper than solar, but this differential is projected by Burgess to be far less by 2040. When translated to a levelised cost of electricity (LCOE) in dollars per megawatt hour ($/Mwh), costs in 2020 are projected to range from as little as around $50 MWh (for biomass steam) to over $300 MWh (for solar PV flat plate). However, by 2040 the cost range is projected to be from around $70 MWh to approximately $190 MWh, with various CCS options being midway in this cost range (Figure 10.3).

Table 10.1: Cost of CO_2 capture for various power plants, demonstrating cost variability and cost rises over the past decade

New plant	Representative cost ($US) per tonne CO_2 avoided	
	2003[1]	2009[2]
Natural gas combined cycle (NGCC)	56	90
Pulverised coal post combustion	59	67
Oxyfuel post combustion	47	51
Integrated gasification combined cycle (IGCC)	25	41

[1] Values from the IPCC 2005
2 Values from Finkenrath 2011

The conclusion to be drawn from all this is that while the cost of all clean energy technologies will add to the cost of electricity, these costs will converge over the next 20-30 years and at this stage there is no basis for ruling CCS or any other clean energy technology out of the mix because of capital cost or on the basis of the levelised cost of electricity. A study by CSIRO has also examined the various technology options for centralised power generation on the basis of cost of electricity per megawatt hour (Australian costs), with costs projected to 2050 (Figure 10.4). The CSIRO study, like the ATSE study illustrated that whatever technology is used, it will add 100% and more to the cost of electricity generation compared to the cost of electricity from conventional coal or gas fired generation (100% cost increase at the generator will mean an additional cost of 25–50% to the home owner). It also suggests that by 2050, solar thermal and hydro will not be competitive in terms of cost of electricity and that biomass is unlikely to be competitive. There is then a range of technologies including geothermal, nuclear, wind and CCS, all with costs within 10–20% of each other and all showing significant cost decreases between now and 2050. Therefore once again there is no cost basis for leaving CCS or most of the other technologies out of the mix, although the CSIRO study suggests that black coal power generation with CCS will be the cheapest clean energy option per MWh for much of the period between now and 2050.

Conclusions

In conclusion, it is extraordinarily difficult to give a direct answer to the question of how much does CCS cost, but the costs arrived at for capture transport and storage of CO_2 and for integrated CCS systems by a range of authors and organisations, and the limited information gleaned from operational projects or projects under construction, are as consistent as might reasonably be expected at this stage of technology development (Tables 10.1 and 10.2). As emphasised repeatedly throughout this chapter, costs can only be established with any degree of confidence on a project-specific basis, but the inherent uncertainties (because of the

Table 10.2: 2002 Cost ranges for the components of a CCS system as applied to a given type of power plant or industrial source[1]

Cost system components	Cost range
Capture from a coal or gas-fired power plant	15–75 US$/t CO_2 net captured
Capture from hydrogen and ammonia production or gas processing	5–55 US$/t CO_2 net captured
Capture from other industrial sources	25–115 US$/t CO_2 net captured
Transportation	1–8 US$/t CO_2 transported
Geological storage	0.5–8 US$/t CO_2 net injected
Geological storage monitoring and verification	0.1–0.3 US$/t CO_2 injected

Adapted from the IPCC Special Report

lack of operational CCS projects) impose a significant extra cost risk which increases the cost estimates, even for a site-specific project.

But even bearing all this in mind, it is realistic to expect that CCS projects in favourable areas will deliver mitigation at around $100 per tonne CO_2 avoided and significantly less than this for 'low hanging fruit' such as production gas separation and some industrial process where, with a nearby storage opportunity, the cost may be $20–$30 per tonne CO_2 avoided or less. Costs will come down for CCS and other clean energy technologies, such as solar and geothermal, in the coming decades, but it is necessary to start building large scale CCS and other clean energy systems within the next few years in order to make this happen. It is also evident that retrofitting of CCS to existing power plants should not be ignored as it may provide a cost-effective option compared to new build in some circumstances. There is no question that CCS, like all clean energy technologies, will add to the cost of electricity. However, in terms of cost, CCS appears to occupy a 'middle of the road' position compared to the range of clean energy technologies and it is therefore essential to have it as part of the mitigation portfolio. The next chapter discusses the question of CCS costs in the broader context of greenhouse policy.

11 THE TECHNOLOGY AND THE POLITICS OF CLEAN ENERGY

Energy lies at the convergence of a range of critical issues including population growth, water supply, poverty reduction, health, resources, food shortages and climate (Figure 11.1). Climate change cannot be viewed in isolation from these issues but at the same time we cannot tackle all the ills of the world in a single, all-encompassing global master plan. Therefore for the moment, let us focus on the fact that climate change is occurring and we need to decrease levels of CO_2

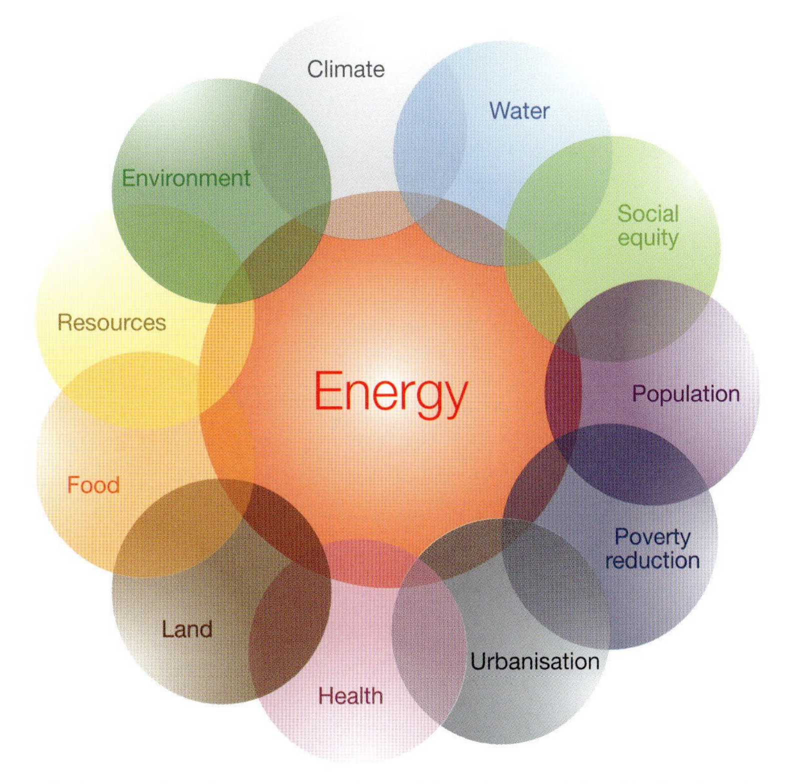

Figure 11.1 Energy lies at the intersection of many economic, social, environmental and technology issues. This interplay, and the need to take a multifaceted approach to energy, issues adds to the challenge of establishing a clean energy policy.

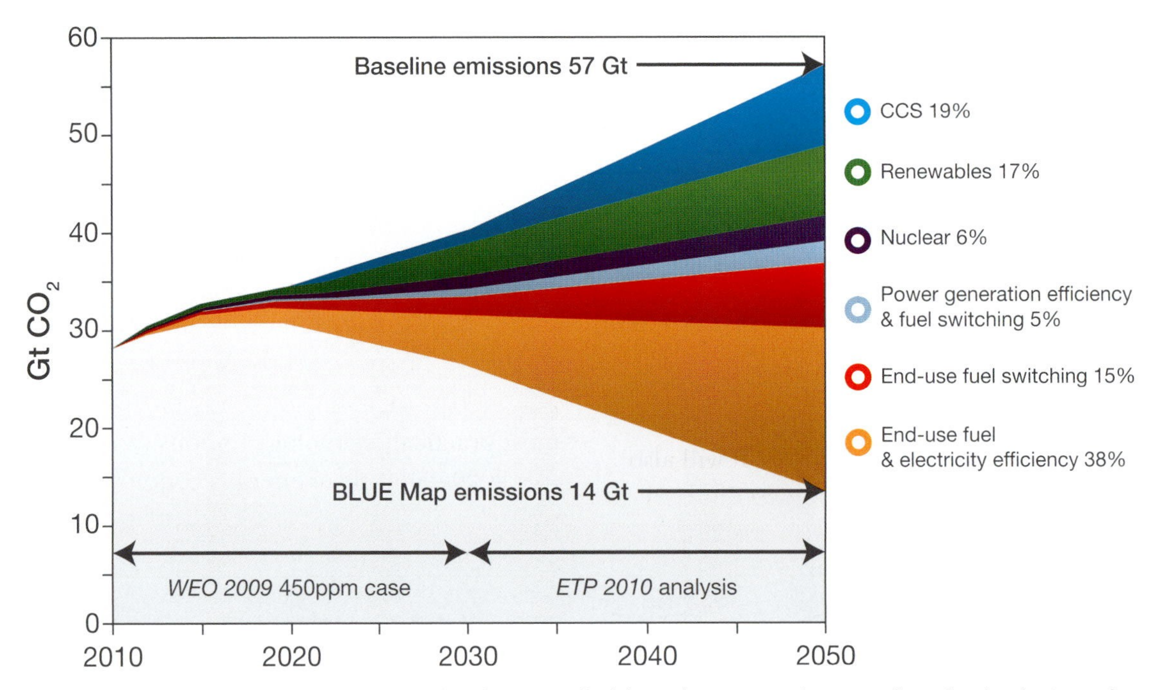

Figure 11.2 There are various versions of the 'wedge diagram'. All of them demonstrate that a number of technologies and actions (each represented by a different coloured wedge) are necessary to achieve a deep cut in emissions by 2050. This diagram, developed for the International Energy Agency (IEA) as part of their so-called Blue Map scenario for the period to 2050, uses the World Energy Outlook (WEO) scenario to 2030 and the Energy Technology Perspectives (ETP) forecast to 2050. Under the BLUE Map scenario, 48 Gt CO_2 are cut from global emissions compared to business as usual. (Image from OECD/IEA 2008)

emissions. The good news is that there are many ways of producing clean energy; the bad news is that for the most part we are not using them! There is no single technology able to provide 'the answer' and a portfolio of responses and technologies is needed (Figure 11.2). There is no ideal clean energy technology for they all have their strengths and weaknesses. For a number, the single biggest problem they face is intermittency of supply, for others the difficulty is one of scale. All forms of clean energy have environmental impacts of one form or another, with some impacts more obvious than others.. There are some clean energy technologies that are here-and-now which will contribute directly to achieving abatement targets for 2020 and there are others not yet deployed which will be absolutely crucial to meeting much more difficult 2050 targets. Some claim that renewable energy can meet all our needs for 24/7 power now or at least within the decade, for example, using solar energy with storage, plus wind power. Others see this as a commendable aspiration but one that is not practical at this stage of technology development.

Future strategies in a carbon constrained world

Many countries have now set targets for emission reductions by 2020 and 2050. For example Australia's 2020 target is a decrease of 5% in emissions compared to 2000, which amounts to an overall decrease of more than 20% compared to business as usual. The 2050 target is set at an even more taxing 80% cut in emissions by 2050. A number of European countries have similar targets in place or proposed. It is useful to consider 2020 and 2050 targets first from the perspective of the clean energy technologies that could deliver them, and second from the perspective of the measures needed to deploy them in time and at the necessary scale.

Using Australia as an example, let us first consider the technologies that will enable the 2020 target to be reached : In essence the choice has already been made for 2020 in that it will have to be based on existing clean energy technologies. These are wind, a modest contribution from solar photovoltaics (PV) and some switching from coal to gas. This will be supported by greater energy efficiency and increased soil carbon sequestration. It will also draw very significantly on carbon credits derived from abatement activities in developing countries. During this period to 2020 there will be no significant abatement from large scale solar or geothermal, or for the most part from CCS. The exception will be CCS applied to gas and liquefied natural gas (LNG) – the Gorgon Project in Western Australia will commence storage of 3–4 million tonnes of CO_2 per annum in 2015 – and perhaps from other gas or industrial projects. The 2020 target will be achieved through existing technologies, the purchase of carbon credits via the Clean Development Mechanism (CDM) or some other device and the use of regulations such as mandatory renewable energy targets (MRET) to drive technology uptake. A domestic carbon price (set at A$23 a tonne CO2) is most unlikely to have any significant impact on technology deployment to 2020. Based on the work of the Productivity Commission the regulation-based abatement to achieve the 2020 target will be extraordinarily expensive.

So, achieving a 2020 target of 5% reduction will be possible, but what will be needed to achieve the much more challenging 2050 emission target? Economic modelling for 2050 by the Australian Treasury assumes 51% of electricity from renewable sources (21% geothermal, 18% wind, 5% solar, 4% hydro, 3% biomass), 31% from coal or gas with CCS and 18% from fossil fuels without CCS. To date, any competitive advantage that Australia has enjoyed in manufacturing, and in the economy more broadly, has rested largely on its ready access to fossil fuels and minerals and low cost energy. With a price placed on emissions, this advantage would be diminished and if a very high price was placed on carbon, the advantage could disappear. This is not to argue for or against a price being placed on carbon, but to point to the necessity to start exploring the most practical technology options in moving to a 2020–2050 clean energy economy, recognising that countries will seek new competitive advantages.

In the case of wind power for example, a country such as Australia has no obvious competitive advantage for deployment; many other parts of the world, for example the Atlantic coastal region of Western Europe, are windier. Similarly Australia has no manufacturing advantage when it comes to wind turbines, which are more likely to be manufactured in China or Europe, where there are established technologies, large domestic markets and a cheaper cost structure. The same applies to wave, tidal and hydro. That is not to say that development and deployment of these technologies should be abandoned by Australia. Where they can be cost effectively deployed, they should be. The argument is often advanced that even if some renewable technologies do not result in massive carbon abatement, one of the key reasons for investing in them is because of the many green jobs that will be created, but it is important to be realistic about this. What green jobs will actually be created? Will they replace highly paid jobs in, say, the mining or fossil fuel industries that may no longer be available in a carbon constrained world? In reality at the present time, many of the so called green jobs are to install solar panels or turbines that are made elsewhere, rather than the high tech jobs that people anticipate. There are obvious exceptions to this. But it is important to retain a sense of realism about the green economy, or we could find that we are chasing a green chimera rather than green jobs.

Many countries are currently heavily dependent on ready access to secure, cost effective and large scale base load electricity generation. To provide that same level of affordable energy certainty, based on their current maturity (as discussed in Chapter 5), the clean energy technologies which have the potential to provide this to a country such as Australia in the period 2020–2050 are solar (probably solar thermal) with storage, geothermal and CCS (with coal, gas, biomass). In the absence of nuclear, it could be appropriate for a national program to concentrate on these three long term clean energy options with clear milestones to ensure that they are realistically assessed and are not pursued irrespective of the cost.

Why are these three technologies highlighted for Australia? In the case of solar, Australia is one of the world's sunniest and driest developed countries, so there is a natural advantage there. But more than that, Australia has an excellent track record for research and innovation in solar energy. That has not necessarily translated into commercial success and therefore any focus on solar must also address the issue of turning an R&D success in solar power into a commercial success. Should the focus be on solar PV or solar thermal? It would be appropriate to reserve judgement on that for the moment and pursue both until, say, 2016, by which time a 'winner' may have emerged. In the case of geothermal, Australia has what appears to be some of the most suitable geology for hot fractured rock (HFR). Added to that, Australia is one of the global leaders in the earth sciences, so the knowledge base is there. Similarly CCS as a focus is justified on the basis of Australia's abundance of fossil fuels, the existence of sufficient CO_2 storage capacity for hundreds of years and a strong science base in chemical engineering and earth sciences, which has grown up around Australia's resource focus. There will be those who will disagree with the choices offered and even with the concept of picking winners to take us beyond 2020. But Australia cannot do everything in the clean energy field. It must seek to play to its obvious strengths in solar, geothermal and CCS and thereby maximise the opportunities for deep cuts in emission and the provision of stable base load power that these can provide. Other countries must similarly evaluate where their natural energy advantages lie and develop their own portfolio to exploit those advantages, sustainable through their own program for clean energy.

Achieving emissions reductions targets

How might such a clean energy development program work, focused on deployment of these new technologies from 2020 onwards? The target for each of them should be delivery of 500 MW of base load electricity by 2020. The vehicle for doing this should not be a competition between technologies and proponents but defined national projects to take each of these technologies forward and provide the basis for final investment decisions on their broader deployment by 2020. The projects should be defined by government and industry working closely with the research community. Once defined in some detail, tenders should be called so that the private sector is responsible for taking them forward and for their delivery at an agreed cost and to an agreed timetable.

Clearly this represents a significant departure from the strategy of using a combination of regulation, mandatory targets, flagships and a low price on carbon, but that strategy is unlikely to deliver the large scale clean energy that is needed when it is needed. The prospect of developing a long term successful mitigation strategy will be greatly enhanced if there is broad political support for these large scale measures aimed at 2020–2050. If efficient and cost effective energy storage systems can be developed by 2020, more abatement may come from wind and solar for 2020–2050, which is why the need for more research and development into energy storage has previously been stressed. If efficient and cost effective energy storage systems can't be achieved by 2020, then it is likely wind will have reached its ceiling by 2020 or earlier and from then on will provide less than 20% of the total mitigation effort. Similarly solar PV will only make a modest contribution by 2020; although solar PV will be important for isolated locations, in the absence of effective power storage it will not be a major contributor. Large scale solar thermal will probably not have become operational by 2020, with 2030–2040 perhaps a more realistic time frame. Geothermal (hot fractured rock – HFR) has the potential to be operational by 2020, although it is unlikely to be making a significant contribution to the electricity network before 2030 and the Australian Treasury modelling that assumes 21% geothermal by 2050 is very questionable at this time. There may be an earlier contribution from hot saline aquifers (HSA), but from 2030 on the main geothermal contribution is likely to be from HFR. By 2020–2030 there could be reconsideration of nuclear.

But perhaps the most significant feature of the period 2020–2030 is that by that time it will be very apparent that a great many countries will continue to have a high level of dependency on fossil fuels and as a consequence, CCS will need to be a critical component of the abatement portfolio. If large scale development of CCS is not taken forward by government and industry, there is no prospect of long term emission targets being attained. This would be such a severe setback to national (and global) abatement that it is vital to ensure that it does not happen. Therefore in the period to 2020 Government and industry will need to have taken forward CCS so that final investment decisions and commercial deployment can happen in a timely manner from 2020 onwards.

In some gas-rich countries, CCS applied to gas processing and gas-fired power generation may become progressively more important than CCS applied to coal, but 'low hanging fruit' such as CCS applied to urea or iron and steel production, may also be important. It is possible that CCS in coal-to-liquid and coal–to-hydrogen projects, coupled with electricity off-take, will be significant by 2025–2030. Australia has the opportunity to exert significant influence in the deployment of CCS beyond 2020 by being an earlier adopter of the technology. As the world's biggest exporter of coal it also has the chance to take on a globally important stewardship role in CCS application to coal fired electricity generation.

CCS in the clean energy mix

With this in mind, the rest of this discussion will focus on CCS and in particular on the steps needed for its wide scale deployment from 2020 onwards. At the present time CCS is not being deployed, or at least not being deployed at anything like the scale needed. There are two reasons for this: technical uncertainty and policy uncertainty. In fact these same issues are faced by all other large scale energy technologies, but the discussion here is restricted to CCS.

The claim is often made that CCS is 'too expensive'. What do we take as our yardstick of 'expensive'? If the yardstick is the current cost of electricity produced by a conventional coal-fired power station, then all clean energy is 'too expensive'. Therefore the issue becomes which is cheaper (or less expensive) in terms of the cost

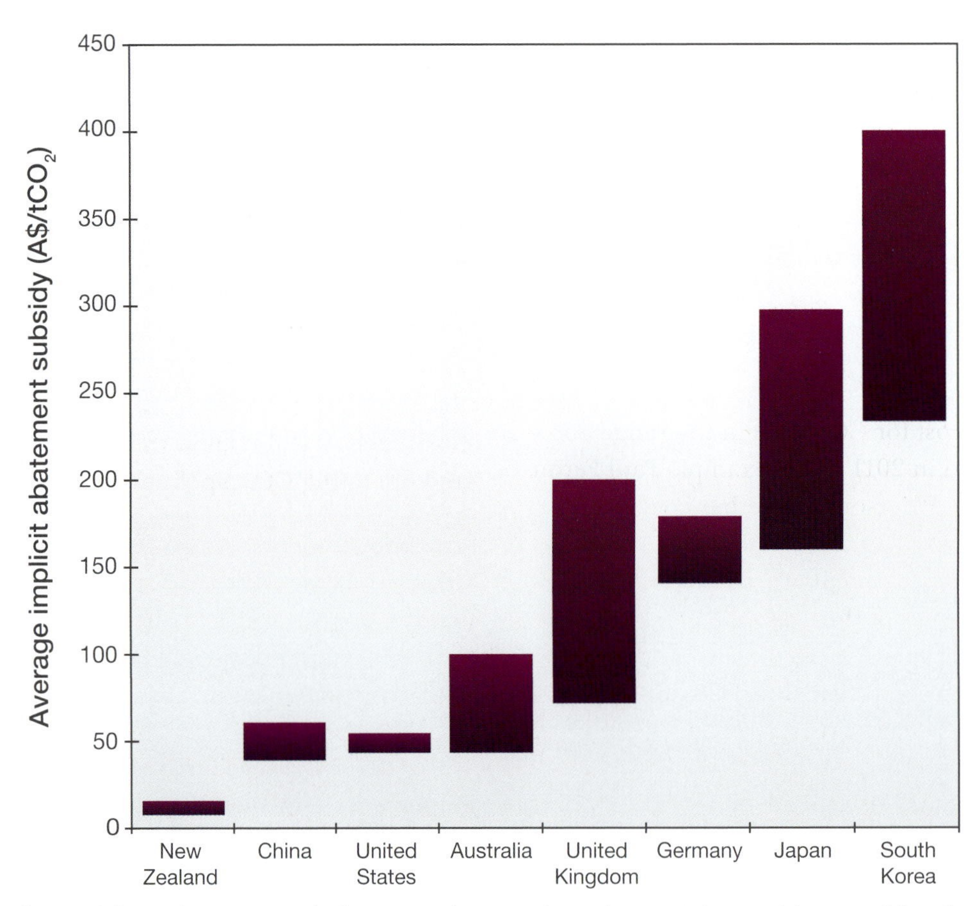

Figure 11.3 The variability in the extent to which countries have sought to abate emissions and the cost of that abatement is illustrated here. (Adapted from Productivity Commission 2011)

per tonne of CO_2 mitigated now, or in 10, 20 or 30 years time? As pointed out in Chapter 10, based on the levelised cost of electricity, CCS is about the same cost as some clean energy technologies and cheaper than others.

Wind and solar PV have been and continue to be installed in many countries whereas CCS is not; does this not then prove that CCS is more expensive than these other clean energy options? The answer is, of course, no. At the present time, in many countries as part of government policy, wind and solar power is heavily subsidised by the taxpayer and/or the consumer, through feed in-tariffs or other government measures. An extraordinarily insightful 2011 report by Australia's Productivity Commission, considered the cost of various clean energy schemes for 10 countries and revealed some staggering costs being incurred to support clean energy technologies (Figure 11.3). For example, Australian schemes used to support solar PV (based on the combined effects of the various funding components) ranged from $431 to $1043 per tonne of CO_2 avoided. Costs in some other countries were even higher. If this cost was to be used as the yard stick, then CCS would be far cheaper than solar PV.

In a 2010 report, the Electric Power Research Institute (quoted by the Productivity Commission) reported the levelised costs of electricity (LCOE) of various sources of electricity in Australia as:

- $78–91 per MWh for coal fired electricity (without CCS)
- $97 per MWh for combined cycle gas turbines (without CCS)
- $150–214 per MWh for wind
- $400–473 per MWh for medium-sized (5 megawatt) solar PV systems.

As pointed out earlier, it is difficult to determine the cost of CCS projects because they are so site dependent, but putting that aside, if the additional cost for CCS were in the range $55–65 proposed in 2011 by for example Paul Feron and Lincoln Paterson or whether coal plus CCS or gas plus CCS were 'around $100 a tonne' as suggested in Chapter 10, then CCS appears to be at the lower end of the cost range for wind power and less than half the cost of 5 MW solar PV. It is important to emphasise that this is not to say we should focus our efforts on CCS to the exclusion of other clean energy technologies. Rather, the point is that indicative costs at this time support the necessity of including CCS in the clean energy mix of many countries.

All large scale technologies including solar, geothermal and CCS, will come down in price as they become more widely deployed. The extent to which this will occur depends on the level of maturity of the technology. Despite this trend, costs projected to 2050 by Ross Garnaut and the Australian Energy Market Operator (AEMO) indicate the ongoing cost competitiveness of CCS. The Intergovernmental Panel on Climate Change (IPCC) Special Report concluded that inclusion of CCS in the portfolio of clean energy options would decrease the costs of mitigation by about one-third compared to the cost of leaving it out of the portfolio. There is no reason to change this assessment. We do not know precisely what the cost of building and operating a large fossil fuel-based power station with CCS will be, because one has yet to be built and operated. But nor have we built a large scale HFR or solar PV or solar thermal power station. The only large scale power systems where we know the construction and operating costs with a high level of confidence are hydro, nuclear and conventional thermal (coal, gas or biomass) power stations. Costs for all other large scale energy systems are speculative to varying degrees. We also know that the existing electricity distribution infrastructure will be compatible with CCS and that CCS does not raise energy security concerns, though obviously there are questions regarding the future cost of fossil fuels. We also know the cost of building and operating CO_2 pipelines, drilling wells and handling sub-surface operations. So we do have the costs of most of the components in a CCS system to a greater extent than for other large scale clean energy systems and from this we know there is no basis for claiming that CCS is 'too expensive'.

The second criticism levelled by the opponents of CCS is that it is 'unproven'. This has been addressed previously but it is worth revisiting it to point out that CCS is 'proven', as demonstrated by:

- the capture activities already underway at many power and industrial plants (although we need to increase their scale by at least one order of magnitude and maybe two)
- the tens of millions of tonnes of CO_2 that are transported by pipeline every year
- the demonstration and industrial scale storage activities, some involving millions of tonnes of CO_2 being injected each year, that have been underway for many years around the world.

CCS is not 'unproven'.

Any perception of CCS being 'risky' is perhaps because it is an unfamiliar technology and CO_2 is regarded as dangerous. But as long as humans have been on earth they have been living above areas where CO_2 is naturally trapped, have bathed in springs charged with CO_2, or have drunk CO_2 mineral waters. Engineered storage of natural gas is used in many parts of the world; transport and injection of tens of millions of tonnes of CO_2 for enhanced oil recovery has been underway for many years and many people already live in close proximity to CO_2 activities. There are many small scale CO_2 capture plants already deployed by the food industry and although capture plants on power stations will be much larger, they will be similar in scale to other industrial operations that are routinely carried out all over the world.

The concept of injecting millions of tonnes of compressed CO_2 into the subsurface for storage is unfamiliar to most people, which is perhaps why individuals and communities consider it 'risky'. An example of this is provided by the community of Barendrecht, in The Netherlands, which has lived above a large natural gas deposit for hundreds of years, but objected to living above that same structure (now depleted of its natural gas) when it was proposed to fill it with non-explosive/non-flammable CO_2. It was perceived as 'risky', yet we accept risk as an integral part of living. In 2010 in a Dutch town near Barendrecht, I was staggered to see a man on a bicycle with a large box in front of it, packed with eight smiling children, 4 or 5 years old, with not a bicycle helmet to be seen! On inquiry, I was told this is not uncommon as a means of transporting groups of young children in The Netherlands. In some countries, such behaviour would be deemed extremely dangerous to the children and the bicyclist would probably be arrested! So the concept of what is 'risky' varies between individuals, between countries and between cultures. The common thread to acceptability is being able to determine that the risk is relatively small and the benefit relatively large.

Perhaps a problem for CCS and the individuals or the community living in the vicinity of a proposed CO_2 storage site is that the benefit is the global benefit of decreased CO_2 in the atmosphere, whereas the perceived impact is on the individual or the community living in the immediate vicinity of the site. This same problem can of course arise for other clean energy technologies such as hydro schemes, wind farms or geothermal projects, so the problem is not unique to CCS. The solution to this apparent disconnect may be to reward the community (or all individuals who are directly or indirectly affected) in a more tangible way rather than relying on the 'feel good factor' of doing something for the global environment. The stronger the link the individual or the community sees between new technology and tangible benefits, the more likely the community will regard a perceived risk as acceptable. Of equal importance is to seek to ensure that individuals and communities accept that a technology (such as CCS) is effectively regulated, is subject to ongoing monitoring, and has a very low risk indeed of CO_2 leaking to the atmosphere. Any risk arising from CCS is lower than that arising from very many other activities that are accepted as part of living. Risk is no basis for leaving CCS out of the clean energy portfolio.

Given that CCS is not expensive compared to other options, not unproven and not risky, what part will it play in the long term energy portfolio? The International Energy Agency (IEA), the IPCC and many other studies conclude that we need all the energy options: in other words, all the wedges (Figure 11.2). The wedges will vary from country to country. Some countries have chosen not to have nuclear power. In other countries, nuclear power is likely to continue to be important. Some countries may choose to have a greater contribution from biomass. Each country will have their own mitigation profile and for some there may be little or no CCS in the mix, but a great many countries will include use of fossil fuels with CCS in the mix because they have access to abundant low cost fossil fuels and favourable geology for storage of CO_2.

It is also important, in pursuing the urgent need to decrease global CO_2 emissions, to not lose sight of the need to also address social equity. For example the World Bank is considering limiting financing of coal-fired power stations for developing countries. This followed criticism of the Bank in 2010 for making a $3.75 billion loan to South Africa to build a coal-fired power station. But over a billion people in the world lack access to electricity and many of them have health problems because of ongoing exposure to smoke from use of solid fuels in the home. Any measure by the World Bank aimed at decreasing global emissions that places an unfair burden on some of the world's poorest people, is inequitable and should have no place in a global strategy for decreasing emissions! Surely a better approach is for the World Bank to recognise that developing countries will seek to use their fossil fuel resources and assist them to use those resources in much cleaner and smarter ways, including through CCS.

An opinion commonly expressed by critics of CCS is that it will 'keep coal going'. There are several aspects to this. First, all the evidence is that coal and fossil fuels more generally will continue to be used with or without CCS (Figure 11.4). A vigorous debate on coal has been underway for some time in Australia because of the country's position as the world's largest exporter of coal. The view of some, is

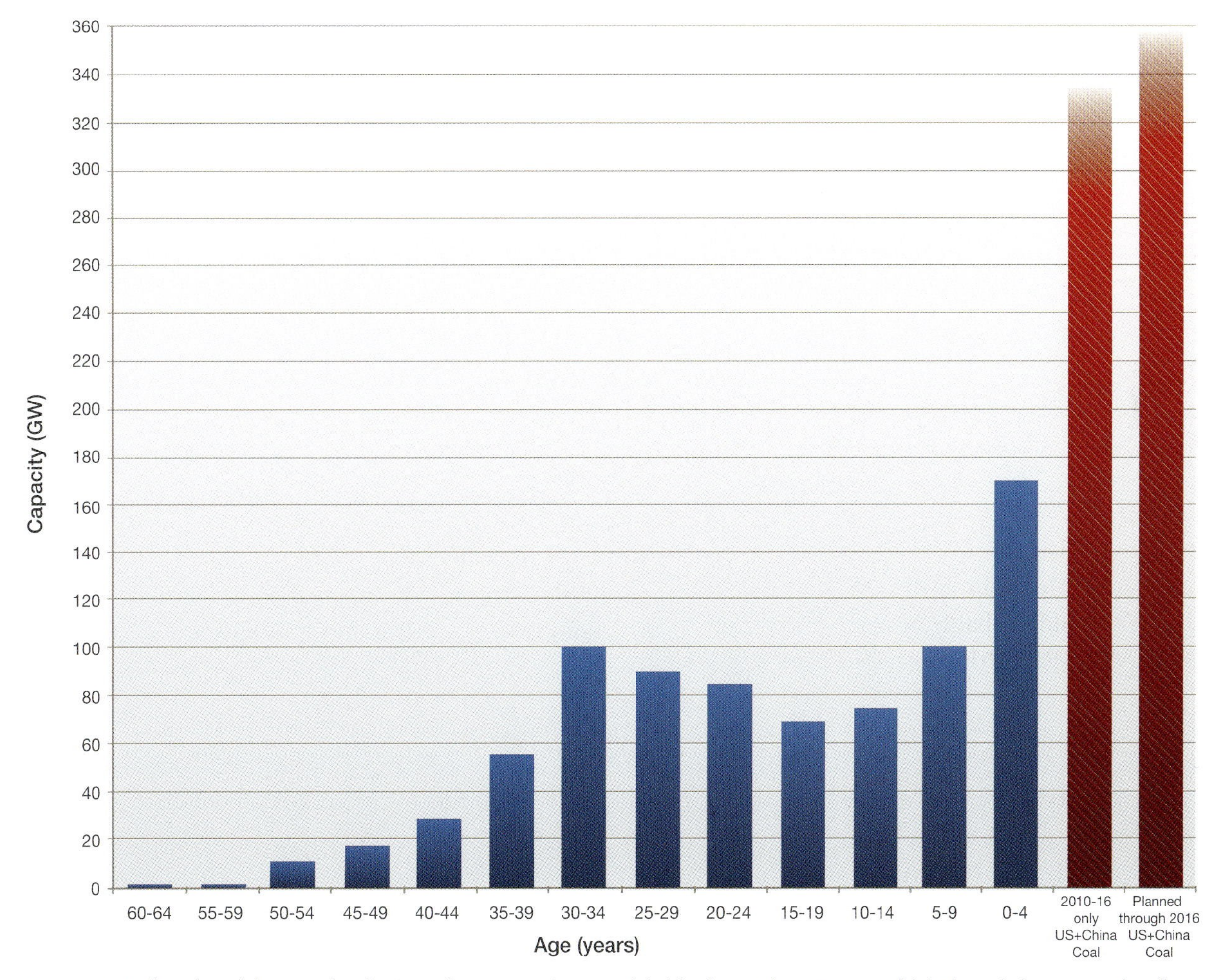

Figure 11.4 This plot of the age distribution of power stations worldwide shows the extent to which the existing generation fleet will continue to emit CO_2 for several decades to come unless CCS is deployed. (Data source: Platts 2008; Shuster 2011)

SOME PROPOSED PROJECTS

ILLINOIS BASIN – DECATUR PROJECT

The Decatur project is a dual phase capture and storage project being conducted by the Midwest Regional Carbon Sequestration Partnership. In the first phase, 1 million tonnes of CO_2 will be captured from the Archer Daniels Midland ethanol plant and injected into the Mt Simon Sandstone at a depth of approximately 2000 metres over a period of 3 years at a rate of 300 000 tonnes per year. For this first phase a well has been drilled and injection is on course to begin in late 2011. An extensive program of site characterisation has been undertaken and plans are in place for a very extensive program of monitoring including seismic monitoring, air, soil and groundwater monitoring and satellite monitoring. For the second phase commencing 2013–2014, a ramped-up capture and storage project will see an additional 2.5 million tonnes of CO_2 stored at a rate of 1 Mtpa.

GREENGEN PROJECT

Greengen is proposed as part of a 650 MW power plant project, located in Tianjin China, that involves coal gasification (integrated gasification combined cycle – IGCC) to produce synfuel and high purity CO_2 that would then be used for enhanced oil recovery. Costing $1 billion, the project will be implemented in stages, beginning with a 250 MW IGCC plant (already under construction) followed by the second 650 MW stage coming online in 2016.

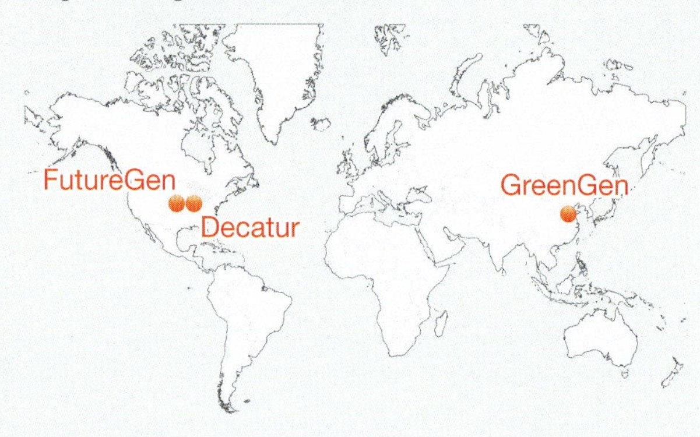

FUTUREGEN PROJECT

The United States CCS flagship project, FutureGen, is proposed as a 200 MW oxyfuel coal-fired power plant that will implement the entire CCS process, capturing and storing 1.3 million tonnes per year when fully operational. It is anticipated that storage will be in the Mount Simon Sandstone at a depth of about 2 km. The project is currently scheduled to begin operation in 2015 with a storage site in Meredosia Illinois. The project is funded by a consortium of seven international companies and the United States Department of Energy.

that those exports should cease. However, although Australia is the world's largest coal exporter, it only produces 6% of the world's coal and it would be relatively easy to replace Australia's coal exports with coal from Indonesia or South America or South Africa, or even more likely, by increasing coal production in coal-rich countries such as China, already the world's largest coal producer. It is also important to be realistic about the fact that coal exports are currently a very important source of income to a number of countries (for example $55 billion a year for Australia) and therefore there is an immediate practical issue of how to replace that income. One view is that it could and should be replaced by renewable technology industries. By way of example, an income of $55 billion could in theory be achieved by manufacturing of the order of 10 000 5-MW wind turbines a year. This is more wind turbines than are built annually in the entire world at present. How realistic is this for

Australia, or South Africa or Indonesia, given that none of these major coal exporting countries manufacture wind turbines at the present time?

An approach based on 'carbon stewardship', which links sales of coal, or liquefied natural gas with facilitation of and support for CCS, would seem to be a more practical way forward in a carbon-constrained world. It is also a more ethical approach than just selling coal and then 'washing your hands' of the consequences. It could also provide some leverage to abatement in the same way that for example Australia only sells uranium to countries that have signed the Nuclear Non-Proliferation Treaty. There is obviously a limit to the leverage that a resource-exporting country can (or should) exert on an importing country but an appropriate course of action, in the absence of an international agreement to limit emissions, is to assist in the development and deployment of the one technology that can enable coal to be used in cleaner and smarter ways. With or without CCS, countries such as India or China will continue to use their existing coal-fired power stations for decades to come. Surely it is better to do all we can to ensure that use continues with CCS?

Part of the problem with the coal-CCS debate perhaps rests with an industry that sought to make the term 'clean coal' synonymous with CCS. This left the coal industry open to some ridicule; magnified by further attempts to use terms such as 'green coal', 'smart coal' or 'newgen coal'. As a colleague in the coal industry has put it, 'The industry does not seem to recognise that no matter what adjective is put in front of the word coal, the problem is still the word 'coal'! So, an attempt to rebadge coal and link it strongly to CCS has led to CCS being incorrectly seen by some as solely about coal. The reality is, as pointed out many times before, is that CCS is applicable to any large scale emission of CO_2, whether coal, gas or biomass-based power, or industrial processes.

In the period 2020–2050, gas will increasingly replace coal in many developed countries and perhaps also in China. The IEA's gas scenario for 2035, set out in a 2011 special report on gas, suggests a 62% rise in the global use of gas compared to 2010 and a rise in coal use of only 11% over the same period. Is there enough gas to meet this demand? The IEA believes there is, with global recoverable reserves of conventional gas equivalent to 120 years at current rates of consumption. While there is no question that gas will provide part of the solution to decreasing CO_2 emissions in the next few years, it will potentially become part of the problem in that it will constitute an ever-larger source of CO_2 emissions.

The emissions from the gas industry are from three sources. The produced gas may be high in CO_2 which needs to be removed so that the gas can be sold into the gas network or sent to the LNG plant. The second source is the LNG plant itself, whether directly or indirectly through its use of power. The third is CO_2 emitted through gas-fired power generation. Fortunately gas is usually produced and processed in areas where there are excellent prospects for CO_2 storage. In addition, the gas companies have more expertise in CCS-type technology than the coal companies and will more readily take up the technology. There is the further advantage that as gas fields are depleted they provide additional storage space for gas. Therefore, depending of course on the price of gas and the stability of that price, the gas industry could be a significant winner in a carbon-constrained world, but to attain that winning position, it will have to embrace CCS. However, gas (through LNG or long distance pipelines) is an internationally traded commodity and the application of CCS to a project will potentially make it uncompetitive compared to a project that does not abate its CO_2 emissions. Obviously an international agreement would help to remove this problem, but even in the absence of such an agreement, CCS is likely to be increasingly applied to gas at all stages.

CCS is applicable to biomass or cogen (coal plus biomass)-fired power stations and the IEA sees this as important during the 2020–2050 period. Bob Williams of Princeton has proposed that a way of more rapidly deploying CCS is through gasification of coal plus biomass to produce synfuels (CO_2 is geologically stored), liquid transportation fuels (a high value product) and emission-free electricity, with the balance between liquid fuels and electricity varying with power needs and economics. This creative proposal warrants further consideration as it does provide a basis for decreasing the normally high carbon footprint of a coal-to-liquids project and may provide one of the building blocks for the future hydrogen economy. There is a need to explore the long term opportunities for hybrid clean energy technologies such as:

- the coordinated use of geological space for geothermal and CCS, or CCS and compressed air storage
- the application of solar power to solvent regeneration in capture plants
- the linking of algal fixation of CO_2 and CCS
- the use of high CO_2 building materials and CCS.

These are not necessarily large scale opportunities but they are nonetheless opportunities to be explored within a clean energy portfolio.

The policy settings

There are no technology 'show stoppers' for CCS, so what are the policy settings that will take CCS forward? Currently initiatives around clean energy and 2020 or 2050 emission targets are being framed within a politically charged atmosphere in many countries, complicated by economic instability and the uncertainty of the climate decision-making process. In the European Union (EU) climate change policy seemed settled with defined targets and emission trading, but the sudden unwillingness of Poland in 2011 to agree to EU emission targets (because of Poland's high level of dependency on fossil fuels) has resulted in heightened uncertainty and a re-evaluation of EU climate policies. In the United States, an unwilling Senate has placed Federal climate measures such as a cap and trade (ETS) scheme in limbo with little prospect of moving from there. There are nonetheless State schemes as well as existing federal regulations that may provide the United States Environmental Protection Authority with levers to decrease greenhouse emissions. In Canada there is no clear agreement on climate measures between the Federal Government and the Provinces, but some Provinces are making their own way with new greenhouse (including CCS) legislation being developed in Alberta, and Saskatchewan is considering a carbon tax.

Globally, much of the policy debate has been about the potential role of a carbon price or tax. Some of the key documents contributing to this debate have been a series of reports by Ross Garnaut, the Productivity Commission and recent carbon legislation. These should be referred to for details of various carbon policies and the factors that influenced those policies, but let us view some of these policy initiatives and pricing carbon through a clean energy technology prism.

For some years governments have pursued a range of climate initiatives, including support for energy efficiency, various forms of support for clean energy R&D and a renewable energy target (RET) backed by feed-in tariffs to subsidise the cost of renewable energy technologies (Figure

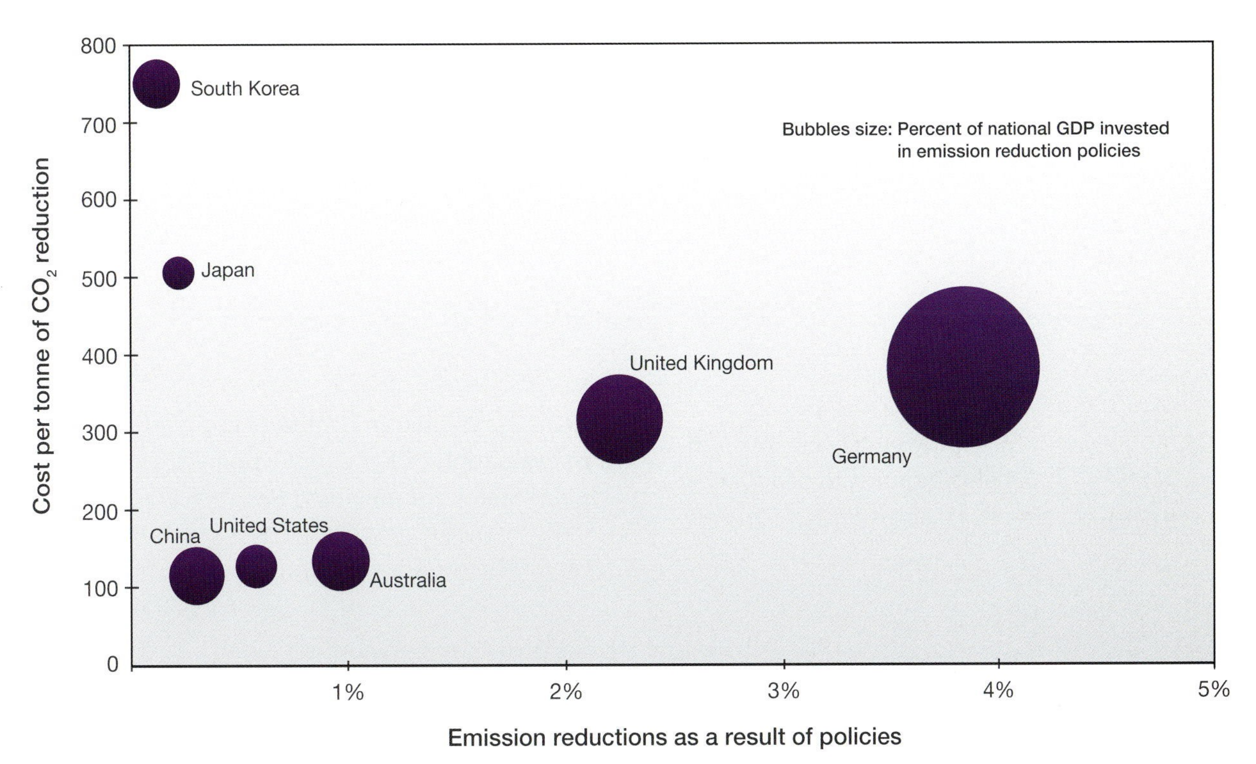

Figure 11.5 The implicit abatement subsidy (in $ per tonne CO_2) varies enormously from country to country. As a proportion of total electricity sector emissions the abatement subsidy for Australia and the United States are about the same; China, Japan and South Korea are lower and the UK and Germany are high. (Adapted from Productivity Commission 2011)

11.5; Table 11.1). A number of countries have also signed the Kyoto Protocol. Many countries have also announced schemes to support development and demonstration of CCS.

In Australia, a $3 billion Flagship Program was announced to support demonstration of large scale clean energy technologies with emphasis on large scale solar and CCS. Separately, more modest support was offered for geothermal and wave energy. On the personal initiative of the then Prime Minister, the Australian Government also established the Global CCS Institute, with an initial commitment of $500 million (now somewhat less) over five years. Following Federal elections (influenced in no small measure by lack of agreement on climate policy) and long and complex negotiations, the minority Labor government developed a package of measures including:

- a carbon tax
- emission targets for 2020 and 2050
- a number of clean energy initiatives, including the establishment of a $10 billion Clean Energy Finance Corporation and support for a carbon farming initiative.

Complementary measures were designed to address problems posed to potentially disadvantaged sections of the population and trade exposed industries, including coal mining and iron and steel. Along with this there are or will be programs supporting green loans, solar cities, smart grids, indigenous carbon farming and the development of clean energy skills. In other words, like many other countries, there is an extraordinarily complex array of initiatives; far too many to consider here and therefore comment will be restricted to those that directly relate to the uptake of clean energy technologies.

Table 11.1 Taxonomy of emissions-reduction policies

Category / Policy
Explicit carbon prices
Emissions trading scheme – cap-and-trade
Emissions trading scheme – baseline and credit
Emissions trading scheme – voluntary
Carbon tax
Subsidies and (other) taxes
Capital subsidy
Feed-in tariff
Tax rebate or credit
Tax exemption
Preferential, low-interest or guaranteed loan
Other subsidy or grant
Fuel or resource tax
Other tax
Direct government expenditure
Government procurement – general
Government procurement – carbon offsets
Government investment – infrastructure
Government investment – environment
Regulatory instruments
Renewable energy target
Renewable energy certificate scheme (REC)
Electricity supply or pricing regulation
Technology standard
Fuel content mandate
Energy efficiency regulation
Mandatory assessment, audit or investment
Synthetic greenhouse gas regulation
Urban or transport planning regulation
Other regulation
Support for research and development (R&D)
R&D – general and demonstration
R&D – deployment and diffusion
Other
Information provision or benchmarking
Labelling scheme
Advertising or educational scheme
Broad target or intergovernmental framework
Voluntary agreement

Source: Productivity Commission 2011

The use of feed-in tariffs and renewable energy targets in many other countries has been notable for two things (Figure 11.6). First it has resulted in a significant uptake of wind power and small scale roof solar PV and second, as pointed out by the Productivity Commission, it has resulted in extraordinarily expensive CO_2 mitigation. A number of renewable programs have been terminated as governments found them to be too costly. There is clearly a need to have full costings for all clean energy technologies, including, among other costs, extra grid costs, power back up costs, remediation costs, and energy outage costs, with the aim of having a level playing field for technology uptake.

If the aim of a mandatory scheme or feed-in tariff is to encourage the development of particular technologies and industries, then this should be the stated aim of the scheme. But if the aim is to meet a particular mitigation target, then it is questionable that there should be explicit or implicit favouring of one technology or one industry over another. As pointed out by the Productivity Commission, the national mitigation strategy has not been well served by some of the existing programs in terms of value for money and its analysis suggests this is also the case for many countries.

What about the effectiveness of support for clean energy R&D, an issue which was not addressed by the Productivity Commission but which will be critical to the achievement of 2050 targets? Let me provide the perspective of a Chief Executive of a research centre with an international perspective. Not surprisingly perhaps, I do think that support for R&D has provided excellent value to the taxpayer across the spectrum of clean energy activities and believe that this could also be said of other countries. CCS research (and clean energy research more generally) has received additional funding in recent years, but in Australia there is a concern that the new funding has been accompanied by the setting up of new organisations, each with their own costly management infrastructure and their own aspirations. The end result can be 'less bang for the buck' but even more importantly it can mean that research teams may not have critical mass. The counter argument is that competition can be a significant driver for taking forward innovation, and this is an argument that has merit.

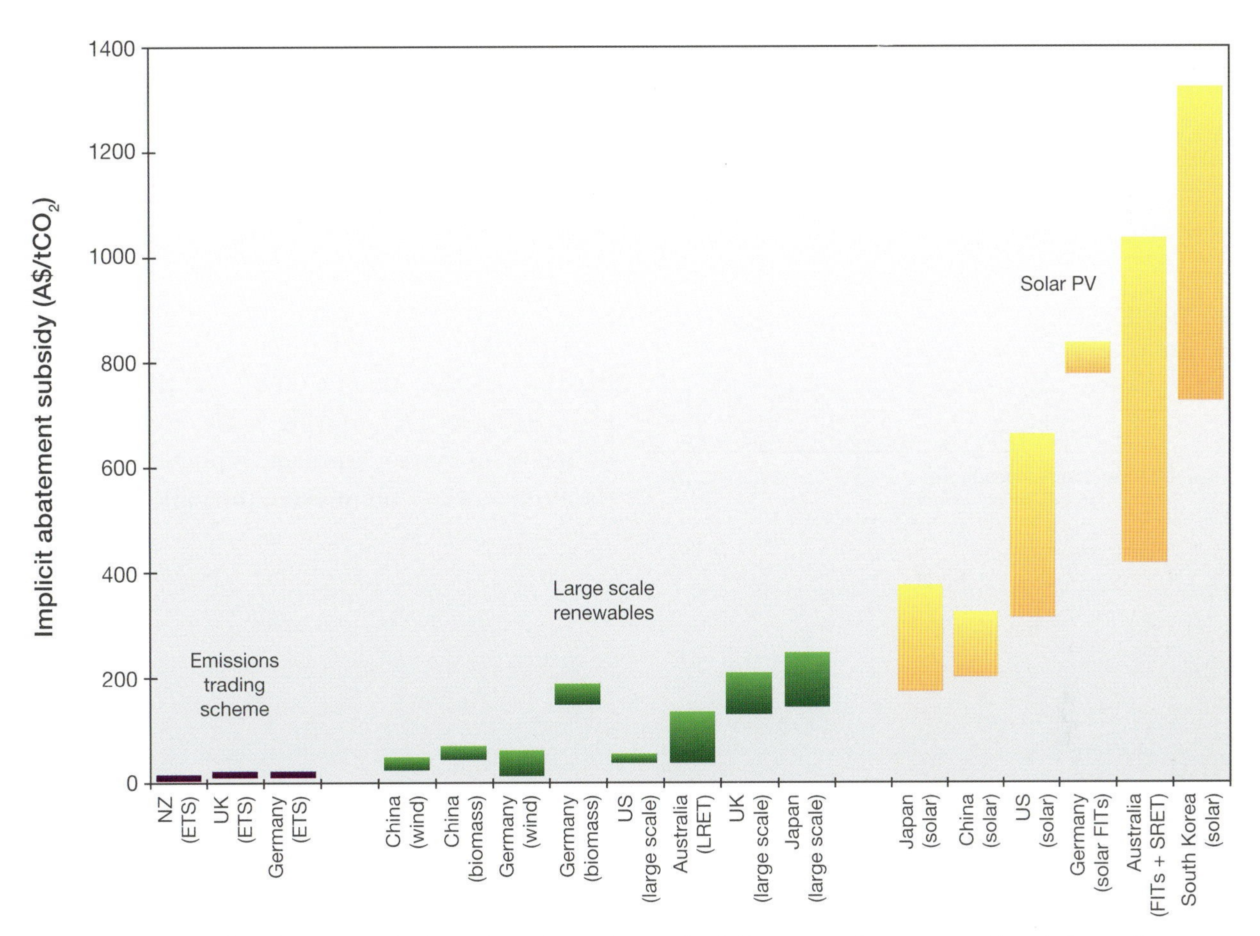

Figure 11.6 The implicit abatement subsidies achieved by various policy initiatives are shown here, with emission trading schemes showing the lowest cost of abatement and feed in tariffs showing the highest (Adapted from Productivity Commission 2011)

However, it is necessary to also bear the magnitude of the task in mind. Developing and deploying cost effective clean energy technologies at the scale required is a massive undertaking – perhaps equivalent to the Manhattan Project or the Apollo Program? In both these examples, a national program was initiated based not on competition but on targeted closely co-ordinated and well funded research collaboration. Competition in these programs came in only to ensure the cheapest and best equipment, nuts and bolts, flanges and gauges et cetera. So it should be with national clean energy research generally, including CCS.

What is needed to deploy CCS? Australia's CCS Flagship Program was announced as a competitive process in May 2009, initially with funding of $1.8 billion, subsequently cut back to $1.5 billion. This was an important start towards getting CCS underway at scale, but unfortunately after two years there has been only modest progress. Some proposals failed to materialise because the quantum of the funding needed was not there from government or industry; others because the expectations of the project proponents were unrealistic. When billions of dollars are involved, it is important that there is not a mismatch between the approach taken to risk by government and that taken by industry. Industry will often spend a great deal of money before deciding that a project is or is not viable; a decision not to proceed is seen by industry not as a 'failure' but as part of the normal commercial process before a final investment decision (FID) is made. However, for a government to spend large amounts of public money and then decide the project is not viable can be seen in the political arena as a failure. As a consequence, governments tend to be more risk averse than companies (and most taxpayers probably prefer it that way) which will mean that governments are slower to make financial decisions.

An issue facing large scale demonstrations, in various countries, is whether having a competition amongst CCS projects is really the best way forward. The UK found that competition was not the best way of taking forward a CCS project. Similarly the Australian experience with a competitive approach has, to date, not been encouraging in that it has been a slow process and has tended to discourage cooperation. Conversely the largely non competitive CCS Regional Partnerships Program in the United States has been very successful. Again using the Apollo or Manhattan analogy, a way forward would be for the government and key stakeholders to collectively decide on the most significant large scale national CCS project (or projects) that could be undertaken and then for industry to work with Government to adequately fund and manage that project or projects. The Global CCS Institute was set up by Australia expressly to accelerate the global up-take of CCS. It has commissioned a number of useful reports on the state of CCS internationally and has provided financial assistance to some projects. But the model was based on the assumption that international funding would flow to the Institute and overall this has not happened. The Global CCS Institute model could be enhanced, in collaboration with stakeholders, and used as a vehicle for developing the 'CCS Apollo model' thereby providing more efficient use of funds set aside for CCS.

The G8 summit (an annual summit of the governments of eight major economies: France, Germany, Italy, Japan, the United Kingdom, the United States, Canada and Russia) aim to have 20 large integrated CCS projects by 2020 is technically feasible, but will not be achieved at the current rate of deployment. The IEA aspiration of 3400 large CCS projects world-wide by 2050 will be impossible to attain unless we start to get things underway now. We need to have front end engineering design (FEED) underway now or in the near future and ensure that some final investment decisions (FID) are being made by 2014–15, preferably sooner, for 20 projects (if the G8 2020 target is to be met), and even earlier to meet the IEA 2050 target.

Despite this, there does not always seem to be the necessary sense of urgency from governments or industry. Indeed the only real sense of urgency arises from scientists and engineers and from some of the concerned NGOs such as the National Resources Defence Council (NRDC) in North America and Bellona in Europe who recognise the magnitude of the task. To put it bluntly, we do not have the luxury of time and rhetoric when it comes to deploying CCS. Given the likelihood of ongoing use of fossil fuels, failure to deploy CCS amounts to continuing with 'business as usual' and will result in atmospheric concentration of CO_2 far in excess of 450 parts per million by 2050.

CCS is currently not moving ahead in many countries, despite the need and the potential benefits and despite the fact that governments around the world have announced of the order of $30–35 billion in support for CCS projects. Unfortunately the rhetoric of announcements does not match the reality of inaction and, for the present, the reality is that probably only 1% of the $35 billion announced has actually been spent and only 5% (or less) actually committed. Overall, governments are not spending the money that needs to be spent to take CCS forward. However, it is not just governments but also industry which is not investing adequately in the large scale demonstration and deployment of CCS. Again the rhetoric of sections of industry regarding available funding for CCS does not seem to accord with the amount of money actually spent. By way of example, in Australia the Coal Association commendably announced a world-first $1 billion fund for CCS in 2005 but regrettably six years on, only $140 million of that billion had been spent. There is an urgent need for industry and government to invest adequately in CCS projects now.

The impact of pricing carbon on clean energy technology uptake

Will putting a price on carbon drive deployment of clean energy technologies? For the most part, the answer is probably no, unless the price is far higher than that being proposed by most countries. A low price can be useful for getting systems in place for the time when a price on carbon will be sufficiently high to drive the necessary technology change. When is that likely to be? No time soon in most countries. That being the case, if a low price is about symbolism or getting systems in place, would it matter if the price on carbon were $10 a tonne rather than, say, $20 or $25 a tonne? From a technical perspective it would not matter as any price in this sort of range is far too low to drive technology change. What would matter, even for a low carbon price, would be if the proceeds from a carbon tax were focused (as they should be) on financially supporting and developing the clean energy technologies that will enable us to reach agreed targets. If this were to be the driver, then the price (or tax) on carbon should reflect very directly the funding required to develop and deploy that clean energy in a timely fashion.

Again returning to the example of Australia, the path proposed for the carbon tax appears to be targeted at a broader tax agenda and what could perhaps be described as social engineering rather than at clean energy engineering. The Australian government has increased funding to improve energy efficiency, enhance soil carbon and take clean energy forward, all positive steps in terms of Australia's response to climate change, but a clear link between the processes of the carbon tax and the investment in clean energy is not explicit.

Equally concerning, in some countries CCS has been put to one side, not included in other 'clean energy' measures. CCS is excluded from the terms of Australia's $10 billion Clean Energy Finance Corporation. There will be some support for CCS through existing measures, but the potential public funding for CCS will be modest (in total around $2 billion) compared to clean energy funding overall, which, with other forms of support, amounts to tens of billions of dollars. It is argued that industry – particularly the coal industry, should provide the funding for CCS. Industry should be providing more funding and this is a valid view, but taking CCS forward cannot be left to industry; it must be a joint endeavour with Government, and it must be adequately funded, at a level comparable to the magnitude of support for renewable energy

Beyond the issue of funding, there is a second and equally important issue, namely that this apparent separation of energy options into 'good' clean energy (usually taken to be renewable energy) and 'bad' clean energy (usually taken to be CCS and nuclear) runs counter to the need to take a holistic approach to energy and CO_2 mitigation. As pointed out time and again in this book, we do not know with certainty what the energy portfolio will look like by 2020 or 2050, but it is highly likely that it will include extensive use of fossil fuels for much, probably all, of that time, and the only way to use fossil fuels and address the issue of their CO_2 emissions is through CCS. The problem posed by the polarisation of clean energy into 'good' and 'bad' is likely to be compounded in government if renewable energy is dealt with by one government department and CCS by another. No doubt the two departments will aim to work together, through interdepartmental committees and so on, but experience in most countries shows that this seldom works well and sometimes does not work at all. In Britain, for example, a single Department takes a comprehensive overview of all the clean energy options. Balkanisation of clean energy policies and programs is the last thing required at this critical time, if sensible climate and energy policies are to be developed.

Policy makers with an unshakeable faith in the market place, may claim that all such problems will be resolved by putting a price on carbon and a key policy initiative of many governments in recent years has been to put a price on carbon using one or other of the available fiscal measures. In Australia, there will be a carbon tax of $23 commencing July 2012 for 3–5 years, rising by increments over that period before moving to a full emissions trading scheme (ETS). Initially the carbon tax will only impact on the top 500 emitters. Other countries have similar schemes in place or proposed, with a price on carbon usually lying in the range $10–30.

Will this impact positively on the uptake of clean energy and particularly CCS? Putting a price on carbon can cost effectively decrease emissions of CO_2 by encouraging behavioural changes and the report of the Productivity Commission demonstrates this. However, international experience to date suggests that initially at least, an ETS will only encourage 'easy' transitions such as greater energy efficiency and perhaps some moves from coal-fired to gas-fired power. It may be seen as a way to drive the transition to wind or solar but the reality is that everywhere, wind and solar uptake has been driven by feed-in tariffs and mandatory renewable energy targets, not by a price on carbon. To deliver most clean energy technologies the price on carbon would need to be very much higher than most countries are prepared to contemplate at present. To propose a price high enough to drive technology change (in excess of $50 a tonne of CO_2) impacting across an entire economy, is not realistic at this time.

To bring about change, it is important there is a very clear link between a price on carbon and the development and deployment of clean energy technologies. This is a point strongly echoed by Garnaut, but governments appear to be less eager to take it up. In most countries, Treasuries will argue against hypothecation of tax revenue to address a particular issue, preferring to have the money in a big bucket from which they can apportion funds as they see fit! The issue of pricing carbon is highly politicised and it would be unrealistic to expect politics and the horse trading that goes along with it, to be taken out of consideration. Nonetheless from a clean energy perspective, it is a concern if a carbon tax is used for addressing issues of social equity or tax reform. Such issues are of course important but it is questionable that using measures supposedly aimed at mitigating CO_2 emissions, is the right way to deal with them.

Does it matter to clean energy technology uptake whether a price is set on carbon through a tax or an ETS? It might be easier to direct a portion of carbon tax revenue towards clean energy deployment than would be the case with an ETS, but no doubt this could be addressed. Conversely would a significant portion of the revenue from a carbon tax go towards its regulation and to supporting an army of public servants, with little actually reaching the technology?

The concern with an ETS approach is exemplified by the companies most strongly in favour of it, such as banks, accounting companies, lawyers, and Google! In other words companies that are not directly responsible for much in the way of emissions, yet see boundless opportunities for making large amounts of money from an ETS. What is there to prevent creative finance houses developing increasingly exotic carbon derivatives and what are then the longer term prospects of it all crashing in a heap? Something as tangible as housing and mortgages provided the basis for complex financial instruments that ultimately proved to be worthless and brought about the GFC. How much scope is there for developing even more exotic and ultimately even more complex financial instruments based on carbon credits derived from say a central African project in a poorly surveyed forest area that few people have seen and which has received uncertain and perhaps quite unreliable carbon auditing? If an ETS is to work, it requires a firm, transparent and effective regulatory regime that can be enforced internationally, and this is certainly not possible in many countries at this time.

There is a fundamental problem with leaving climate policy and the uptake of clean energy technology to the market. A market works fine when there is a clear choice: Should I buy this television which is cheaper or that television which has more options? Should I buy this big house in a downmarket suburb or this small house in an upmarket suburb? But what is on offer at present in the case of clean energy technologies? There is this technology, which will work but which won't be ready for large scale deployment until about 2025, and we don't know exactly how much it will cost. Or we have this other technology which needs more research and which might (or might not) be better, but it too won't be ready for a number of years and we won't know exactly how much it will cost for another five years? A market will work now if the choices are there now. A carbon market and a price on carbon alone will not progress longer term large scale clean energy technologies of the type (solar, geothermal or CCS) that we will need to make deep cuts in emissions between 2020 and 2050. Complementary measures will also be needed to achieve this.

There needs to be a clear and strong policy directly linking a carbon price to clean energy development, otherwise there is no real carbon policy, just another speculative market. Those with greater faith in the market will argue that the market can factor in the risks arising from technology uncertainty and that any speculative behaviour can be avoided through appropriate regulation. The same claims were made before the Global Financial Crisis! Just as there is no single technology to provide the answer to climate change, so there is no single financial instrument that will provide the answer to climate change. Financial instruments and a carbon market may ultimately provide part of the mechanism to accelerate the deployment of CCS and other clean energy technologies, but not for a number of years.

If a price on carbon is only part of the answer, what are the other measures? Governments might include subsidies, direct government investment and regulatory instruments. Some of these are already in place and will drive technology uptake to 2020 but there are few that will take us into the critical 2020–2050 period. Governments do not have unlimited financial resources and consequently there will be some hard choices to be made within the abatement portfolio, which is why for the period beyond 2020, it is proposed that large scale solar, geothermal and CCS be the clean energy focus and that also we plan for the likelihood of ongoing use of fossil fuels.

In most countries politics, rather than public policy have dominated much of the climate debate with only limited recognition of the fact that no amount of legislation, or fiscal measures, carbon trading or economic modelling will result in any abatement of emissions whatsoever, unless effective clean energy technologies are available, when they are needed. This point appears to have been lost in the complexity of the debate. Therefore it is useful to conclude by summarising the key points relating to clean energy technologies.

Conclusions

Meeting the emission target for 2020 will be through existing mature relatively small scale technologies and their deployment will be driven largely by regulation, not by a price on carbon. Wind and solar PV will be prominent in the mix and switching from coal to gas will be a feature of some countries. Some of the abatement measures to 2020 will prove to be expensive and inadequate to meet national emissions targets. The application of a price on carbon will have no significant impact on meeting 2020 emission targets other than through the device of buying international carbon credits. This will have the benefit of helping some developing countries but will do little for the development and deployment of large scale clean energy technologies and may even delay their deployment in some countries. A carbon price should be set primarily to provide the level of financial support necessary to get critical clean energy technologies developed and deployed. Politically and also socially, it may be necessary to compensate disadvantaged groups to the extent of actual price rises resulting from a carbon price. The deployment of clean energy will result in significantly higher costs for electricity and it is unrealistic to pretend otherwise, but a direct link between that higher cost and technology deployment may be preferable to a carbon price or tax that bears little or no relationship to clean technology uptake. Preferably, the carbon price should be very directly related to clean energy technology development and deployment.

The period to 2020 should be the time when key large scale clean energy technologies, particularly solar, geothermal and CCS, are developed and tested at the scale of delivering 500 MW of base load power. Governments and industry will need to stop announcing how much money they will spend and actually start spending it on projects. It will also require a rethink of the competitive approach currently in vogue for CCS. An 'Apollo Project' for large scale and reliable clean energy, with CCS a key component of the portfolio, should be set up. This also provides time for rethinking the role of nuclear in providing zero emission power beyond 2020.

The period 2020–2050 will be critical for taking measure to ensure that global atmospheric concentrations of CO_2 stay below 'dangerous levels', taken here to be 450 ppm. None of the climate policy measures currently in place will be adequate to the task. Part of the problem revolves around unrealistic expectations regarding existing renewable technologies, expectations that are driven more by dogma and hyperbole than a realistic assessment of what is likely. Improved energy storage could be a 'game changer' and there should be greater R&D investment in this. However, fossil fuels will continue to be a major feature of the energy scene until at least 2050 and therefore we must plan for this, through the application of CCS.

Beyond 2020, CCS will be an essential technology for coal, gas and biomass-fired electricity generation and for a range of industrial processes (Figure 11.7). It also has the potential to be part of new hybrid technologies. It is therefore vital that it is included in the clean energy portfolio. Measures that separate CCS from other clean energy technologies may be politically expedient but are technologically inappropriate and will not serve to take the most appropriate clean energy portfolio forward. The ongoing use of fossil fuels will pose a technological challenge and the only large scale technology that can deal with the related emissions is CCS. Contrary to some claims, CCS is proven, cost-competitive and safe and it is a critical component of the long term clean energy portfolio. But just as there is no single technology solution, so there is no single policy or fiscal or economic solution. A price on carbon will do little unless it is accompanied by

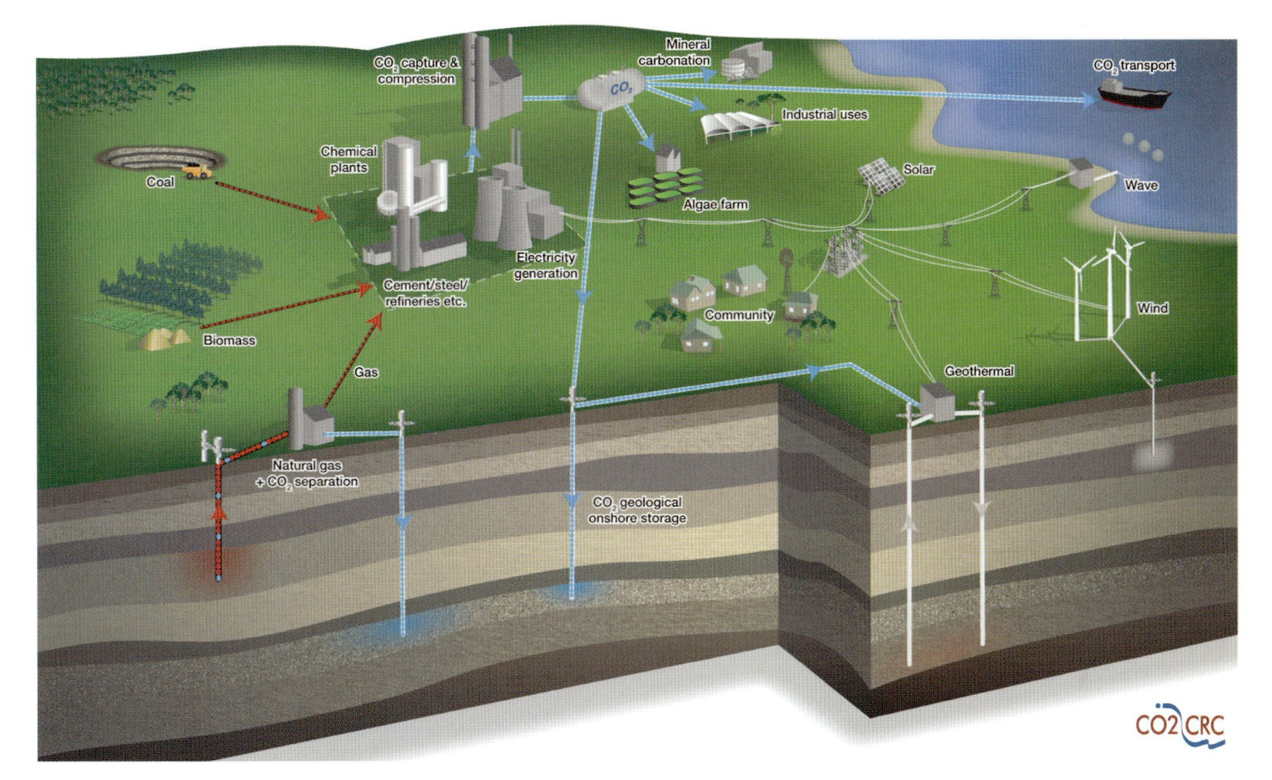

Figure 11.7 An energy systems approach must be taken to the deployment of clean energy beyond 2020. Meeting long term emission targets will require scientific creativity, a willingness to cross conventional energy boundaries to develop hybrid technologies and a realistic level of funding. In this representation of energy sources and systems that could be in place by 2020–2030, it is assumed coal will still be a significant energy source for a great many countries but that gas will be a major, and perhaps the major fossil energy source by that time. In addition there will be greater use of biomass by itself but more likely in co-generation systems. CCS will be applied to all these fuel sources and systems. Polygeneration of chemicals and production of liquid fuels and hydrogen will be a feature of some of the energy – chemical installations. There may be some algal sequestration and other beneficial uses of CO_2: while these will provide some cost offsets, they will make only a modest contribution to emission reduction. Geothermal power (hot fractured rocks) will be making a contribution to base load power generation and hybrid CCS-geothermal technology (using CO_2 as a thermal transfer fluid) may be a new development with significant potential. CO_2-EOR (enhanced oil recovery) will be applied in suitable geological settings, but existing transport fluids will be progressively replaced by electric transport with some use of hydrogen beyond 2020–2030. Wind power will have reached its peak by 2020 or sooner, in the absence of power storage. However in this representation, it is suggested that compressed air storage in suitable porous-permeable rocks may be making some inroads. The contribution from solar is shown as modest in the absence of storage, but this would change significantly by 2020–2030 if there are major breakthroughs in storage technology. Wave energy will make a small but locally useful contribution. Changed farming and forest practices, along with increased energy efficiency, will provide important abatement contributions. In this depiction of the energy scene, it is assumed that nuclear will not be part of the clean energy mix, but for many countries, nuclear power would lie in the centre of this figure.

complementary measures to develop the technologies that are needed to make a market approach work. Without technology, a carbon tax or an ETS is a potentially very expensive and ultimately empty gesture. Without market discipline, unrealistically expensive and ultimately impractical clean energy solutions will be offered by researchers. It is therefore time for scientists, industrialists, economists and politicians to get together in a more meaningful way. Making deep cuts to emissions is far too important to leave solely to politicians, or economists or industrialists or scientists. We are all in this together.

Finally, it is important to restate the fact all credible energy projections indicate fossil fuels will continue to be used for decades to come. CCS is the only technology we have for avoiding the greenhouse consequences of that use. Therefore countries that are major users of fossil fuels and especially countries that are both major users and producers of fossil fuels, such as Australia, Canada, China, Russia, UK and USA, have a particular need (and obligation) to ensure that CCS is deployed soon. If we are to have a solution to the greenhouse problem, CCS must be part of national and global clean energy portfolios.

ACRONYMS

ARI	Advanced Resources Institute
AEMO	Australian Energy Market Operator
ASU	air separation unit
ATSE	Australian Academy of Technological Sciences and Engineering
BGS	British Geological Survey
BMR	Bureau of Mineral Resources
CBM	coal bed methane
CCS	carbon capture and storage
COP	Conference of the Parties
CSIRO	Commonwealth Scientific and Industrial Research Organisation
CCSD	The Cooperative Research Centre for Coal in Sustainable Development
CTL	coal to liquids
DSA	deep saline aquifer
ECBM	enhanced coal bed methane
EGR	enhanced gas recovery
EOR	enhanced oil recovery
EPA	Environmental Protection Agency
EPRI	Electric Power Research Institute
ESA	electrical swing adsorption
GCCSI	Global Carbon Capture and Storage Institute
GHG	greenhouse gas
FEED	Front-end engineering design
FID	final investment decisions
FIT	feed-in tariff
GHG	greenhouse gas
GTL	gas to liquids
GW	Gigawatt
GWP	global warming potential
HFR	hot fractured rock
HSA	hot saline aquifer
IEA	International Energy Agency
IEAGHG	International Energy Agency Greenhouse Gas Program
IGCC	integrated gasification combined cycle
IPCC	Intergovernmental Panel on Climate Change
ISCG	in situ coal gasification
LCOE	levelised cost of electricity
LNG	liquefied natural gas
LPG	liquefied petroleum gas
LRET	large scale renewable energy target
LULUCF	land use, land use change and forestry
MIT	Massachusets Institute of Technology
Mt	million tonnes
MW	megawatt

NETL	National Energy Technology Laboratory
NEV	net energy value
NGCC	natural gas combined cycle
NRDC	National Resources Defence Council
OECD	Organization for Economic Co-operation and Development
OTEC	ocean thermal energy conversion
PCC	post combustion capture
PIG	pipeline inspection gauge
ppb	parts per billion
ppm	parts per million
PSA	pressure swing adsorption
PV	photovoltaics (solar)
RCSP	Regional Carbon Sequestration Partnership
REC	renewable energy certificate
RET	renewable energy target
TCF	trillion cubic feet
TSA	temperature swing adsorption
UN	United Nations
UNFCCC	United Nations Framework Convention on Climate Change
VSA	vacuum swing adsorption

ADDITIONAL GENERAL READING

Australian Academy of Science, 2010. *The Science of Climate Change: Questions and answers.* Australian Academy of Science, Canberra

Burgess, J, 2010. *Low Carbon Energy: Evaluation of new energy technology choices for electric power generation in Australia.* Australian Academy of Technological Sciences and Engineering (ATSE), Melbourne

Burgess, J, 2011. *New Power Cost Comparisons: Levelised cost of electricity for a range of new power generating technologies.* Australian Academy of Technological Sciences and Engineering (ATSE), Melbourne

CAETS Working Group, 2010. Deployment of low-Emissions technologies for electric power generation in response to climate change. CAETS, Melbourne. Available from http://www.caets.org/cms/7122/9933.aspx

Climate Commission, 2011. The critical decade. Department of Climate Change and Energy Efficiency. Available from http://climatecommission.gov.au/wp-content/uploads/4108-CC-Science-WEB_3-June.pdf

Cruz, J, 2008. *Ocean Wave Energy: Current status and future perspectives.* Springer, Berlin

CSIRO, 2008. Soil carbon: The basics. CSIRO, Available from http://www.csiro.au/resources/soil-carbon.html

Edenhofer, O., et al., 2011. IPCC Special Report on Renewable Energy Sources and Climate Change Mitigation. Cambridge University Press, Cambridge. Available from http://srren.ipcc-wg3.de/

Garnaut, R, 2011. *Garnaut Climate Change Review – Update 2011 – Australia in the Response to Climate Change – Summary.* Cambridge University Press, New York. Available from http://www.garnautreview.org.au/update-2011/garnaut-review-2011.html

InterAcademy Council (Coordinating Author: Shapiro, H.), 2010. *Climate Change Assessments: Review of the processes and procedures of the IPCC.* InterAcademy Council, Amsterdam. Available from http://www.ipcc.ch/pdf/IAC_report/IAC%20Report.pdf

IPCC, 2005. *IPCC Special Report on Carbon Dioxide Capture and Storage. Prepared by Working Group III of the Intergovernmental Panel on Climate Change* (Eds Metz, B., Davidson, O., de Coninck, H. C., Loos, M. and Meyer, L. A.). Cambridge University Press, Cambridge (UK) and NY

IPCC, 2007. *Climate change 2007: Synthesis report. Contribution of working groups I, II and III to the fourth assessment report of the Intergovernmental Panel on Climate Change* (Eds, R. K. Pachauri and A. Reisinger). IPCC, Geneva, Switzerland.

MacKay, D., 2009. *Sustainable energy – Without the Hot Air.* UIT, Cambridge, UK. Available from http://www.withouthotair.com/download.html

Manwell, J., 2002. *Wind Energy Explained: Theory, design and application.* Wiley, Chichester, New York

Milne, J., Cameron, J., Page, L., Benson, S. and Pakrasi, H., 2009. *Report from Workshop on Biological Capture and Utilization of CO2.* Workshop held at the Charles F. Knight Center, Washington University in St. Louis, September 1-2, 2009. The Global Climate and Energy Project (GCEP), Stanford, CA

Neelin, J., 2011. *Climate Change and Climate Modelling.* Cambridge University Press, Cambridge

Nelson, V., 2011. *Introduction to Renewable Energy.* CRC Press, Florida

OECD/IEA, 2008. *Energy Technology Perspectives.* International Energy Agency, Paris France. Available from http://www.iea.org/textbase/nppdf/free/2008/etp2008.pdf

OECD/IEA, 2010. *Energy Technology Perspectives.* International Energy Agency, Paris France

Pacala, S. and Socolow, R., 2004. Stabilization Wedges: Solving the climate problem for the next 50 years with current technologies. *Science.* Vol. 305 No. 5686, pp. 968-972

Porter, J. and Phillips, P. (eds.), 2007. *Public Science in Liberal Democracy.* University of Toronto Press, Toronto

Productivity Commission, 2011. *Carbon Emission Policies in Key Economies.* Research Report, Productivity Commission, Canberra

Schneider, S., 2003. *Abrupt Non-linear Climate Change, Irreversibility and Surprise.* OECD, Paris, France

Seligman, P., 2010. *Australian Sustainable Energy – by the Numbers.* Melbourne Energy Institute, Parkville, Victoria. Available from http://energy.unimelb.edu.au/index.php?page=ozsebtn

Stanford University Global Climate & Energy Project, 2006. An assessment of solar energy conversion technologies and research opportunities. Technical assessment report. Available from http://gcep.stanford.edu/pdfs/assessments/solar_assessment.pdf

Stern, N, 2007. *The Economics of Climate Change: The Stern review.* Cambridge University Press, Cambridge (UK) and New York. Available from http://webarchive.nationalarchives.gov.uk/+/http://www.hm-treasury.gov.uk/independent_reviews/stern_review_economics_climate_change/stern_review_report.cfm

Tester, J., Drake, E., Driscoll, M., Golay, M. and Peters, W., 2005. *Sustainable Energy: Choosing among options.* MIT Press, Cambridge, Massachusetts

The Royal Society, 2009. Geoengineering the climate: science, governance and uncertainty. Available from http://royalsociety.org/Geoengineering-the-climate/

U.S. Department of Energy and National Energy Technology Laboratory, 2010. Carbon sequestration atlas of the United States and Canada – third edition. Available from http://www.netl.doe.gov/technologies/carbon_seq/refshelf/atlasIII/index.html

US EPA, 2010. Carbon sequestration in agriculture and forestry. Available from http://www.epa.gov/sequestration/faq.html

World Resources Institute (WRI), 2008. *Guidelines for Carbon Dioxide Capture, Transport, and Storage.* World Resources Institute, Washington DC. Available from http://www.wri.org/publication/ccs-guidelines

REFERENCES TO DATA SOURCES

Chapter 1

Climate Commission, 2011. The critical decade. Department of Climate Change and Energy Efficiency. Available from http://climatecommission.gov.au/wp-content/uploads/4108-CC-Science-WEB_3-June.pdf

Garnaut, R., 2011. Garnaut climate change review – Update 2011 – Australia in the response to climate change – Summary. Garnaut Climate Change Review. Available from http://www.garnautreview.org.au/update-2011/garnaut-review-2011.html

House of Commons Science and Technology Committee, 2010. The disclosure of climate data from the Climatic Research Unit at the University of East Anglia. *Eighth Report of Session 2009-10 of UK Parliament*. Available from http://www.publications.parliament.uk/pa/cm200910/cmselect/cmsctech/387/387i.pdf

IEA, 2011. Prospect of limiting the global increase in temperature to 2C is getting bleaker. [Online Press Release, 30 May 2011]. Available from http://www.iea.org/index_info.asp?id=1959

InterAcademy Council (Coordinating Author: Shapiro, H), 2010. Climate change assessments – review of the processes and procedures of the IPCC. InterAcademy Council, Amsterdam Available from http://www.ipcc.ch/pdf/IAC_report/IAC%20Report.pdf

Lomborg, B., 2007. *Cool It: The sceptical environmentalist's guide to global warming*. Alfred A. Knopf Publishing, New York

MacKay, D., 2009. *Sustainable energy – Without the Hot Air*. Cambridge, England: UIT. Available from http://www.withouthotair.com/download.html

Meyer, L., 2010. Assessing an IPCC assessment. An analysis of statements on projected regional impacts in the 2007 report. Netherlands Environmental Assessment Agency (PBL). Available from http://www.pbl.nl/sites/default/files/cms/publicaties/500216002.pdf

Russell, M., 2010. The independent climate change e-mails review. Review prepared for the University of East Anglia. Available from http://www.cce-review.org/pdf/FINAL%20REPORT.pdf

Seligman, P., 2010. Australian sustainable energy – by the numbers. Melbourne Energy Institute. Available from http://energy.unimelb.edu.au/index.php?page=ozsebtn

Stern, N., 2007. *The Economics of Climate Change: The Stern review*. Cambridge University Press, Cambridge (UK), New York. Available from http://webarchive.nationalarchives.gov.uk/+/http://www.hm-treasury.gov.uk/independent_reviews/stern_review_economics_climate_change/stern_review_report.cfm

United Nations, 1992. United Nations Framework Convention on Climate Change. United Nations. Available from http://unfccc.int/essential_background/convention/background/items/2853.php

United Nations, 1997. Kyoto Protocol to the UNFCCC. United Nations. Available from http://unfccc.int/resource/docs/convkp/kpeng.pdf

Zinser, T., 2011. Response to Sen. James Inhofe's request to OIG to examine issues related to internet posting of email exchanges taken from the Climatic Research Unit of the University of East Anglia. Office of Inspector General, US Dept of Commerce. Available from http://www.

oig.doc.gov/Pages/Response-to-Sen.-James-Inhofe%27s-Request-to-OIG-to-Examine-Issues-Related-to-Internet-Posting-of-Email-Exchanges-Taken-from-.aspx

Chapter 2

Arrhenius, S., 1896. On the influence of carbonic acid in the air upon the temperature of the ground. *Philosophical Magazine and Journal of Science. Vol. 41, S 5., pp. 237-276*

Blasing, T., 2011. Recent greenhouse gas concentrations. *DOI: 10.3334/CDIAC/atg.032.* Available from http://cdiac.ornl.gov/pns/current_ghg.html

Brohan, P., Kennedy, J.J., Harris, I, Tett, S. F. B, Jones, P. D. 2006: Uncertainty estimates in regional and global observed temperature changes: a new dataset from 1850. *Journal of Geophysical Research. Vol. 111, D12106*

Bureau of Meteorology, 2010. State of the climate. Available from http://www.bom.gov.au/inside/eiab/State-of-climate-2010-updated.pdf

Church, J., White, N., Hunter, J., Lambeck, K., 2008. A post-IPCC AR4 update on sea-level rise. Antarctic climate & ecosystems Cooperative Research Centre; Report No. MB01_080911. Available from http://www.cmar.csiro.au/sealevel/downloads/797655_16br01_slr_080911.pdf

Cook, P., and Shergold, J, 1979. *Proterozoic-Cambrian Phosphorites.* Australian National University Press, Canberra

Cook, P., and Shergold, J., (eds), 1986. *Phosphate Deposits of the World: Volume 1 Proterozoic and Cambrian phosphorates.* Cambridge University Press, Cambridge (UK) and NY

Crutzen, P. J., and Stoermer, E. F., 2000. The 'Anthropocene'. *Global Change Newsletter. Vol 41, pp. 17-18.* Available from http://www.mpch-mainz.mpg.de/~air/anthropocene/

Essex, C. and McKitrick, R., 2002. Taken by Storm: The troubled science, policy and politics of global warming. Key Porter Books, Ontario, Canada

Flint, R. F, 1971. *Glacial and Quaternary Geology.* John Wiley and Sons, NY

Frank, D. C., Esper, J., Raible, C. C., Büntgen, U., Trouet, V., Stocker, B., and Joos, F. 2010. Ensemble reconstruction constraints on the global carbon cycle sensitivity to climate. *Nature. Vol. 463, No. 7280,* DOI:10.1038/nature08769

Garnaut, R., 2011. Garnaut climate change review – Update 2011 – Australia in the response to climate change – Summary. Garnaut Climate Change Review. Available from http://www.garnautreview.org.au/update-2011/garnaut-review-2011.html

Gutro, R., 2008. What's the difference between weather and climate? Available from http://www.nasa.gov/mission_pages/noaa-n/climate/climate_weather.html

Hays, J, Imbrie, J, Shackleton, N. 1976. Variations in the Earth's orbit: pacemaker of the ice ages. *Science. Vol. 194, No. 4270. pp. 1121-1132*

Idnurm, M. and Cook, P. J, 1980. Palaeomagnetism of beach ridges in South Australia and the Milankovitch theory of ice ages. *Nature. Vol 2867, No. 5774, pp. 699-702*

Idso, S., Kimball, B., Anderson, M., Mauney, J. 1987. Effects of atmospheric CO_2 enrichment on plant growth: the interactive role of air temperature. *Agriculture, Ecosystems & Environment. Vol 20, No. 1, pp. 1-10*

Intergovernmental Panel on Climate Change (IPCC), 2007. Summary for Policymakers. In *Climate Change 2007: The physical science basis. contribution of working group I to the fourth assessment report of the Intergovernmental Panel on Climate Change.* (Eds S. Solomon, D. Qin, M. Manning, Z. Chen, M. Marquis, K.B. Averyt, M.

Tignor and H.L. Miller). Cambridge University Press, Cambridge (UK) and NY

IPCC, 2007. *Climate Change 2007: The physical science basis. Contribution of Working Group I to the fourth assessment report of the Intergovernmental Panel on Climate Change.* (Eds S. Solomon, D. Qin, M. Manning, Z. Chen, M. Marquis, K.B. Averyt, M. Tignor and H.L. Miller). Cambridge University Press, Cambridge (UK) and NY

Jouzel et al., 2007. 800KYr δD data and temperature reconstruction. Available from http://www.ncdc.noaa.gov/paleo/icecore/antarctica/domec/domec_epica_data.html

Lisiecki, L. E., and Raymo, M. E. 2005, A Pliocene-Pleistocene stack of 57 globally distributed benthic $\delta^{18}O$ records, *Paleoceanography. Vol. 20,* PA1003

Lüthi et al., 2008. 800KYr CO_2 data. Available from http://www.ncdc.noaa.gov/paleo/icecore/antarctica/domec/domec_epica_data.html

Mann, M. E., Bradley, R. S., Hughes, M. K., 1998. Global-scale temperature patterns and climate forcing over the past six centuries. *Nature. Vol. 392, pp. 779-787*

Mann, M. E., Bradley, R. S., and Hughes, M. K., 1999. Northern hemisphere temperatures during the past millennium: inferences, uncertainties, and limitations. *Geophysical Research Letters. Vol. 26, pp. 759-762*

Mann, M. E., Zhang, Z., Hughes, M. K., Bradley, R. S., Miller, K., Rutherford, R.S and Ni, F., 2008. Proxy-based reconstructions of hemispheric and global surface temperature variations over the past two millennia. *Proceedings of the National Academy of Sciences Vol. 105, No. 36, pp. 13252-13257*

Mann, M. E., Zhang, Z., Rutherford, S., Bradley, R. S., Hughes, M. K., Shindell, D., Ammann, C., Faluvegi, G., and Ni, F., 2009. Global signatures and dynamical origins of the little ice age and medieval climate anomaly. *Science. Vol. 326, pp. 1256-1260*

Mcintyre, S. and McKitrick, R., 2005. Hockey sticks, principal components, and spurious significance. *Geophysical Research Letters. Vol. 32, No. 3*

Meinhausen, M., Meinhausen, N., Hare, W., Raper, S., Frieler, K., Knutti, R., Frame, D., Allen, M., 2009. Greenhouse-gas emission targets for limiting global warming to 2C. *Nature. Vol. 458, pp. 1158-1162*

Miller, K. et al., 2005. The Phanerozoic record of global sea-level. *Science. Vol. 310, pp. 1293-1298*

Park, R., Epstein, S., 1961. Metabolic fractionation of C13 & C12 in plants. *Plant Physiology, Vol. 36, No. 2, pp. 133-138*

Peacock, J., 2007. Million year plus ice core project (MY+): Drilling an Antarctic ice core over a million years old. Australian Government Fact Sheet. Available from http://www.gg.gov.au/res/File/PDFs/Millionyearplusicecoreproject.pdf

Pearman, G., 2011. Personal communication regarding the discrepancy between northern and southern hemisphere CO_2 concentrations

Petit J. R., Jouzel J., Raynaud D., Barkov N. I., Barnola J. M., Basile I., Bender M., Chappellaz J., Davis J., Delaygue G., Delmotte M., Kotlyakov V. M., Legrand M., Lipenkov V., Lorius C., Pépin L., Ritz C., Saltzman E., Stievenard M, 1999. Climate and atmospheric history of the past 420,000 years from the Vostok Ice Core, Antarctica. *Nature. Vol. 399, pp.429-436*

Pittock, B., 2009. Climate change – the science, impacts and solutions. CSIRO PUBLISHING, Australia

Royer, D., Berner, R., Montañez, I., Tabor, N., Beerling, D., 2004. CO_2 as a primary driver of Phanerozoic climate. *Geological Society of America Today. July 2004, Vol 14, No 3, pp. 4-10*

Schneider, S., Rosencranz, A., D., M., and Kuntz-Duriseti, K., 2009. *Climate Change Science and Policy.* Island Press, Washington

Steinfield, H., Gerber, P., Wassenaar, T., Castel, V., Rosales, M., and de Haan, C., 2006. *Livestock's Long Shadow.* Food and Agricultural Organization of the UN, Rome, Italy. Available from ftp://ftp.fao.org/docrep/fao/010/a0701e/a0701e00.pdf

Stern, N., 2007. *The Economics of Climate change: The Stern review.* Cambridge University Press, Cambridge (UK) and NY. Available from http://webarchive.nationalarchives.gov.uk/+/http://www.hm-treasury.gov.uk/independent_reviews/stern_review_economics_climate_change/stern_review_report.cfm

Trudinger, C., Enting, I., Etheridge, D., Francey, R., Rayner, P., 2005. The carbon cycle over the past 10,000 years inferred from the inversion of ice core data. In *A History of Atmospheric CO_2 and its Effects on Plants, Animals and Ecosystems,* (Eds J. Ehleringer, T. Cerling and M. Dearing). Springer

United States Bureau of Land Management, 2006. Salt Creek phases iii/iv environmental assessment: appendix C. Howell Petroleum. (#WYO60-EA06-18) Casper Field Office: Dept of the Interior.

Vail P. R. et al., 1977. Seismic stratigraphy and global changes of sea level: Part 3: relative changes of sea level from costal onlap. In Payton, C. E., Seismic stratigraphy – applications to hydrocardbon exploration. Memo 26. American Association of Petroleum Geologists, Tulsa

Watson, R., 2000. Land Use, Land-use Change, and Forestry. Cambridge University Press, Cambridge (UK)

Weidenbach K., 2008. Rock star; the story of Reg Sprigg – outback legend. *East Street Publications*

Zachos, J, Pagani, M, Sloan, L, Thomas, E, Billups, K, 2001. Trends, rhythms, and aberrations in global climate 65 ma to present. *Science. Vol. 292 (5517), pp. 686-693*

Chapter 3

Agyei, Y., 1998. Deforestation in Sub-Saharan Africa. *African Technology Forum. Vol 8.1*

Boden, T., Marland, G., Andres, B., 2011. National CO_2 emissions from fossil-fuel burning, cement manufacture and gas flaring 1751-2008. Carbon Dioxide Information Analysis Center, Oak Ridge National Laboratory; Oak Ridge, Tennessee. Available online at http://cdiac.ornl.gov/ftp/ndp030/nation.1751_2008.ems

BP, 2011. BP statistical review of world energy June 2011. BP, London. Available from www.bp.com/statisticalreview

FAO, 2011. State of the World's forests. Food and agricultural organization of the UN, Rome, Italy. Available from http://www.fao.org/docrep/013/i2000e/i2000e00.htm

Freese, B., 2003. *Coal: a human history.* Perseus Publishing, Cambridge, Massachusetts

Gorham, R., 2002. Air pollution from ground transportation. United Nations. Available from http://www.un.org/esa/gite/csd/gorham.pdf

Helm, D., 2008. Climate-change policy: why has so little been achieved? *Oxford Review of Economic Policy. Vol. 24, No. 2, pp. 211-238*

Howarth, R., Santoro, R., Ingraffea, A., 2011. Methane and the greenhouse-gas footprint of natural gas from shale formations. *Climate Change.* DOI:10.1007/s10584-011-0061-5

Hubbert M. K., 1956. Nuclear energy and the fossil fuels. American Petrol Institute Drilling & Production Practice. Proceedings of the Spring Meeting, San Antonio, Texas. *pp. 7-25*

Intergovernmental Panel on Climate Change, 2003. Good practice guidance for land use, land-use change and forestry. (Eds J. Penman, M.

Gytarsky, T. Hiraishi, T. Krug, D. Kruger, R. Pipatti, L. Buendia, K. Miwa, T. Ngara, K. Tanabe, W. Wagner, W.). Institute for Global Environmental Strategies, Hayama, Japan. Available from http://www.ipcc-nggip.iges.or.jp/public/gpglulucf/gpglulucf.html

International Energy Agency (IEA), 2009. CO2 emissions from fuel combustion: Highlights. OECD/IEA. Available from http://ccsl.iccip.net/co2highlights.pdf

Kigomo, B., 2003. Forests and woodland degradation in dryland Africa: A case for urgent global attention. Kenya Forestry Research Institute. Available from http://www.fao.org/DOCREP/ARTICLE/WFC/XII/0169-B3.HTM

Malthus, T.R., 1789. *An essay on the principle of population.* J Johnson, London. Reprinted 1959.

Meadows, D. 1972. *The Limits to Growth.* Universe Books, New York

OECD/IEA, 2010. World Energy Outlook. International Energy Agency, Paris, France

OECD/IEA, 2011. Country statistics: Share of total primary energy supply in 2008. IEA. Available from http://www.iea.org/stats/graphsearch.asp

Pena, N., Bird, N., Frieden, D., Zanchi, G., 2010. Conquering space and time: The Challenge of emissions from land use change. *CIFOR infobrief.* Available from www.cifor.cgiar.org/publications/pdf_files/infobrief/3269-infobrief.pdf

Rasheed, S. 1996. The challenges of sustainable development in 1990s and beyond. *Sustainable Development Beyond Rhetoric - Africa Environment. Environment Studies and Regional Planning Bulletin. Vol X, No.1-2*

Stanton, E., Ackerman, F., Sheeran, K., 2010. Why do state emissions differ so widely? E3 Network. Available from http://www.e3network.org/papers/Why_do_state_emissions_differ_so_widely.pdf

Chapter 4

Australian Energy Market Operator (AEMO), 2011. Pricing event reports – January 2011. Available from http://www.aemo.com.au/reports/pricing_jan.html

Burgess, J., 2011. New power cost comparisons: Levelised cost of electricity for a range of new power generating technologies. Australian Academy of Technological Sciences and Engineering (ATSE), Melbourne

Department of the Environment, Water, Heritage and the Arts, 2008. Energy use in the Australian residential sector 1986-2020. Commonwealth of Australia. Available from http://www.energyrating.gov.au/library/details2008-energy-use-aust-res-sector.html

Geoscience Australia. Interpreted temperature at 5 km depth. Available at http://www.ga.gov.au/energy/projects/geothermal-energy.html

Gerardi, W., and Nsair A., 2009. Comparative costs of electricity generation technologies. Report to AGEA. Available from www.agea.org.au/media/docs/mma_comparative_costs_report_2.pdf

Heaton, E., Dohleman, F., and Long, S., 2008. Meeting US biofuel goals with less land: the potential of Miscanthus. *Global Change Biology. Vol 14. pp. 1-15*

Institute for Energy Research (IER), 2009. Levelized cost of new generating technologies. Available from http://www.instituteforenergyresearch.org/wp-content/uploads/2009/05/levelized-cost-of-new-generating-technologies.pdf

Institute of Public Utilities, 2011. Energy Information Administration electricity generation estimates (2011). Michigan State University. Available from http://www.energytransition.msu.edu/documents/ipu_eia_electricity_generation_estimates_2011.pdf

Joskow, P., 2011. Comparing the costs of intermittent and dispatchable electricity generating technologies. *American Economic Review. Vol. 101, No. 3, pp. 238-241*

OECD/IAEA, 2008. *Uranium 2007: Resources, production and demand.* A joint report by the OECD Nuclear Energy Agency and the International Atomic Energy Agency, OECD, Paris

OECD/IEA, 2010. *World Energy Outlook.* International Energy Agency, Paris, France

Pimentel, D., 2003. Ethanol fuels: energy balance, economics and environmental impacts are negative. *Natural Resources Research. Vol. 12, No. 2, pp. 127-134*

RED Electrica De Espana, 2011. Wind power energy, the main source of electricity in March. [press release]. Available from http://www.ree.es/ingles/sala_prensa/web/notas_detalle.aspx?id_nota=180

Sandia National Laboratory, 2010. Electric power industry needs for grid-scale storage applications. Report sponsored by US Department of Energy. Available from http://www.oe.energy.gov/DocumentsandMedia/Utility_12-30-10_FINAL_lowres.pdf

Seimens AG, 2009. Online monitoring of polysilicon production in photovoltaic industry. Answers for Industry. Available from http://www.automation.siemens.com/mcms/solar-industry/en/polysilicon-production/Documents/Case_Study_EN.pdf

Seligsohn, D., Heilmayr, R., Tan, X., Weischer, L., 2009. China, the United States and the climate change challenge. World Resource Institute, Washington D.C.. Available online at http://pdf.wri.org/china_united_states_climate_change_challenge.pdf

Shapouri, H., Duffield, J., Wang, M., 2002. The energy balance of corn ethanol: an update. USDA Agricultural economic report No. 813. Available from http://www.transportation.anl.gov/pdfs/AF/265.pdf

Strategic energy technologies information system (SETIS), 2009. Hydropower. European Commission. Available from http://setis.ec.europa.eu/technologies/Hydropower

US Dept of Agriculture, 2009. 2007 Census of agriculture. US Summary and state data: Geographic Area Series. Vol. 1, Part 51. Available from http://www.agcensus.usda.gov/Publications/2007/Full_Report/usv1.pdf

US Department of Energy/EERE. Estimated temperatures at depths of 6km. Available at http://teeic.anl.gov/er/geothermal/restech/dist/index.cfm

Chapter 5

3TIER, 2011. Global solar irradiance map. 3TIER resource maps. Available at http://www.3tier.com/en/support/resource-maps/

3TIER, 2011. Global wind speed map. 3TIER resource maps. Available at http://www.3tier.com/en/support/resource-maps/

Anderson, E., Arundale, R., Maughan, M., Oladeinde, A., Wycislo, A. and Voigt, T., 2011. Growth and agronomy of *Miscanthus giganteus* for biomass production. *Biofuels. Vol. 2, No. 2, pp. 167-183*

Engelman, R., 2011. An end to population growth: why family planning is key to a sustainable future. *Solutions. Vol. 2, No. 3, pp. 32-41.* Available from http://www.thesolutionsjournal.com/node/919

Garnaut, R., 2008. *The Garnaut Climate Change Review.* Cambridge University Press, Cambridge (UK). Available at http://www.garnautreview.org.au/2008-review.html

Glassman, D., Wucker, M., Isaacman, T., Champilou, C., 2011. The water energy nexus. The World Policy Institute in partnership with

EBG Capital. Available from http://www.worldpolicy.org/sites/default/files/policy_papers/THE%20WATER-ENERGY%20NEXUS_0.pdf

IEA, 2009. Technology roadmap – carbon capture and storage. International Energy Agency. Available from http://www.iea.org/roadmaps/ccs_roadmap.asp

IPCC, 2007. Climate change 2007: Synthesis report. Contribution of working groups I, II and III to the fourth assessment report of the Intergovernmental Panel on Climate Change. (Eds R. K. Pachauri and A. Reisinger, A.). IPCC, Geneva, Switzerland

IPCC, 2011. Summary for Policymakers. In *IPCC Special Report on Renewable Energy Sources and Climate Change Mitigation.* [Eds O. Edenhofer, R. Pich, Y. Madruga, K. Sokona, P. Seyboth, S. Matschoss, T. Kadner, P. Zwickel, G. Eickemeier, S. Hansen, C. Schlömer von Stechow). Cambridge University Press, Cambridge (UK) and NY

Mercer-Blackman, V., Samiei, H., Cheng, K., 2007. Biofuel demand pushes up food prices. *Survey Magazine.* IMF Research Department. Available from http://www.imf.org/external/pubs/ft/survey/so/2007/RES1017A.htm

Milbrandt, A., Overend, R., 2008. Survey of biomass resource assessments and assessment capabilities in APEC economies. Report for the APEC Energy Working Group. Available from http://www.nrel.gov/docs/fy09osti/43710.pdf

NASA, 2001. Measuring solar insolation. Visible earth: A catalogue of NASA images Available at http://visibleearth.nasa.gov/view_rec.php?id=1683

OECD/IEA, 2008. Energy technology perspectives. International Energy Agency, Paris France. Available online at http://www.iea.org/textbase/nppdf/free/2008/etp2008.pdf

OECD/IEA, 2010. World Energy Outlook. International Energy Agency, Paris, France

Pacala, S., and Socolow, R., 2004. Stabilization wedges: Solving the climate problem for the next 50 years with current technologies. *Science, Vol. 305, No. 5686, pp. 968-972*

Roy, S., Pacala, S., and Walko, R., 2004. Can large wind farms affect local meterology? *Journal of Geophysical Research, Vol. 109.* Available from http://www.atmos.illinois.edu/~sbroy/publ/jgr2004.pdf

Soder, L., Hofmann, L., Orths, A., Holttinen, H., Wan, Y., Tuohy, A., 2006. Experience from wind integration in some high penetration areas. Report prepared for IEA Wind task 25. Available from http://www.ieawind.org/AnnexXXV/PDF/Soder/Soeder%20et%20al%20IEEE%20Paper.pdf

US Census Bureau, 2011. Online international database. Available from http://www.census.gov/population/international/data/idb/informationGateway.php

U.S. Department of Energy, 2006. Energy demands on water resources: report to Congress on interdependency of energy and water. *Sandia National Laboratories [online].* Available from http://www.sandia.gov/energy-water/congress_report.htm

United Nations, 2004. *World Population to 2300.* United Nations Department of Economic and Social Affairs, NY. Available from http://www.un.org/esa/population/publications/longrange2/WorldPop2300final.pdf

Walsh, B, 2011. Why biofuels help push up world food prices. *TimeScience.* Available from http://www.time.com/time/health/article/0,8599,2048885,00.html

Chapter 6

Boden, T. A., Marland, G. and Andres, R. J., 2011. *Global, Regional, and National Fossil-Fuel CO_2 Emissions*. Carbon Dioxide Information Analysis Center, Oak Ridge National Laboratory, U.S. Department of Energy, Oak Ridge, Tennesee, USA. DOI: 10.3334/CDIAC/00001_V2011. Available at http://cdiac.ornl.gov/trends/emis/tre_prc.html

Boonpoke, A., Chiarakorn, S., Laosiripojana, N., Towprayoon S., Chidthaisong, A., 2011. Synthesis of activated carbon and MCM-41 from bagasse and rice husk and their carbon dioxide adsorption capacity. *Journal of Sustainable Energy & Environment. Vol. 2, pp. 77-81.* Available at http://www.jseejournal.com/JSEE%202011/JSEE%20Vol%202%20Issue%202/12.%20Synthesis%20of%20activated%20carbon_pp.77-81.pdf

Dakota Gasification Company. The greatest CO_2 story ever told. Available at http://www.dakotagas.com/CO2_Capture_and_Storage/index.html

Department of Climate Change, 2009. National Inventory Report 2007. Commonwealth of Australia. Available at http://www.climatechange.gov.au/publications/greenhouse-acctg/~/media/publications/greenhouse-acctg/national-industry-report-vol-1-complete.ashx

Foss, M., 2004. Interstate natural gas – quality specifications and interchangibility. Report produced for the Center for Energy Economics, University of Texas, Austin, Texas. Available at http://www.beg.utexas.edu/energyecon/lng/documents/CEE_Interstate_Natural_Gas_Quality_Specifications_and_Interchangeability.pdf

Ho, M., and Wiley, D., 2011. Implementing CO_2 capture at power plants – retrofit or new build? CO2CRC Report No. RPT11-2946, CO2CRC, Canberra

IEA, 2009. Cement technology roadmap 2009: carbon emissions reductions up to 2050. International Energy Agency. Available from http://www.iea.org/papers/2009/Cement_Roadmap.pdf

IEA GHG, 2011. Retrofitting CO_2 capture to existing power plants. IEAGHG Report 2011-02

IEA News Centre. Pulverized Coal Combustion. IEA Clean Coal Centre. Available at http://www.iea-coal.org.uk/site/2010/database-section/ccts/pulverised-coal-combustion-pcc?

IPCC, 2005. *IPCC Special Report on Carbon Dioxide Capture and Storage. Prepared by Working Group III of the Intergovernmental Panel on Climate Change* (Eds Metz, B., Davidson, O., de Coninck, H. C., Loos, M. and Meyer, L. A.). Cambridge University Press, Cambridge (UK) and NY

Mills, S., 2010. Coal use in the new economies of China, India and South Africa. IEA Clean Coal Centre Report, London. Summary available at http://www.iea-coal.org.uk/publishor/system/component_view.asp?LogDocId=82270&PhyDocID=7433

OECD/IEA, 2005. Energy policies of IEA countries: Australia 2005 review. International Energy Agency, Paris, France

Specker, S., Phillips, J., Dillon, D., 2009. The potential growing role of post combustion CO_2 capture retrofits in early commercial applications of CCS to coal-fired power plants. MIT Coal Retrofit Symposium, Cambridge, Massachusetts, March 2009

Taylor, H., 1997. *Cement Chemistry*. Thomas Telford Publishing, London

Walter, K., 2007. Fire in the hole. *Science and Technology Review, Lawrence Livermore National Laboratory, Issue: April 2007.* Available at https://www.llnl.gov/str/April07/pdfs/04_07.2.pdf

Walker, L., 1999. Underground coal gasification: a clean coal technology ready for development. *The Australian Coal Review, October 1999.* Available at http://www.cougarenergy.com.au/pdf/AustCoalReviewPaperOct1999.pdf

Chapter 7

Ciferno, J., Litynski, J., Plasynski, S., Murphy, J., Vaux, G., Munson, R., Marano, J., 2010. DOE/NETL Carbon dioxide capture and storage RD&D roadmap. Report prepared for the US Department of Energy. Available from http://www.netl.doe.gov/technologies/carbon_seq/refshelf/CCSRoadmap.pdf

Coleman, D., 2009. Transport infrastructure rationale for carbon dioxide capture and storage in the European Union to 2050. *Energy Procedia, Vol. 1, No. 1, pp. 1673-1681*

CSLF, 2009. CO_2 Transportation – Is it safe and reliable?. *CSLF in Focus Series.* Available from http://www.cslforum.org/publications/documents/CSLF_inFocus_CO2Transport.pdf

Interagency Taskforce on CCS, 2010. Report of the interagency taskforce on carbon capture and storage. Report produced for Presidential Memorandum. Available from http://www.epa.gov/climatechange/policy/ccs_task_force.html

IPCC, 2007. *Climate change 2007: Synthesis report. Contribution of working groups I, II and III to the fourth assessment report of the Intergovernmental Panel on Climate Change* (Eds, R. K. Pachauri and A. Reisinger). IPCC, Geneva, Switzerland

KinderMorgan, 2011. Kindermorgan CO_2. Available from http://www.kindermorgan.com/business/co2/default.cfm

Parfomak, P., and Folger, P., 2007. Carbon Dioxide (CO_2) pipelines for carbon sequestration: emerging policy issues. Congressional research service report for Congress. Available from http://ncseonline.org/nle/crsreports/07may/rl33971.pdf

Rubin, E., 2008. CO_2 capture and transport. *Elements. Vol. 4, No. 5., pp. 311-317.* Available from http://web.mit.edu/mitei/docs/reports/rubin-capture.pdf

Seevam, P., Race, J., and Downie, M., 2008. Carbon dioxide impurities and their effects on CO_2 pipelines. *The Australian Pipeliner. April 2008.* Available from http://pipeliner.com.au/news/carbon_dioxide_impurities_adn_their_effects_on_co2_pipelines/011846

Seiersten, M., Kongshaug, K., 2005. Materials selection for capture, compression, transport and injection of CO_2. In *Carbon Dioxide Capture for Storage in Deep Geologic Formations - Results from the CO2 Capture Project Capture and Separation of Carbon Dioxide from Combustion Sources.* (ed. David C. Thomas). Elsevier

Watt, J., 2010. Lessons from the US: Experience in carbon dioxide pipelines. *The Australian Pipeliner. October 2010.* Available from http://pipeliner.com.au/news/lessons_from_the_us_experience_in_carbon_dioxide_pipelines/043633/

World Resources Institute (WRI), 2008. *Guidelines for Carbon Dioxide Capture, Transport, and Storage.* World Resources Institute, Washington DC: WRI. Available from http://www.wri.org/publication/ccs-guidelines

Chapter 8

BP, 2011. BP statistical review of world energy June 2011. Available from www.bp.com/statisticalreview

Bradshaw, B., Simon, G., Bradshaw, J., and Mackie, V., 2005. GEODISC research: carbon dioxide sequestration potential of Australia's coal basins. CO2CRC Report No. RPT05-0011, Canberra, ACT

BSCSP, 2009. Basalt pilot factsheet. Available from http://www.bigskyco2.org/sites/default/files/documents/basalt_fatsheet.pdf

Carbon Storage Mapping Taskforce, 2009. National carbon mapping and infrastructure plan – Australia. Department of Resources, Energy and Tourism, Canberra. Available from http://www.ret.gov.au/resources/Documents/Programs/CS%20Taskforce.pdf

ChevronTexaco's Rangely Oil Field Operations. Colorado School of Mines Fact Sheet. Available from http://emfi.mines.edu/emfi2005/ChevronTexaco.pdf

CO2CRC, 2008. Storage capacity estimation, site selection and characterisation for CO_2 storage projects. CO2CRC Report No. RPT08-1001, Canberra, ACT

CSLF (Carbon Sequestration Leadership Forum), 2005. A taskforce for review and development of standards with regards to storage capacity measurement. CSLF-T- 2005-9 15, August 2005, Available from http://www.cslforum.org/documents/Taskforce_Storage_Capacity_Estimation_Version_2.pdf

Dahowski, R., Li, X., Davidson, C., Wei, N., and Dooley, J., 2009. Regional opportunities for carbon dioxide capture and storage in China. Report for the U.S. Department of Energy. Available from www.zeroemissionsplatform.eu/downloads/491.html

Dahowski, R. T., Li, X., Davidson, C. L., Wei, N., Dooley, J. J. and Gentile R. H., 2009. Early assessment of carbon dioxide capture and storage potential in China. In 8th Annual Conference on Carbon Capture & Sequestration. Pittsburgh, Pennsylvania, 5 May 2009. Available from http://www.cslforum.org/publications/documents/PNWD_SA_8600.pdf

EU Geocapacity, 2009. Assessing European capacity for geological storage of carbon dioxide. Geocapacity Final Report. Available from http://www.geology.cz/geocapacity/publications

Forsythe, J., 2009. CCS operating flexibility experience from In Salah. Presentation to IEAGHG workshop on Operating flexibility of power plants with CCS, Imperial College London November, 2009. Available from http://www.ieaghg.org/docs/flexibility%20workshop/07_Flexibility%20workshop%20Forsyth.pdf

Global CCS Institute and Parsons Brinckerhoff, 2011. Accelerating the uptake of CCS: industrial use of captured carbon dioxide. Global CCS Institute. Available from http://www.globalccsinstitute.com/resources/publications/accelerating-uptake-ccs-industrial-use-captured-carbon-dioxide

Godec, M., 2011. Global technology roadmap for CCS in industry: Sectoral assessment CO_2 enhanced oil recovery. Report for United Nations Industrial Development Organization; prepared by Advanced Resources International Inc. Available from http://cdn.globalccsinstitute.com/sites/default/files/publication_20110505_sector-assess-eor.pdf

Herzog, H., 2002. Carbon sequestration via mineral carbonation: Overview and assessment. MIT Laboratory for Energy and the Environment: Technology Assessment. Available from http://sequestration.mit.edu/pdf/carbonates.pdf

IPCC, 2005. *IPCC Special Report on Carbon Dioxide Capture and Storage. Prepared by Working Group III of the Intergovernmental Panel on Climate Change.* (Eds B. Metz, O. Davidson, H. de Coninck, M. Loos and L. Meyer). Cambridge University Press, Cambridge (UK) and NY

Kuuskraa, V., G, 2008. Maximizing oil recovery efficiency and sequestration of CO_2 with "next generation" CO_2-EOR technology. Presentation to the 2nd Petrobras International Seminar on CO_2 Capture and Geological Storage, Brazil.

Available from http://www.adv-res.com/pdf/V_Kuuskraa%20Petrobras%20CO2%20SEP%2008.pdf

Matter, J. M and Kelemen, PB. 2008. In situ carbonation of peridotite for CO2 storage. *PNAS. Vol. 105, No. 45, pp. 17295-17300*

Matter, J. M. and Kelemen, P. B., 2009. Permanent storage of carbon dioxide in geological reservoirs by mineral carbonation. *Nature Geoscience. Vol. 2, No. 12, pp. 837-841*

McGrail, B. P., Schaef, H. T., Ho, A. M., Chien, Y. and Dooley J. J., 2006. Potential for carbon dioxide sequestration in flood basalts. *Journal of Geophysical Research. Vol. 111, B12201*

McPherson, B., and Lichtner, P., 2001. CO_2 sequestration in deep aquifers. Paper presented at the First National Conference on Carbon Sequestration, Washington DC. Available from http://www.netl.doe.gov/publications/proceedings/01/carbon_seq/7a2.pdf

Pedersen, T., 2008. Results of the EU GeoCapacity project. *GEO ENeRGY. No. 18.* Available from http://www.energnet.eu/GeoEnergy_18.pdf

Plasynski, S., Brickett, L., and Preston, C., 2008. Weyburn Carbon Dioxide Sequestration Project. US Department of Energy: Project Facts. Available from http://www.netl.doe.gov/publications/factsheets/project/Proj282.pdf

Reeves, S., Taillefert, A., Pekot, L., and Clarkson, C., 2003. The Allison unit CO_2 – ECBM pilot: a reservoir modelling study. Report prepared for the US Department of Energy by Advanced Resources International and Burlington Resources. Available from http://www.coal-seq.com/Proceedings2003/The%20Alllison%20Unit%20CO2rec-resubmitted%20120103.pdf

Sigurðardóttir, H. CarbFix: CO_2 fixation into basalts. *Annual Status Report 2009.* Available from http://www.or.is/media/PDF/CarbFix_StatusReport2009_09062010.pdf

TNO, 2007. K12-B, CO_2 storage and enhanced gas recovery. [TNO informational handout]. Available from http://www.tno.nl/downloads/357benol.pdf

U.S. Department of Energy and National Energy Technology Laboratory, 2010. Carbon sequestration atlas of the United States and Canada – third edition. Available from http://www.netl.doe.gov/technologies/carbon_seq/refshelf/atlasIII/index.html

Vangkilde-Pederson, T. et al., 2009. WP2 report storage capacity. EU GeoCapacity. Available from http://www.geology.cz/geocapacity/publications

Chapter 9

Benson, S., 2006. Carbon dioxide capture and storage: assessment of risks from storage of carbon dioxide in deep underground geological formations. Earth Sciences Division, Lawrence Berkley National Laboratory. Available from http://southwestcarbonpartnership.org/_Resources/PDF/GeologicalStorageRiskAssessmentV1Final.pdf

European Commission, 2011. CO_2 storage life cycle risk management framework. European Commission. Implementation of directive 2009/31/EC on the geological storage of carbon dioxide. Available from http://ec.europa.eu/clima/policies/lowcarbon/ccs_implementation_en.htm

Friedmann, J, and Benson, S., 2009. Carbon sequestration risks and hazards: What we know and what we don't know. Presentation given to NRDC Workshop in New York City. Available from http://docs.nrdc.org/globalwarming/files/glo_10062101d.pdf

Gaviot. Underground natural gas storage facility, 2010. *Enagas.* Available from http://www.enagas.

com/cs/Satellite?blobcol=urldata&blobheader=application%2Fpdf&blobkey=id&blobtable=MungoBlobs&blobwhere=1146251976537&ssbinary=true

Goff, F., Love, S., Warren, R., Counce, D., Obenholzner, J., Siebe, C., and Schmidt, S., 2001. Passive infrared remote sensing evidence for large intermittent CO_2 emissions at Popocatepetl volcano, Mexico. *Chemical Geology. Vol. 177, No. 1-2, pp. 133-156*

IEA, 2011. Carbon capture and storage – legal and regulatory review. Edition 2. International Energy Agency. Available from http://www.iea.org/ccs/legal/review.asp

Knowledge Networks, 2009. Field report of the carbon sequestration survey. Massachusetts Institute of Technology. Available from http://sequestration.mit.edu/research/survey2009.html

Material Safety Data Sheet (MSDS) for Carbon Dioxide. 2003. BOC Gases. Available from http://www.bocsds.com/uk/sds/industrial/carbon_dioxide.pdf

Nelson, L. 2000. Carbon dioxide poisoning. *Emergency Medicine. Vol. 32, No. 5, pp. 36-38.* Available from http://www.emedmag.com/html/pre/tox/0500.asp

Schneider, A., Stark M., and Littmann, W, 2002. Erdgasspeicher Berlin-Methoden der Betriebsführung. [Berlin Natural Gas Storage - Methods of Management.] *Erdöl Erdgas Kohle* [*Oil Gas Carbon*], *Vol. 118*

Chapter 10

Allinson, G., Cinar, Y., Hou, W. and Neal, P. R., 2009. The Costs of CO_2 transport and injection in Australia. CO2Tech Consultancy Report for the Department of Resources, Energy and Tourism. Available from http://www.ret.gov.au/resources/Documents/Programs/cst/CO2Tech%20-%20The%20Costs%20of%20CO2%20Transport%20and%20Injection%20in%20Australia.pdf

Allinson, W. G., Cinar, Y., Neal, P. R., Kaldi, J. and Paterson, L., 2010. CO_2 storage capacity – combining geology, engineering and economics. In SPE Asia Pacific Oil Gas Conference, Adelaide, South Australia, 18-20 October 2010. Society of Petroleum Engineers, SPE Paper 133804

Burgess, J., 2011. New power cost comparisons: Levelised cost of electricity for a range of new power generating technologies. Australian Academy of Technological Sciences and Engineering (ATSE), Melbourne

CAETS Working Group, 2010. Deployment of low-emissions technologies for electric power generation in response to climate change. CAETS. Available from http://www.caets.org/cms/7122/9933.aspx

Chevron, 2010. Company submission. Strategic Energy Initiative: Issues Paper. Western Australian Office of Energy. Available from http://www.energy.wa.gov.au/cproot/1833/2/Chevron.pdf

Dahowski, R et al., 2009. A preliminary cost curve assessment of carbon dioxide capture and storage potential in China. *Energy Procedia, Vol. 1, No. 10, pp. 2849-2856*

Daley, J. and Edis, T., 2011. Learning the hard way: Australia's policies to reduce emissions. Grattan Institute. Available from http://www.grattan.edu.au/pub_page/077_report_energy_learning_the_hard_way.html

Feron, P. and Paterson, L., 2011. Reducing the costs of CO_2 capture and storage (CCS). CSIRO PUBLISHING. Available from http://www.csiro.au/files/files/p10pa.pdf

Finkenrath, M., 2011. Cost and performance of carbon dioxide capture from power generation. International Energy Agency, Paris. Available from http://www.iea.org/publications/free_new_desc.asp?pubs_ID=2355

IEA, 2009. Technology roadmap – carbon capture and storage. International Energy Agency. Available from http://www.iea.org/roadmaps/ccs_roadmap.asp

IEA, 2010. CO_2 capture & storage. *IEA ETSAP – Technology Brief, Vol. E14 (October 2010).* Available from http://www.etsap.org/E-techDS/PDF/E14_%20CCS%20draft%20oct2010_%20GS-gc_OK.pdf

IEA, 2011. *Climate and Electricity Annual* 2011 - Data and Analyses. International Energy Agency, Paris

McKinsey & Company, 2008. Carbon capture & storage: assessing the economics. McKinsey & Company. Available from http://ww1.mckinsey.com/clientservice/electricpowernaturalgas/thinking.asp [y 2011]

Mourits, F., 2008. Overview of the IEAGHG Weyburn-Midale CO_2 monitoring and storage project. Presentation to the Workshop on Capture and Sequestration of CO_2. Mexico D.F, 10 July 2008. Available from http://www.cslforum.org/publications/documents/11_MouritsWeyburnMexico2008.pd

U.S. Department of Energy and National Energy Technology Laboratory, 2010. Carbon sequestration atlas of the United States and Canada – Third edition. Available from http://www.netl.doe.gov/technologies/carbon_seq/refshelf/atlasIII/index.html

Chapter 11

APH (Australian Parliament House), 2011. Feed-in tariffs. *Canberra.* Available from www.aph.gov.au/library/Pubs/ClimateChange/governance/domestic/national/feed.htm

Daley, J. and Edis, T, 2011. Learning the hard way: Australia's policies to reduce emissions. Grattan Institute Report. No. 2011-2

Department of Resources, Energy and Tourism, 2011. Carbon capture and storage flagships program. Available at http://www.ret.gov.au/Department/archive/cei/ccsfp/Pages/default.aspx

Electric Power Resource Institute (EPRI) (Coordinating Author: Booras, G), 2010. Australian Electricity generation technology costs – Reference case 2010. Report prepared for the Australian Government Department of Resources Energy and Tourism. Available from http://www.ret.gov.au/energy/Documents/AEGTC%202010.pdf

Feenstra, C., Mikunda, T., and Brunsting, S., 2010. What happened in Barendrecht. Report for the International Comparison of Public Outreach Practices Associated with Large Scale CCS Projects. Available at http://www.csiro.au/files/files/pybx.pdf

Feron, P., and Patterson, L., 2011. Reducing the cost of CO_2 capture and storage (CCS). Report produced for the 2011 Garnaut Review: CSIRO Energy Technology. Available at http://www.garnautreview.org.au/update-2011/commissioned-work/reducing-costs-CO2-capture-storage.pdf

IEA, 2009. Technology roadmap – carbon capture and storage. International Energy Agency. Available from http://www.iea.org/roadmaps/ccs_roadmap.asp

International Energy Agency, 2011. Are we entering a golden age of gas? International Energy Agency. Paris, France. Available at http://www.iea.org/weo/docs/weo2011/WEO2011_GoldenAgeofGasReport.pdf

OECD/IEA, 2008. Energy technology perspectives. International Energy Agency, Paris France. Available online at http://www.iea.org/textbase/nppdf/free/2008/etp2008.pdf

Platts, 2008. UDI World electric power plants database. Platts, UDI Products Group, Washington D.C

Productivity Commission, 2011. *Carbon Emission Policies in Key Economies.* Research Report, Productivity Commmision, Canberra

Shuster, E, 2011. Tracking new coal-fired power plants. Energy Analysis publication from the National Energy Laboratory (NETL). Available from http://netl.doe.gov/energy-analyses/refshelf/PubDetails.aspx?Action=View&PubId=194

World Bank, 2011. The World Bank and energy: focusing on access and expanding renewables. World Bank. Available at http://web.worldbank.org/WBSITE/EXTERNAL/TOPICS/EXTSDNET/0,,contentMDK:22951717~menuPK:64885113~pagePK:7278667~piPK:64911824~theSitePK:5929282,00.html

Wright, M., and Hearps, P., 2010. *Australian Sustainable Energy: Zero Carbon Australia Stationary Energy Plan.* University of Melbourne: Energy Research Institute, Carlton, Victoria. Available at http://media.beyondzeroemissions.org/ZCA2020_Stationary_Energy_Report_v1.pdf

INDEX